CRISPR Crops

Aftab Ahmad • Sultan Habibullah Khan •
Zulqurnain Khan
Editors

CRISPR Crops

The Future of Food Security

Springer

Editors
Aftab Ahmad
Center for Advanced Studies in Agriculture and Food Security (CASAFS)
University of Agriculture
Faisalabad, Pakistan

Sultan Habibullah Khan
Centre of Agricultural Biochemistry and Biotechnology (CABB)
University of Agriculture
Faisalabad, Pakistan

Center for Advanced Studies in Agriculture and Food Security (CASAFS)
University of Agriculture
Faisalabad, Pakistan

Zulqurnain Khan
Institute of Plant Breeding and Biotechnology
MNS University of Agriculture
Multan, Pakistan

ISBN 978-981-15-7141-1 ISBN 978-981-15-7142-8 (eBook)
https://doi.org/10.1007/978-981-15-7142-8

This Springer imprint is published by the registered company Springer Nature Singapore Pte Ltd.
The registered company address is: 152 Beach Road, #21-01/04 Gateway East, Singapore 189721, Singapore

Preface

This volume is a piece of commitment and dedication with voluntary and extensive collaboration among various parties. We wish to express our sincere gratitude to all the authors and contributors who strived to make this venture fruitful. The genome-editing research work has been emphasized in 2012 by taking a start with TALEs and TALENs. Many reviews and research articles were published focusing on the application of the technology for solving problems in economically important crops. We would like to extend our gratitude to Prof. Caixia Gao and Prof. Weicai Yang, IGDB, Chinese Academy of Sciences, Beijing, China and Professor Yinong Yang, Pennsylvania State University, USA for being our collaborators and part of the workshop on CRISPR-based genome editing in Pakistan. The share of the country rapporteurs and our students cannot be forgotten. Our students made a lot of effort in compiling the final manuscript. We hope that the readers will benefit from this book by getting the latest review and discussion on the subject matter. The way in which the book has been composed is very comprehensive and reader-friendly making concepts palpable. A big proportion of the credit goes to the publisher for their support in publishing the idea. This book is an endeavor toward understanding the concept, application, and regulation of the CRISPR technology in the field of crop sciences. We hope that along with the agricultural scientific community, scientists, researchers, and students of other fields will also found this book beneficial to understand the technology and its safe usage.

At the end, once again, we are thankful to all of the persons who have floated, wrote, composed, and published the idea in the shape of a book for the scientific community.

Faisalabad, Pakistan — Aftab Ahmad
Faisalabad, Pakistan — Sultan Habibullah Khan
Multan, Pakistan — Zulqurnain Khan
January, 2020

Contents

About the Editors

Aftab Ahmad is currently working as Assistant Professor at Department of Biochemistry, US-Pakistan Center for Advanced Studies in Agriculture and Food Security (CAS-AFS), University of Agriculture Faisalabad (UAF), Pakistan. His research interests include developing virus-resistant and salt-tolerant plants using CRISPR/Cas technology. He has published more than 15 research articles in international peer-reviewed journals including the Journal of Biomedical Science, Scientific Reports-Nature, PLOS One, Plant Molecular Biology Reporter, Molecules, Journal of Cereal Science, and Genetic Resources and Crop Evolution. He pursued his PhD and postdocs in Plant Molecular Biology from University of Shizuoka, Japan under Center of Excellence (COE, Twenty-First Century) program. He also served as visiting scientist in Department of Plant Sciences, University of California, Davis, USA. He has been teaching biochemistry and molecular biology at UAF and his research focuses in the area of CRISPR/Cas-based genome editing in plants.

Sultan Habibullah Khan received his PhD from the University of Manchester, England in the field of Plant Sciences in 2004. Since his graduation from UK, he has been working as a faculty at the University of Agriculture Faisalabad (UAF). His postdoctoral experience also includes teaching and research at the University of California, Berkeley and Sultan Qaboos University, Oman. Currently, he is working as the Deputy Chief of Party at the US-Pakistan Center for Advanced Studies in Agriculture and Food Security (CAS-AFS) and a Tenured Associate Professor of Plant Molecular Sciences at the Centre of Agricultural Biochemistry and Biotechnology (CABB), UAF. For the last 20 years, he has been engaged in the fields of plant biotechnology, breeding, genetics, and molecular biology. He is a recipient of national/international scholarships, Research Productivity Award (2014), and over 40 international publications with impact factor.

Zulqurnain Khan is currently working as Assistant Professor in the Institute of Plant Breeding and Biotechnology (IPBB), Muhammad Nawaz Sharif University of Agriculture, Multan (MNSUAM), Pakistan. He earned his PhD in Biotechnology from University of Agriculture Faisalabad (UAF), Pakistan and his research focuses on CRISPR/Cas9 system for the development of resistance against begomoviruses using model plants. He has published one book chapter, three review articles, and three research articles in international peer-reviewed journals including the Journal of Biomedical Sciences, Plant Molecular Biology Reporter, and International Journal of Agriculture and Biology. He also worked as a visiting researcher with a group of Professor Caixia Gao at the Institute of Genetics and Developmental Biology (IGDB), Chinese Academy of Sciences (CAS), Beijing, China.

Abbreviations

ABEs	Adenine base editors
BLESS	Breaks labeling and enrichment on streptavidin and sequencing
Bt	*Bacillus thuringiensis*
Cas	CRISPR-associated protein
CBEs	Cytosine base editors
CBP	Cartagena biosafety protocol
CDE	CRISPR-mediated directed evolution
CEO	CRISPR-edited organisms
CRE	Cis-regulatory elements
CRISPR	Clustered regularly interspaced short palindromic repeats
CRISPRi	Cluster interspaced specific short palindromic repeats interference
crRNAs	CRISPR ribonucleic acid
CTLs	Complex trait loci
DSB	Double strand break
dsRNA	Double-stranded RNA
EPSPS	5-Enolpyruvylshikimate-3-phosphate
EU	European Union
FAO	Food and Agriculture Organization
FSANZ	Food Standards Australia New Zealand
GDP	Gross domestic product
GE	Genome editing
GMOs	Genetically modified organisms
gRNAs	Guide RNA
HDR	Homology directed repair
HRM	High resolution melt
HTGTS	High-throughput genome-wide translocation sequencing
ICPs	Insecticidal crystalline proteins
IDLV	Integrase-deficient lentiviral vectors
indel	Insertion or deletion
IR	Insect-resistant
KI	Knock in
KO	Knockout
LEA	Late embryogenesis abundant

NHEJ	Non-homologous end joining
NPBT	New plant-breeding technique
PAGE	Polyacrylamide gel electrophoresis
PAM	Protospacer adjacent motif
PEG	Polyethylene glycol
PNT	Plant novel traits
rDNA	Recombinant DNA
RGEN	RNA-guided engineered nucleases
RISC	RNAi-induced silencing complex
RNAi	RNA interference
RNPs	Ribonucleoproteins
SDNs	Site-directed nucleases
siRNA	Short interfering RNA
SIT	Sterile insect technique
TALENs	Transcription activator-like effector nucleases
TDNA	Transfer DNA
TET	Ten-eleven translocation
TRV	Tobacco rattle virus
UAS	Upstream activating system
US	United States
WHO	World Health Organization
ZFNs	Zinc finger nucleases

CRISPR/Cas System: An Introduction

1

Nayla Munawar and Aftab Ahmad

Abstract

The process of altering genetic code to modify DNA precisely and efficiently within a cell is known as genome editing. All techniques used for changing a gene in a cell are based on cutting of double-stranded DNA through DNA endonucleases at specific site. DNA nucleases used for genome editing are composed of programmable DNA binding domains to bind with target sequence followed by DNA cleavage domain to break double-stranded DNA that excites error prone non-homologous end joining (NHEJ) or homology-directed repair (HDR) at specific genome locations. Programmable nucleases enabled researchers to manipulate practically any genomic sequence, provided opportunities to create cell lines and animal models to study human diseases, and also promoting new possibilities of treating human diseases by gene therapy. Four genome editing technologies, meganucleases, zinc-finger nucleases (ZFNs), transcription activator-like effector nucleases (TALENs), and clustered regularly interspaced short palindromic repeats (CRISPR/Cas9) based on the above-mentioned principle have been developed so far and have been used successfully for the correction of disease-causing mutations, for addition of therapeutic gene and the deletion of specific genes from the specific sites in the genome. Genome editing revolutionized the scientific world and have intense impact on various fields of biotechnology, such as biopharmaceutical production, agriculture, creation of transgenic organisms and cell lines, regulation and function of the genome, etc. All four genome editing techniques, their history, principles,

N. Munawar (✉)
Department of Chemistry, United Arab Emirates University, Al-Ain, UAE

A. Ahmad
Center for Advanced Studies in Agriculture and Food Security (CASAFS), University of Agriculture, Faisalabad, Pakistan
e-mail: aftab.ahmad@uaf.edu.pk

A. Ahmad et al. (eds.), *CRISPR Crops*,
https://doi.org/10.1007/978-981-15-7142-8_1

advantages, disadvantages, and applications, with a special focus on CRISPR/Cas, have been discussed in this chapter.

1.1 Genome Editing: An Overview

The alteration of specific DNA sequences in the cellular genome, referred to as genome editing, has gained popularity in recent years. Genome editing can alter the phenotypic characteristics of cells.

The discovery of the DNA double-helix structure in the mid-twentieth century and then an improved understanding of the transfer of genetic material from parents to off-spring have led researchers to conclude that minor changes in the DNA sequence can cause disease. The term "genome" refers to the complete DNA sequence of an organism and can be altered by intrinsic factors such as DNA replication errors during cell division or environmental factors such as UV radiation from the sun. Moreover, DNA mutations can be acquired from parents and can be present in virtually every cell in the body. Most DNA variations do not result in an abnormal protein and are called "polymorphisms." These DNA polymorphisms can lead to diverse phenotypic characteristics, such as differences in hair color and type, eye color, and blood type, and do not affect the health of an individual. However, some DNA variations can increase the risk for certain disorders and may cause serious health issues.

The identification of disease-causing "DNA errors" has led researchers to develop strategies to prevent or reverse genetic diseases. On the basis of this notion, the concept of gene editing or trait alteration was introduced in the late 1980s and later led to the discovery of different genome editing technologies. The first genome editing study to target specific genes by homologous recombination (HR) in mouse embryonic stem cells to generate knockin and knockout cells was reported by Capecchi et al. in 1989. While the process was revolutionary, it was highly inefficient as it required extensive selection and screening to identify the one-in-a-million cell expressing the modified gene.

Although, the recombination frequency was low because of the low rate of HR in mammalian cells, this proof-of-concept study suggested that it was possible to target a specific gene with precision. The method to achieve targeted DNA modifications using a sequence-specific nuclease was first introduced in the 1990s. The Cre-lox system comprising a site-specific DNA endonuclease Cre, which recognizes the 34-nt locus *loxP*, to generate double-stranded DNA breaks (DSBs) and facilitate subsequent recombination at these sites was developed in the early 1990s. This technology led the successful knockout of any desired gene and was used as the most effective gene editing tool during the time to develop transgenic mouse models. While the Cre-lox system provided an opportunity to control gene expression more easily than HR, the efficiency of the system was low because of the large genetic distance between the *loxP* sites.

After developing this successful genome editing technique, the search to develop a reliable method to generate a DSB at any desired location for targeted editing began. Subsequently, zinc-finger nucleases (ZFNs) consisting of a DNA binding domain and a nuclease domain, were used for the first time to incorporate changes in the genome of *Drosophila melanogaster* fruit fly in 2002 and to repair a mutation in the human genome in 2005. Despite being time-consuming and laborious, this technique is being employed in a growing number of applications currently. Almost a decade ago, another genome editing technique called transcription activator-like effector nucleases (TALENs) was developed in 2010 and successfully used for genome modification in more than 40 different organisms and cell types. However, limitations of both ZFNs and TALENs propelled the search for alternative gene targeting technologies. In 2012, the most efficient and widely used genome editing technology of clustered regularly interspaced short palindromic repeats (CRISPR)/ Cas associated system 9 (CRISPR/Cas9) was discovered and proved to be the most reliable genome editing technique so far.

1.2 Techniques for Genome Editing

The introduction of site-specific DSBs is a prerequisite for genome editing. Hence, by using the intrinsic DNA repair machinery of cellular organisms, tools that generate DNA DSBs can be used to precisely alter the genome. After the success of Cre-lox DNA endonuclease-based gene editing technology in 1992, researchers began to identify nucleases (enzymes that cleave nucleic acids) that can generate DNA DSBs as well as methods to direct these nucleases to specific regions of the genome.

Eventually, four genome editing techniques were established over the following 20 years. These techniques use designer/programmable nucleases to generate DSBs and employ the cellular DNA repair mechanism to precisely modify genomic sequences (Voytas 2013). The foremost characteristics of a successful genome editing technique are (1) target selectivity (nucleotide sequences, epigenetic markers), (2) precision for the target, and (3) regulation of its activity. The following four major classes of customizable DNA binding proteins currently exist for genome editing, and their popularity, applicability, effectiveness, and pitfalls can be determined on the basis of the above-mentioned factors.

1. Meganucleases
2. ZFNs
3. TALENs
4. CRISPR/Cas

1.2.1 Meganucleases

Meganucleases (MNs), also known as homing nucleases, are derived from microbial mobile genetic elements (Smith et al. 2006). These are sequence-specific endonucleases with a large restriction site (14–25 nt) and are classified into five families based on their sequences and structure motifs: (1) LAGLIDADG, (2) GIY-YIG, (3) HNH, (4) His-Cys box, and (5) PD-(D/E)XK (Orlowski et al. 2007; Zhao et al. 2007). The LAGLIDADG homing nuclease (LHE family) is the most abundant form and is found in Archaea, Bacteria, Eukarya, and their respective viruses (Stoddard 2005). These meganucleases have been extensively studied and are often termed as "selfish genetic elements" because their role in the host is not yet known. The LHE family endonucleases exist in both homo-dimeric and monomeric forms and can bind to target DNA sequences of up to 24 nt in length. Because meganucleases can recognize specific DNA sequences through protein–DNA interactions, efforts were made to manipulate LAGLIDADG protein–DNA interactions. However, several studies have shown that related proteins can use different subsets of residues to recognize similar DNA sequences (Chevalier et al. 2003; Lucas et al. 2001). Hence, it was difficult to re-engineer the specificity of these meganucleases. Nevertheless, site-specific cleavage of a maize genomic sequence (Gao et al. 2010) and two human genes *XPC* and *RAG1* (Smith et al. 2006; Arnould et al. 2007) by engineered meganucleases has been reported in literature.

Despite being the only nuclease naturally evolved for genome editing, it has been difficult to modify meganucleases to specifically recognize target DNA sequences. Thus, they have not achieved widespread acceptance as a genome editing technique. Moreover, the high probability of off-target binding is another limitation of this technique (Argast et al. 1998). However, a minor advantage of meganucleases is their small size (approximately 40 kDa). They can be used in viral vectors where short coding sequences are required. Nevertheless, because of their inability to predict and modify target DNA sequences, meganucleases are less preferred than other genome editing techniques, including ZFNs and TALENs, which are more flexible with regard to re-targeting the reagent to different sequences using a distinct DNA-cutting domain that can be non-specific and a specific programmable DNA binding domain.

1.2.2 Zinc-Finger Nucleases (ZFNs)

ZFNs, a genome editing tool, is based on eukaryotic transcription factors (Urnov et al. 2010; Miller et al. 2011). Zinc fingers are small peptide domains in which a zinc ion is coordinated in a tetrahedral geometry to provide structural stability. These domains bind to specific DNA sequences in a precise manner. The currently used ZFNs consist of restriction endonucleases and a zinc-finger DNA binding domain to target specific DNA sequences within the genome. The natural type IIS restriction enzyme *Fok*I was first discovered by Chandrasegaran (Chandrasegaran and Carroll 2016), and consists of distinct DNA binding and cleavage domains. Researchers

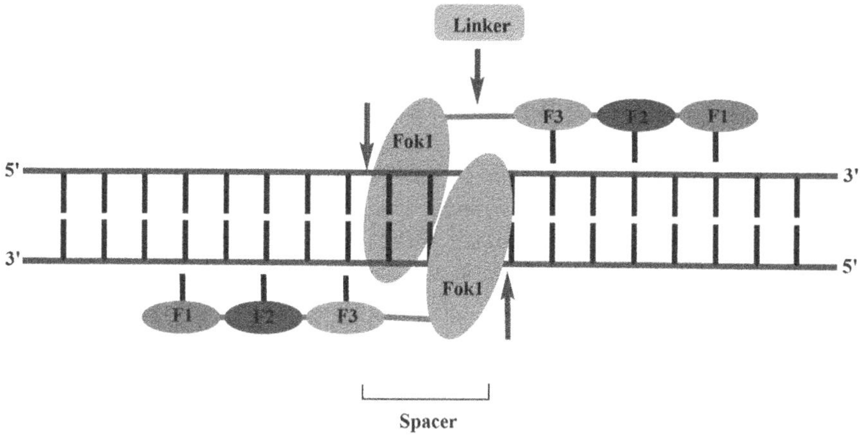

Fig. 1.1 Mechanism of ZFNs: An illustration of a zinc-finger nuclease (ZFN). A ZFN consists of three to six zinc-finger protein monomers and the *Fok*I restriction enzyme, which cleaves DNA upon dimerization. Each ZFP recognizes a 3-nt target DNA sequence

cleaved the DNA binding domain of *Fok*I and fused its DNA cleavage domain with the zinc-finger motif to construct a distinct DNA recognition and cleavage tool called ZFN (Kim et al. 1996). The association of the zinc-finger DNA binding domain with the DNA cleavage domain enables the enzymatic machinery to target a unique locus in the genome and provoke endogenous DNA repair. The mechanism of action of this technique is shown in Fig. 1.1. The DNA binding domain identifies and binds to a specific DNA sequence. Thereafter, *Fok*I is recruited to the DNA sequence to induce a DSB. To cleave a specific site in the genome, a pair of ZFNs is designed to recognize two sequences flanking the site, one on the forward strand and the other on the reverse strand. Each DNA binding domain of ZFN identifies three nucleotides on either side of the DNA strand. Upon binding to the DNA strand, the *Fok*I domains dimerize and cleave the DNA at the site to generate a DSB, which could be repaired either by non-homologous end joining (NHEJ) or homology-directed repair (HDR) mechanism (Fig. 1.5).

Because each ZFN can recognize a 3-nt DNA sequence, it is possible to assemble 6–7 unique ZFNs to specifically target 18–21 nt in the genome. Hence, sequence-specific chimeric nucleases can be constructed feasibly. Because of its ease of application and high specificity, this technique has gained widespread popularity and has been adopted by many researchers for different applications ranging from genome modification to gene therapy in both animals and plants (Townsend et al. 2009; Shukla et al. 2009; Urnov et al. 2010; Mashimo et al. 2010; Kim et al. 2009). Theoretically, it is possible to target any gene in any organism using this technique. However, the assembly of functional ZFPs with the desired DNA binding specificity is a major limitation of this technique because it requires an extensive screening process. The binding of zinc-finger domains to an extended stretch of nucleotides with high affinity is difficult to achieve (Ramirez et al. 2008). Moreover, the requirement to assemble zinc fingers in new combinations for each new target DNA sequence made it difficult for non-specialists to routinely engineer ZFNs. To overcome this difficulty, an academic consortium developed an open source library of zinc-finger components and protocols to perform screening to identify ZFNs that bind with high affinity to a desired sequence. Nonetheless, it can take months for non-specialists to construct optimized ZFNs (Maeder et al. 2008). Taken together, the construction of site-specific zinc-finger modules is a rate-limiting step in utilizing this technology for genome editing. It is partially due to this reason, ZFNs were not excessively used for genome editing by the scientific community.

1.2.3 Transcription Activator-Like Effector Nucleases (TALENs)

Although, artificially designed ZFNs generated enormous excitement in the field of genome editing, the discovery of TALENs in 2011 was recognized in "Nature Methods" as the "method of precise genome editing." TALENs are based on transcription activator-like effector proteins from plant pathogenic bacteria of *Xanthomonas* spp. (Christian et al. 2010; Miller et al. 2011; Boch et al. 2009; Moscou and Bogdanove 2009). The endogenous function of these proteins is to activate specific host genes by mimicking eukaryotic transcription factors to facilitate the growth of the pathogen (Schornack et al. 2006). These effector proteins contain a DNA binding domain consisting of monomers, with each monomer bound to one nucleotide in the target DNA sequence (Römer et al. 2007). The monomers are tandem repeats of 34 amino acids with two highly variable amino acids at positions 12 and 13 in each repeat. These residues have been termed as repeat variable di-residues (RVDs) and are responsible for DNA specificity of TALE proteins. The DNA specificity of TAL effector proteins can be modulated by manipulating these RVDs. To synthesize artificial DNA binding modules of TALE proteins, monomers containing RVDs, such as Asn and Ile (NI), Asn and Gly (NG), two Asn (NN), and His and Asp (HD) for binding the nucleotides A, T, G, and C, respectively, have been utilized (Nemudryi et al. 2014). The construction of artificial TALENs and their mode of action resemble that of ZFNs (Fig. 1.2). However, unlike ZFNs, the DNA binding domain of TALENs recognize only one nucleotide, thereby enhancing the site specificity with fewer off-target effects (Chandrasegaran and Carroll 2016). Moreover, the chimeric fusion of the *Fok*I nuclease domain with a combination of TALE modules resulted in a robust and effective programmable genome editing tool.

TALENs offer two distinct advantages for genome editing over ZFNs: (1) no requirement of selection or directed evolution to engineer TALE arrays, resulting in a considerable reduction in the amount of time and experience needed to assemble a functional nuclease, (2) provide higher specificity and reduced toxicity than ZFNs because of their increased affinity for target DNA (Meckler et al. 2013) or a greater energetic penalty for association with base mismatches (Mussolino et al. 2014). On the other hand, the construction of a site-specific DNA binding module is challenging because of high similarity between TALE recognition sequences; this restricts

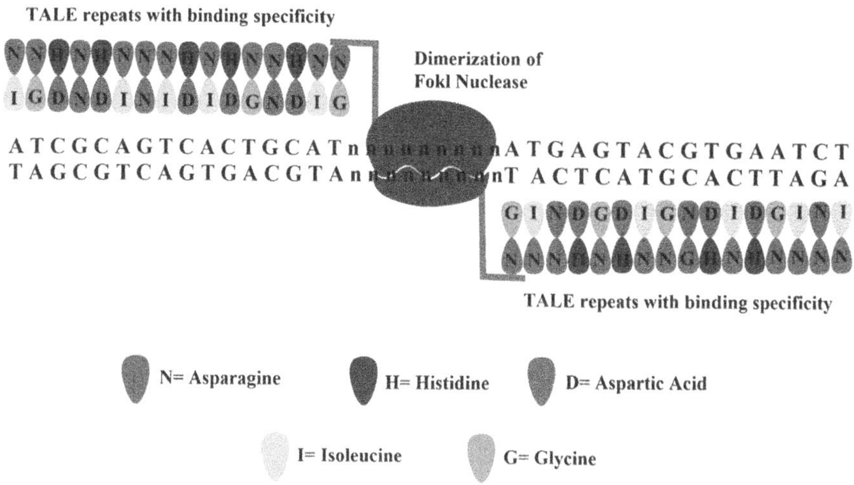

Fig. 1.2 Mechanism of action of TALENs: An illustration of a transcription activator-like effector nuclease (TALEN) showing left and right monomers of TALE proteins and the *Fok*I restriction enzyme, which cleaves DNA upon dimerization. Each TALE protein recognizes a single DNA base pair

the wide use of TALENs in genome engineering. Moreover, the requirement of the presence of thymine (T) at start of the TALEN binding sequence, specific target-site length, and a spacer between two TALEN arms for the formation of the *Fok*I dimer negatively affect its routine application. Attempts are undertaken to reduce the limitations of TALENs and improve their efficacy by adapting approaches, such as using *Fok*I variants and nickases to reduce off-target binding (Miller et al. 2011), using adenoviral vectors for TALEN delivery to difficult-to-transfect cell types (Holkers et al. 2014; Maggio et al. 2016), and constructing a library of TALENs targeting 18,740 human protein-coding genes to facilitate new studies (Kim et al. 2013). However, the repetitive sequences in TALENs render the construction of novel TALE arrays labor extensive and costly. While the engineering of modular DNA binding proteins is challenging, the discovery of the CRISPR-Cas methodology with a novel mode of DNA recognition simplifies the development of custom nucleases.

1.2.4 Clustered Regularly Interspaced Short Palindromic Repeats (CRISPR)/CRISPR-Associated Protein 9 (Cas9) System

Most recently, the CRISPR/Cas system has been reported as an efficient and facile alternative of ZFNs and TALENs for inducing targeted genetic modifications. The microbial adaptive immune system CRISPR can target any genome location of choice using a synthetic short guide RNA (sgRNA) (Jinek et al. 2012).

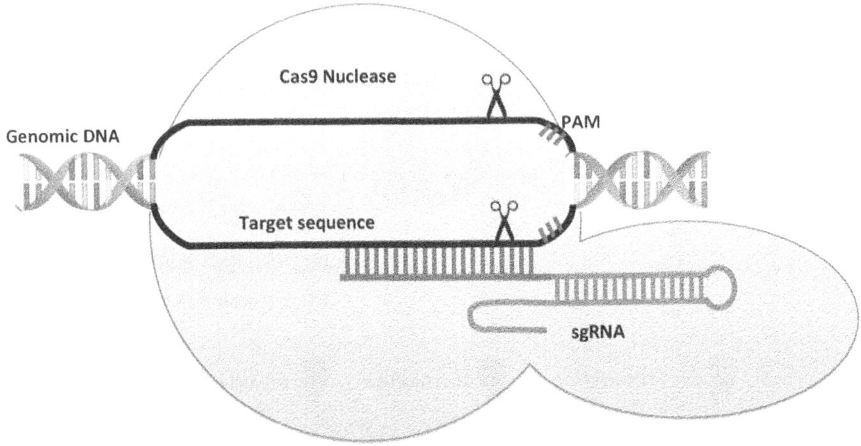

Fig. 1.3 General structure of CRISPR-Cas locus with all components. Cas9 indicated in blue color. Protospacer adjacent motif (PAM) "NGG" for Cas9 shown in red color. Pairing of 20 nt of sgRNA (purple color) with target site in genome

The general structure of the CRISPR-Cas locus is shown in Fig. 1.3. The locus harbors an array of short repetitive sequences (repeats) interspaced by unique, short non-repetitive sequences (spacers) that originate from mobile genetic elements (MGEs), such as bacteriophages and transposons, followed by an AT-rich leader sequence, and a set of diverse CRISPR-associated *cas* genes that encode Cas proteins (Barrangou et al. 2007; Haft et al. 2005; Makarova et al. 2006; Mojica et al. 2005).

1.2.5 Mode of Action

The general mechanism of action of the CRISPR/Cas adaptive immunity system, regardless of its different types, is divided into three steps: (1) adaptation, (2) expression and maturation, and (3) interference.

These three stages are shown in Fig. 1.4. In adaptation, distinct short DNA fragments from the invading pathogen (known as protospacers) are recognized by the Cas protein and integrated into CRISPR repeats as a new spacer, which then serves as a genetic record to develop immunological memory in the host and enables it to recognize same invading pathogens in the future. Several CRISPR/Cas systems require a short 3–5 nt protospacer adjacent motif (PAM) in the target DNA to acquire the protospacer (Deveau et al. 2008; Shah et al. 2013; Bolotin et al. 2005).

Expression and maturation involves the transcription of the CRISPR array into a precursor-CRISPR RNA (pre-RNA), which is further processed to yield short mature CRISPR RNA (crRNA) through endo-nucleolytic cleavage and contain the memorized sequences of the invading pathogen (Carte et al. 2008; Haurwitz et al.

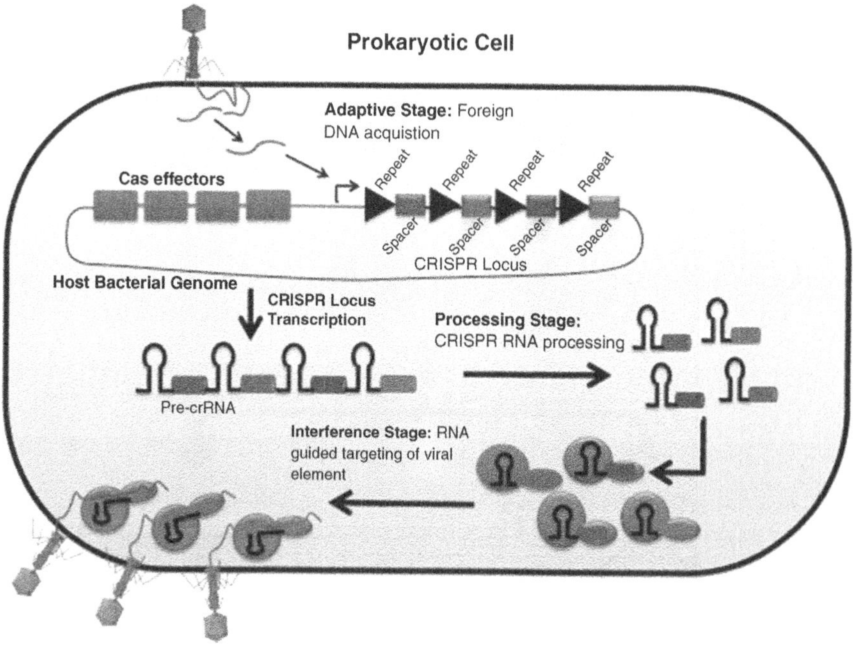

Fig. 1.4 General mechanism of action of the CRISPR-Cas system. Adaptive stage: repeats and spacer are expressed as pre-crRNAs. Processing stage: pre-crRNAs are processed into mature crRNA.Interference stage: mature crRNA along with Cas nuclease (Cas9) binds with target site in phage genome and creates DSBs

2010). Each crRNA contains a single spacer (short RNA segment, complementary to the DNA sequence of the foreign genetic element) at its 5′ end and a CRISPR repeat sequence at its 3′ end. The mature crRNA associates with one or more Cas effector proteins to form an active Cas-crRNA effector complex.

During the interference stage, this effector complex scans and targets the foreign nucleic acids in the cell. The crRNA part of the complex functions as a guide to recognize target DNA using the specific PAM sequence upstream or downstream of the protospacer (Fonfara et al. 2016; Zetsche et al. 2015; Fineran et al. 2014; Jiang et al. 2013; Semenova et al. 2011; Westra et al. 2013) through Watson–Crick base pairing. The successful recognition of the target DNA leads to its cleavage and destruction by Cas nuclease.

In recent years, diverse classes of the CRISPR immune system based on different effector Cas proteins and PAM recognition sequences have been identified in various microorganisms. These classes have been described in detail below in Sect. 1.3.2.

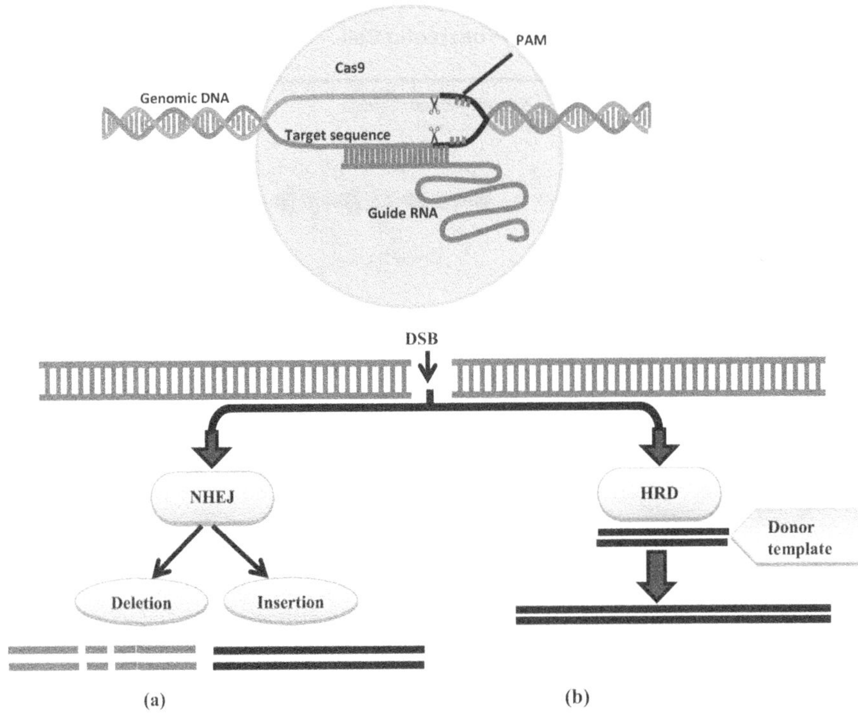

Fig. 1.5 Mechanisms of gene editing: DNA double strand breaks (DSBs) can be repaired by two mechanisms: (**a**) Non-homologous end joining (NHEJ), which occurs in the absence of a donor template and introduces small base insertions or deletions that can result in gene disruption. (**b**) Homology-directed repair (HDR), which occurs in the presence of a homologous donor template

1.3 Molecular Mechanism of Genome Editing

All modern genome editing techniques rely on the principle of inducing a site-specific DNA DSB that can be subsequently repaired using inherent cellular repair mechanisms, such as HDR or the error prone NHEJ pathway (Shalem et al. 2015). These two pathways of DNA repair in the cell permit the insertion, deletion, or mutation of nucleotides in the target DNA, resulting in incorporating desired changes in the genome. Although the NHEJ repair mechanism is error prone and often results in variable lengths of insertion and deletion (indel) variants, it can be used for gene knockout (Fig. 1.5a). On the other hand, the HDR mechanism is error-free because it involves recombination between homologous DNA sequences obtained from the undamaged chromatid of the cell or if engineered, can use an extrinsic homologous donor DNA template to alter the genome (Fig. 1.5b).

In plants, DSBs are mainly repaired by NHEJ. In this process, several enzymes are used to re-anneal the broken ends of DSBs without the requirement of a homologous DNA template (Puchta 2005). NHEJ occurs throughout the cell cycle in higher eukaryotes and exhibits low fidelity repair. Because of its error prone nature, NHEJ repair often leads to the addition or deletion of nucleotides and causes alterations at targeted DNA DSB sites. In many cases, the indels introduced in exons during NHEJ repair can lead to missense or nonsense mutations, and in some cases, result in a complete loss of gene function. A majority of the published articles on plant genome editing has used the NHEJ pathway to knockout genes (Mladenov and Iliakis 2011).

In the error-free HDR pathway, a homologous sequence serves as a template to repair the DSBs. The HDR pathway can be used to precisely modify nucleotide sequences and perform gene replacement or insertion at target loci in the presence of an exogenous donor DNA as a repair template. Unlike the NHEJ pathway, the HDR pathway occurs mainly during the S and G2 phases of the cell cycle and has low efficiency in higher eukaryotes. Only one in a million treated cells undergoes the desired genome modification (Capecchi 1989). However, the balance between HDR and NHEJ pathways can be shifted towards HDR by inactivating some components of the NHEJ pathway (Beumer et al. 2008, 2013; Bozas et al. 2009). The knockout of DNA ligase IV increased HDR frequency from 20% to 65% in *Drosophila*. This result suggests the possibility of limiting a specific type of repair mechanism in the cell; however, a common method to enhance HDR does not exist because of different cell types and overall complexity of the repair mechanism in the cell (Chandrasegaran and Carroll 2016).

DSBs are resolved by the NHEJ pathway in the absence of an intrinsic donor DNA sequence. This may result in insertion or deletion (indels) and ultimately gene knockout. On the other hand, DSBs are repaired by the HDR pathway in the presence of an intrinsic or extrinsic donor DNA sequence, ultimately resulting in gene knockin (Bibikova et al. 2002). Overall, the combination of these two repair mechanisms is effective for modifying the eukaryotic genome (Hsu et al. 2014). Genome editing systems use sequence-specific nucleases (SSNs) that consist of a DNA binding domain to provide sequence specificity and a nuclease domain to introduce DNA DSBs at the targeted site (Puchta 2005). These DSBs can be repaired by either of the above-mentioned repair mechanisms.

1.3.1 A Bacterial and Archaeal Immune System as a Genome Editing Tool in Eukaryotes

All living organisms are continuously exposed to infectious agents such as viruses and transposons. Consequently, the host organisms have developed two mechanisms to defend themselves from these pathogens: (1) innate immunity and (2) adaptive immunity. Innate immunity is naturally present in the genetic system of the host and is responsible for the recognition of the invading pathogen and provides non-specific protection. On the other hand, adaptive immunity is characterized by mounting a specific immune response against the invading pathogen. The immune system memorizes the antigenic properties of the invading pathogen to provide protection to the host upon re-exposure. Adaptive immunity was presumed to be a characteristic only of eukaryotes until the previous decade. However, the discovery of CRISPR loci and CRISPR-associated proteins (Cas) in both bacteria and archaea demonstrated the presence of the adaptive immune system in prokaryotes as well (Mojica et al. 2000).

Approximately 50% and 90% of the sequenced genomes of bacteria and archaea, respectively, possess adaptive immunity in the form of the CRISPR/Cas system (Van der Oost et al. 2009). This system consists of a "memory chip" and "DNA scissors" in the form of the CRISPR loci. As mentioned in Sect. 1.2.4, the CRISPR loci store the immunological memory of an invading pathogen in the form of a "spacer," a short DNA sequence of the invading pathogen. The second essential component is the variable Cas gene cassette. This cassette is typically located adjacent to a CRISPR locus and produces Cas proteins, which are essential elements of the adaptive immunity system. The defensive role of CRISPR/Cas against pathogens has been well established since its discovery (Barrangou et al. 2007; Brouns et al. 2008; Sapranauskas et al. 2011). Moreover, CRISPR/Cas proteins are involved in other cellular functions such as gene regulation, inhibition of biofilm formation, multicellular development, and virulence (Sampson et al. 2013; Zegans et al. 2009; Vercoe et al. 2013). However, the precise under lying mechanisms of these functions remain unclear (Rath et al. 2015). After identifying the CRISPR/Cas system as an adaptive immune mechanism in various microorganisms, several attempts were undertaken by researchers to understand the mechanism of action of these modules in the cell (Marraffini and Sontheimer 2008; Plagens et al. 2015). These studies identified of distinct classes of the CRISPR/Cas system (Haft et al. 2005) and paved the way for these systems to be used an effective genome editing technology (Jinek et al. 2012; Cong et al. 2013; Mali et al. 2013).

1.3.2 Classification of the CRISPR/Cas System

1.3.2.1 Classification

The CRISPR-Cas modules (loci) have been identified as an as adaptive immune mechanism in archaea and bacteria and provide sequence-specific protection against foreign DNA or RNA. The CRISPR/Cas system is classified into two broad classes based on variability in the CRISPR loci corresponding to these modules in different organisms: "Class 1" and "Class 2," which are further subdivided into types and subtypes on the basis of comparative genomic analysis, structure and biochemical activities of CRISPR components (Koonin et al. 2017), domain architecture of effector proteins, and PAM sequence identification. The major difference between the two classes lies in the type of the crRNA-effector complexes. In class 1 systems, the crRNA-effector complex exists as a multi-subunit complex, whereas in class 2 systems, a single protein such as Cas9 performs all functions of the effector complex. Moreover, class 2 systems are less complex than class 1 systems.

The remarkable variability in the genomic architecture of CRISPR-Cas loci because of the constant threat to adaptive immunity from viruses and foreign plasmids poses a major challenge for the consistent classification of CRISPR/Cas systems. However, a consistent classification is essential for a robust characterization of CRISPR-Cas loci in new genomes and to advance research in the field. A strategy to combine signature genes with elements of the *Cas* loci was proposed in 2015 to assign nearly all of the detected CRISPR-Cas loci to specific subtypes. This classification strategy is compatible with the results of sequence-based clustering of crRNA-effector complexes and can be adopted for the classification of newly discovered CRISPR/Cas variants from new genomes.

The class 1 CRISPR/Cas system is categorized into three types: type I, type III, and type IV, which are further divided into subtypes. Similarly, the class 2 CRISPR/Cas system is categorized into three types, which are further divided into subtypes (Fig. 1.6). The two principle modules of the CRISPR/Cas system include the adaptation module, which is responsible for spacer acquisition and the effector module, which involves pre-crRNA processing, target recognition, and cleavage. The adaptation module is largely uniform across the CRISPR/Cas system and consists of endonucleases and the structural subunits Cas1 and Cas2 (Amitai and Sorek 2016). However, the variability of effector modules is high between CRISPR types and subtypes.

Within class 1, type I and type III CRISPR/Cas systems are present most commonly in archaea and less frequently in bacteria. Type IV CRISPR/Cas system is rare and lacks the adaptation module. The type I effector complex (CRISPR-associated complex for antiviral defense) consists of a 61-nt mature crRNA and five Cas proteins with an irregular stoichiometry.

Type I and type III CRISPR/Cas systems possess a complex architecture, with a long backbone of repeat-associated mysterious proteins (RAMPs), including Cas5 and Cas7. These proteins contain an RNA recognition motif (RRM) fold and additional large and small subunits (Zhao et al. 2014; Van Der Oost et al. 2014; Jackson et al. 2014; Jackson and Wiedenheft 2015; Hochstrasser et al. 2014, 2016; Staals et al. 2013, 2014). The effector complexes, which accommodate the guide RNA, consist of one Cas5 and multiple Cas7 subunits. The Cas5 subunit binds to 5′-crRNA and interacts with the large subunit of Cas8 in type I and Cas10 in type III. Cas6 is loosely associated with the effector complex and functions as a repeat-specific RNase in pre-crRNA processing (Charpentier et al. 2015; Niewoehner and Jinek 2016). Therefore, the type I CRISPR/Cas system utilizes Cas5 or Cas6 for pre-processing of crRNA; further cleavage requires Cas3 nucleases, cascade, and crRNA for interference (Khan 2019). Different subunits of the type I system require distinct PAM sequences for target acquisition and recognition (Fig. 1.6). Type I-A system found in *Sulfolobus islandicus* and *Sulfolobus solfataricus* requires a 5′-CNN-3′ PAM motif for interference (Gudbergsdottir et al. 2011; Lillestøl et al. 2009). Type I-B system established in haloarchaea recognizes six distinct PAMs (Fig. 1.6). Type I-C system—characterized by a Cas5-dependent crRNA maturation pathway (Nam et al. 2012)—identifies an "NTTC" consensus PAM sequence in *Bacillus halodurans* (Sorek et al. 2013). Type I-E effector complex from *E. coli* and *Thermobifida fusca* bind to 5′-AAG-3′ PAM sequence (Xiao et al. 2017). Type I-F effector complex from *Pseudomonas aeruginosa* consists of four Cas proteins instead of five protein found in type I-E, and targets foreign DNA using a PAM

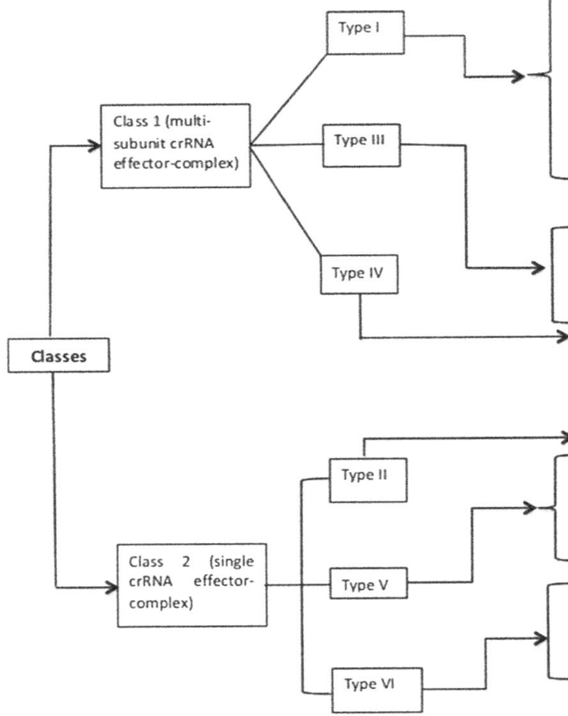

Subtype	Effectors/signature genes	PAM/PFS
I-A	Cas8a2, Cas5	5′-CCN- 3′
I-B	Cas8b	5′-TTC- 3′, 5′-ACT- 3′, 5′ -TAA- 3′, 5′ -TAT- 3′, 5′-TAG- 3′, 5′-CAC- 3′
I-C	Cas8c	5′-NTTC- 3′
I-U	Cas8u2	NA
I-D	Cas10d	NA
I-E	Cse1, Cse2	5′-AAG- 3′ (21 PAM sequences allow acquisition of new spacer)
I-F	Csy1, Csy2, Csy3, Cas6f	Two consecutive G-C base pairs
I-Fv	Csy1, Csy2	

Subtype	Effectors/signature genes	PAM/PFS
III-A	Cas10, Cas5, Cas7,Cas6	NR*
III-B	Cas10, Cas5, Cas7, Cas6	NR
III-Bv	Cas10, Cas5, Cas7, Cas6	NR
III-C	Cas10, Cas5, Cas7	NR
III-D	Cas10, Cas5, Cas7	NR
IV	DinG (Csf4)	NR

Subtype	Effectors/signature genes	PAM/PFS
II-A,B,C	Cas9	5′-NGG- 3′
V-A	Cas12a (cpf1)	5′-TTTN- 3′
V-B	Cas12b (c2c1)	5′-TTT- 3′, 5′-TTA- 3′, 5′-TTC- 3′
V-C	Cas12c (c2c3)	5′-AT rich PAM
V-D	CasY (Cas12d)	5′-AT rich PAM
V-E	CasX (Cas12e)	5′-TTCN- 3′
VI-A	Cas13a	3′ PFS: non-G
VI-B1	Cas13b	5′ PFS: non-C; 3′ PFS: NAN/NNA
VI-B2	Cas13b	5′ PFS: non-C; 3′ PFS: NAN/NNA
VI-C	Cas13c	NA

Fig. 1.6 Experimentally characterized classes and types of CRISPR system consisting of different effectors and PAM sequences. For each CRISPR/Cas subtype, signature genes and PAM/PFS sequences are shown. NR is representing "not required" and NA indicating "not available"

element consisting of two consecutive G-C base pairs (Rollins et al. 2015). Type I-Fv is a variant of type I-F and was discovered in *Shewanella putrefaciens*. The PAM recognition elements are absent in this variant due to the lack of small and large cascade subunits (Rollins et al. 2015).

Type III CRISPR system is largely found in archaea and some bacteria. It has been divided into four subtypes based on the differences in their adaptation, interference, and recognition strategies. Whereas type III-A system possesses an adaptation module, type III-B, -C, and -D systems lack the adaptation genes; hence, these systems depend on the other systems to incorporate new spacer. Similar to the type I system, interference in the type III system is performed by a crRNA-guided multiprotein complex called Csm/Cmr complex (Rath et al. 2015). The Csm complex is present in A and D subtypes, while the Crm complex is present in B and C subtypes. Similar to the type I system, the type III system uses Cas6 for crRNA processing. Because of the scarcity of structural data, PAM recognition sequence for many subtypes of the class 1 CRISPR system is not yet known. A rare type IV CRISPR system includes an elementary CRISPR-Cas locus that lacks the adaptation domain

The effector module in the class 2 CRISPR/Cas system consists of a single, large, multidomain protein. Hence, the CRISPR-Cas loci are organized more uniformly in the class 2 system than in class 1 system. The class 2 system is also categorized into three types: II, V, and VI. Type II is the most studied and well-characterized system among all types of class 2 systems and consists of the effector Cas9 endonuclease that is widely used in genome editing. In this system, the recruitment of Cas9 to target DNA is regulated by crRNA. However, the mechanism by which RNase III, trans-activating RNA (tracrRNA), and protein factors perform the 5′ end processing of crRNA is not yet determined.

The type V CRISPR/Cas system consists of a single effector protein Cas12 (Dong et al. 2016; Yang et al. 2016; Yamano et al. 2016), and is categorized into five known subtypes A-E and a predicted U subtype. Subtype A of the type V (type V-A) CRISPR system consists of Cas12a (Cpf1), an active RNA-guided endonuclease that does not require additional tracrRNA for target cleavage (Zetsche et al. 2015). This subtype uses the "TTN" PAM sequence for target recognition (Gleditzsch et al. 2019). The type V-B system utilizes both tracrRNAs and crRNAs to target dsDNA using a T-rich PAM sequence such as TTT, TTA, or TTC. Other variant systems such as type-VD and -VE consist of CasY (Cas12d) and CasX (Cas12e) signature proteins, respectively. As opposed to CasY, CasX requires tracrRNA for interference. Moreover, while CasY employs a "TA" PAM sequence for target recognition, CasX employs a 5′-TTCN-3′ PAM sequence (Burstein et al. 2017).

All subtypes of the type IV system consist of HEPN domains (higher eukaryotes and prokaryotes nucleotide-binding domains). These subtypes are predicted to possess RNase activity instead of DNA cleavage activity. Moreover, the activity of the type VI system relies on the presence of a protospacer flanking site (PFS), a PAM analogue for RNA targets (Gleditzsch et al. 2019). Type VI-A and -B systems rely on Cas13a and Cas13b effectors, respectively, for their interference activity.

The diversity of the CRISPR-Cas systems in the cell is caused by the presence of novel effector proteins and new molecular mechanisms. Bacteria and archaea possess different CRISPR systems. Type II systems are exclusively found in bacteria, whereas type III systems are more common in archaea. These findings have helped researchers to utilize the unique characteristics of CRISPR systems to design novel and robust technologies for genetic modifications.

Cas12 is frequently used for genome editing due to its relatively small size, no requirement for tracrRNA, and asymmetric cleavage sites. Strategies to expand the range of Cas12 targets by designing different protein variants with different PAM specificities are recently being tested (Gao et al. 2017).

1.3.3 History of CRISPR/Cas

The CRISPR/Cas system was developed over 20 years, across 12 cities, in 9 countries (Fig. 1.7). Briefly, this system was first reported by Franscisco Mojica in 1993, who began his doctoral studies at University of Alicante in the Mediterranean port of Santa Pola on Spain's Costa Blanca in 1989. Mojica found multiple copies of a near-perfect, roughly palindromic, repeat sequence of 30 nt, separated by spacers of approximately 36 nt that did not resemble any family of repeats known in

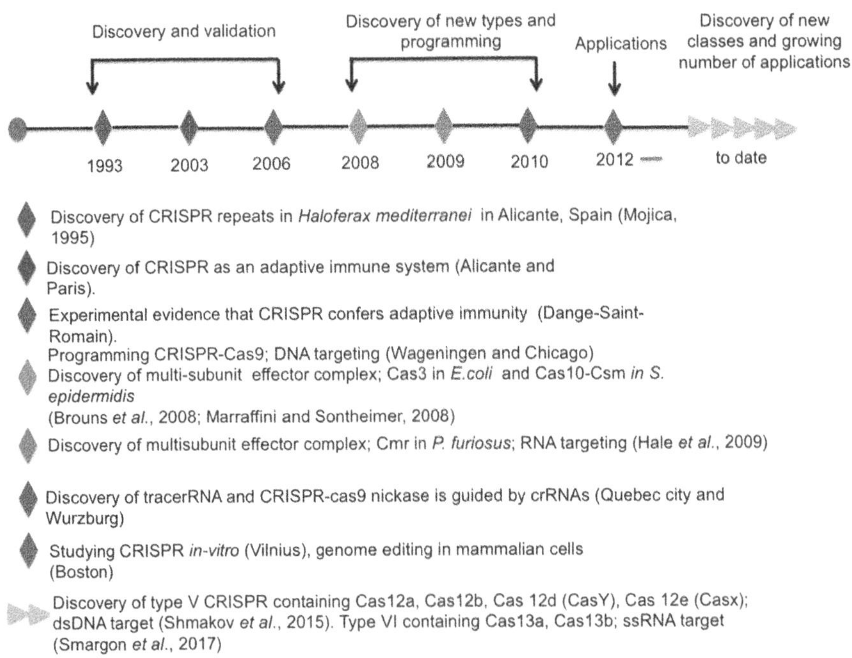

Fig. 1.7 Timeline of CRISPR transformation from nature to technology

microbes, in a DNA fragment from *Haloferax mediterranei,* a halophilic microorganism. These repeats were named as "clustered regularly interspaced palindromic repeats" or CRISPR in 2002 (Mojica and Garrett 2013). In 2003, after performing a detailed bioinformatics analysis of these repeats, Mojica realized that the CRISPR loci belonged to the adaptive immune system that protects microbes against infections from invading pathogens (Mojica et al. 2005; Bolotin et al. 2005).

Horvath et al. in 2005 confirmed the correlation between CRISPR activity and adaptive immune response in microorganisms. They used a phage-sensitive *Streptococcus thermophilus* strain to study the mechanism of phage resistance in this bacterium. They found that the resistant strains acquired phage-derived sequence at their CRIPSR loci, resulting in enhanced immunity against subsequent infection (Barrangou et al. 2007). They also studied the role of Cas7 and Cas9 proteins. Cas7 was not found to be involved in eliciting an immune response; however, it was required to generate new spacers and repeats. On the other hand Cas9 was found to possess nuclease activity and play an active role in adaptive immunity (Cas9 is referred to as Cas5 in CRISPR-related publications prior to 2012).

Later on, van der Oost et al. in 2008 studied the CRISPR system of *E. coli* by inserting it in another *E. coli* strain lacking its endogenous CRISPR system. They characterized the most complex CRISPR system (class 1, type I) that contains Cas3 instead of Cas9. They also reported that CRISPR arrays are transcribed to small crRNAs that consist of individual spacers to guide Cas nuclease activity.

In the same year, it was demonstrated that Cas possesses DNA nuclease activity, but not RNA cleavage activity, in type III-A CRISPR system from *Staphylococcus epidermidis* (Marraffini and Sontheimer 2008). However, type III-B CRISPR system from *Pyrococcus furious* revealed the presence of crRNA-directed RNA cleavage activity in addition to DNA nuclease activity (Hale et al. 2009, 2012). The importance of the PAM sequence for spacer acquisition and nuclease activity of type II Cas9 was demonstrated by Deveau et al. (2008). More importantly, they demonstrated that type I and II CRISPR systems prevent self-targeting by preventing the binding of PAM sequences to direct repeat sequences. However, a mismatch between the 5′ end of crRNA and target DNA is required for interference in type III CRISPR systems (Marraffini and Sontheimer 2010).

By 2010, studies on the functional mechanism of the native type II CRISPR system revealed that only Cas9 possesses DNA cleavage activity (Garneau et al. 2010). Moreover, it was found that non-coding tracrRNAs hybridize with crRNA to facilitate RNA-guided targeting of Cas9. This finding facilitated the engineering of a simple RNA-programmable DNA endonuclease for genome editing. Moreover, the dual RNA hybrid structure of Cas9 and endogenous RNAase III is required for the formation of mature crRNA from the CRISPR array primary transcript (Deltcheva et al. 2011). These studies demonstrated that at least three components (Cas9, mature crRNA, and tracrRNA) are essential for the reconstitution of the type II CRISPR system. In 2011, the successful reconstitution of CRISPR interference of type II CRISPR locus obtained from *Streptococcus thermophiles* and transformed into *E. coli* validated the transferable properties of the CRISPR system (Sapranauskas et al. 2011). Until 2012, - in-vitro DNA cleavage using purified Cas9 from

Streptococcus thermophiles or *Streptococcus pyogenes* guided by crRNA provided a strong basis for utilizing type II CRISPR system for genome editing. In 2013, a few research groups, Cong et al. (2013) and Mali et al. (2013) successfully engineered the type II CRISPR system from *Streptococcus thermophiles* and *Streptococcus pyogenes* for genome editing in eukaryotic cells.

Since then, researchers have reported the use of CRISPR-based genome editing in organisms such as yeast, fruit fly, nematode, mouse, zebrafish, and monkey. Moreover, new types of CRISPR systems have been discovered, their biology has been elucidated in detail, the existing technology for genome editing has been improved (Barrangou and Marraffini 2014; Hsu et al. 2014; Van Der Oost et al. 2014; Sander and Joung 2014; Jiang and Marraffini 2015; Sternberg and Doudna 2015; Wright et al. 2016; Smargon et al. 2017).

1.4 CRISPR: Experimental Design and Considerations

Scientific advancement requires continuous development of novel tools and technologies. Protein-based nuclease systems such as meganucleases, ZFNs, and TALENs have been used for DNA sequence editing; however, their restricted application have led researchers to design more convenient gene editing strategies to modify DNA sequences. The discovery of the CRISPR/Cas system and the understanding of its physiological role in prokaryotic cells paved the way to modify the innate genome editing mechanisms into targeted genome editing technology. Different classes and types of the CRISPR system require different components to target specific DNA sequences using different mechanisms. Among all types of natural CRISPR systems, the type II CRISPR system is the best re-engineered genome editing system in eukaryotes. The type II CRISPR/Cas9 system consists of two components: Cas9 protein that harbor the DNA binding and cleavage domains, and crRNA to guide Cas9 protein to target DNA with the help of tracrRNA. The crRNA and tracrRNA duplex work synergistically to dock Cas9 protein onto the target DNA. However, Cas9 can recognize the target DNA only in the presence of a short (2–5 nt) PAM sequence adjacent to the target DNA. Cas9 obtained from different sources can identify different PAM sequences, as described in the previous section. The ease of operation and the following beneficial characteristics of the type II CRISPR system facilitated its use as a robust, specific, and simple genome editing technology.

1. crRNA and tracrRNA can be combined into a single guide RNA (sgRNA) (Jinek et al. 2012).
2. Unlike Cas3, which degrades the target DNA, Cas9 produces a single DNA DSB, a prerequisite for gene editing.
3. Cas9 can be re-targeted by designing a sgRNA sequence.
4. Cas9 nuclease domain can be inactivated while retaining its target specificity, thereby allowing researchers to regulate the transcription of a desired gene.

The CRISPR/Cas system can be transformed into cells using conventional methods as well as microinjection. The generation of DNA DSBs by Cas nuclease triggers the activation of DNA repair pathways in the cell. The NHEJ pathway repairs DSBs in a non-template directed manner, resulting in the addition or deletion (indels) of nucleotides in the target DNA. On the other hand, precise nucleotide sequence modifications, gene replacement, and gene insertion at target loci is achieved by HDR in the presence of a homologous template.

Knowledge regarding different CRISPR systems and cellular DNA repair mechanisms has been extensively utilized to design the most popular gene editing technology—CRISPR-Cas—today a billion-dollar industry.

1.5 CRISPR/Cas: Beyond Genome Editing

CRISPR/Cas9 is the most widely used genome editing technique in both medical and agricultural fields. In contrast to ZFNs and TALENs where target DNA specificity and nuclease activity rely on the fusion of the sequence-specific DNA binding domain and the nuclease domain of the protein, the CRISPR system uses Cas9 protein to target any desired DNA sequence using a sequence-specific guide RNA (sgRNA), which is a hybrid of the naturally occurring tracrRNA: crRNA duplex. The use of sgRNA has simplified the application of the CRISPR system for genome editing. The application of ZFNs and TALENs for genome editing is restricted because of their complex re-engineering requirement for each target. However, the simple design of CRISPR/Cas9 has facilitated its use for genome editing in virtually all commonly studied eukaryotes, ranging from yeast to plants and from zebrafish to humans (Terns and Terns 2014; Sampson and Weiss 2014).

The application of the CRISPR technology is not restricted to genome editing. Owing to the programmable targeting capability of catalytically inactive Cas9, dCas9-based techniques have been used for several other genome manipulating applications. The Cas9 molecular scissors have been engineered into a versatile delivery tool by deactivating its catalytic activity through D10A and H804A substitutions. The deactivated Cas9 (dCas9)/sgRNA system can deliver different gene regulatory components to a specific target DNA, thereby acting as a shuttle without disrupting the double-stranded DNA. Therefore, dCas9/sgRNA can be used as a powerful tool for the transcriptional control of target genes. The CRISPRi (CRISPR interference) technique has been developed to knockdown gene expression with the aid of dCas9 (Qi et al. 2013). The fusion of dCas9 with a strong repressor complex such as Kruppel-associated box (KRAB) improves gene repression to a higher degree than dCas9 alone (Gilbert et al. 2013). Conversely, the fusion of dCas9 with the transcriptional activation domains VP16 and VP64 generated a gene activation platform. The fusion of dCas9 with a tripartite transcriptional complex, which is composed of VP64, P65, and Rta (VPR) proteins, induced gene expression (Chavez et al. 2015).

The programmable capacity of dCas9 has also been exploited for epigenome editing by recruiting various epigenetic writers and erasers to specific loci.

Understanding the role of epigenetic markers in gene regulation (expression and repression) is important in chromatin biology. The fusion of dCas9 with epigenetic modifier complexes provides a tool to specifically manipulate the expression of target loci. Although dCas9 lacks nuclease activity, it can act as a DNA binding domain and interact with a transcriptional repressor or activator domain to achieve CRISPR/dCas9-mediated epigenome editing. Therefore, epigenome altering tools in the form of CRISPR-dCas9 can be used to characterize regulatory elements and cellular states.

The three-dimensional structure of the genome is thought to play an important role in gene regulation and cell differentiation (Lanctôt et al. 2007; Schneider and Grosschedl 2007; Dixon et al. 2015; Peric-Hupkes et al. 2010). The unavailability of tools to visualize sequence-specific genome dynamics poses a challenge to understand the correlation between the linear DNA sequence and its three-dimensional organization. However, fluorescently tagged dCas9-based live cell imaging has enabled the study of gene organization and its cellular dynamics. The evolutionary dynamics of coding and non-coding sequences in human cells has been studied using a dCas9/green fluorescent protein (GFP) fusion protein (Chen et al. 2013). Moreover, multicolor fluorescent dCas9-based genome imaging tools were used by Ma et al. (2015) to track the dynamics of multi-genomic loci simultaneously (Ma et al. 2015). The use of different dCas9 proteins, each with a distinct PAM recognition sequence, tagged with different fluorophores provides a tool to visualize sequence-specific genomic dynamics and enhances the utility of CRISPR-based imaging to study complex chromosomal architecture and nuclear organization. Fluorescently tagged dCas9 is a powerful alternative to DNA-FISH (fluorescent in situ hybridization) for live cell imaging (Chen et al. 2013).

The discovery of RNA-targeting Cas proteins opened up an exciting possibility for using these proteins to target RNA sequences. RNA-targeting Cas9 from *Francisella novicida* does not require a PAM sequence for its nuclease activity; moreover, this protein has been successfully used to target hepatitis C virus in eukaryotic cells (Price et al. 2015). A Cas9-based RNA manipulating tool can provide a significant opportunity to track the real-time dynamics of RNA processing in the cell (Nelles et al. 2015).

Inducible Cas9 is required for the controlled application of the CRISPR/Cas system in cells to determine their function in cells. The structural analysis of Cas9 revealed the presence of two domains of the Cas protein, which could be reassembled through chemically inducible small molecules, such as ABI and PYL or light-inducible dimerization domains CIB1 and CRY2 (Hsu et al. 2014). Chemical- and light-inducible TALENs have been previously constructed (Konermann et al. 2013). A similar approach can be adopted to engineer an inducible Cas9 system to expand its application.

A heat shock inducible CRISPR/Cas9 (HS-CRISPR/Cas9) system was recently reported by Nandy et al. (2019). In this system, off-target effects during gene editing can be prevented by heat-dependent regulation of Cas9 expression. Some of the off-target effects have been shown to be caused by high doses of sgRNA:Cas9 in human, maize, and rice cells (Hsu et al. 2014; Hu et al. 2018; Pattanayak et al. 2013;

Svitashev et al. 2016). The HS-CRISPR/Cas9 system not only reduced off-target mutations, but also increased the mutation rate by 50–63% compared to the constitutively overexpressing CRISPR/Cas9 lines (Nandy et al. 2019).

1.6 Applications of the CRISPR/Cas System

The efficacy of the CRISPR/Cas9 technology was determined in model plant systems of *Arabidopsis*, rice, sorghum, and tobacco (Feng et al. 2013; Mao et al. 2013; Miao et al. 2013; Nekrasov et al. 2013; Shan et al. 2013; Xie and Yang 2013; Li et al. 2013; Jiang et al. 2013) after its successful genome editing application in animals and bacteria (Cong et al. 2013; Hwang et al. 2013). Currently, this method is extensively used in various plant species as a genome editing tool (detailed description is provided in the subsequent chapters). Several vectors containing the Cas9 and sgRNA expression cassettes are available from the plasmid repository Addgene (http://www.addgene.org). Plasmids can be transfected into cells using conventional transformation methods. Alternatively, in-vitro synthesized Cas9 mRNA and sgRNA can be microinjected or transfected into animal embryos and plant protoplasts. However, this microinjection method can be used only for selected plant species because of low regeneration capacity of protoplasts in most plant species.

The Cas9/sgRNA system is considered to be the primary choice for gene knockout in plants. Several gene knockout studies have been conducted in maize (Svitashev et al. 2015; Liang et al. 2014), soybean (Li et al. 2015; Jacobs et al. 2015), wheat (Wang et al. 2014), tomato (Brooks et al. 2014), and barley (Lawrenson et al. 2015) over the past 3–4 years. Feng et al. (2014) and Zhao et al. (2014) have demonstrated that Cas9/sgRNA knockout target genes with variable efficiency and edited genes can be propagated to the next generation of *Arabidopsis* and rice (Feng et al. 2014). The production of stable gene knockout mutants encouraged researchers to utilize CRISPR/Cas technology for other plant studies. Cas9/sgRNA-mediated site-directed mutagenesis and gene knockin has been reported in rice and *Arabidopsis* protoplasts (Li et al. 2013; Shan et al. 2013), and has been shown to be immensely beneficial in precision crop breeding. Transcriptional regulation by the catalytically deactivated dCas9/sgRNA system has been verified using the agroinfiltration transient expression system in *Nicotiana benthamiana* leaves (Piatek et al. 2015). Targeted gene expression or suppression enables researchers to improve crops by enhancing the expression of specific traits in plants. Moreover, the transcriptional reprogramming ability of dCas9-based CRISPR system can redirect the plant metabolome to produce valuable nutrients or bio-agents by regulating the expression of multiple genes.

The use of the CRISPR/Cas system has become increasingly relevant in several areas of medicine. In addition to its use for diagnostic purposes (Groenen et al. 1993), CRISPR/Cas system has been recently used as a direct antimicrobial tool.

Because of the high specificity of the CRISPR system, an artificial CRISPR array was designed to selectively kill pathogenic bacteria by targeting their antibiotic resistance gene and allowing non-pathogenic strains to survive. Consequently, the pathogenic strains were eliminated from the bacterial population (Bikard et al. 2014; Citorik et al. 2014; Gomaa et al. 2014). Therapeutic methods for treating genetic disorders based on Cas9 engineering are increasingly being studied. The CRISPR/Cas9 system has been used to prevent monogenic recessive disorders caused by loss-of-function mutations, such as cystic fibrosis and sickle-cell anemia (Schwank et al. 2013; Xie et al. 2014).

Cancer immunotherapy has been extensively studied in medical research in recent years. Of note, adoptive T-cell immunotherapy has been successfully used to treat cases of lymphoma, leukemia, and melanoma (Kennedy et al. 2014). CRISPR-based T-cell therapeutic systems have been reported by Schumann et al. (2015). CRISPR T-cell therapies hold great promise for several diseases including cancer, HIV, and autoimmune diseases (Huang et al. 2018).

CRISPR-based "gene drive" is a powerful technique that uses the CRISPR system to transfer a desirable genetic trait into an entire population (or species). This technique can be used to efficiently transfer a specific allele to an offspring, which under natural circumstances have a 50% chance to get transferred in offspring. CRISPR-based gene drive has been utilized by researchers to spread a gene allele that confers a parasite-resistant phenotype in mosquitos into a population, in a non-Mendelian manner (Gantz et al. 2015; Hammond et al. 2016). This approach can be utilized against malarial diseases. However, social and ethical barriers can hinder the application of this technique.

In recent years, CRISPR/Cas technology has gained attention in plant breeding industries, pharmaceutical companies, as well as the research community because of its simplicity and quick turnaround time. This technology will continue to be used in clinical, biological, and agricultural research for various purposes (Fig. 1.8). However, several concerns need to be addressed for the widespread and safe application of this technology. The following strategies can be used to prevent off-target cleavage by Cas9 and improve target specificity:

1. Injection of purified Cas9 protein directly into a cell instead of expressing a recombinant protein in the target cell (Ramakrishna et al. 2014; Kim et al. 2014).
2. Use of two sgRNAs that target both strands of the target sequence in combination with a DNA-nicking variant of Cas9 (Ran et al. 2013).
3. Truncation of sgRNA by 2 or 3 nucleotides has been shown to improve target specificity (Fu et al. 2014, 2014).
4. Addition of two guanine nucleotides to the 5′ end of DNA, immediately downstream of the complementary target region of sgRNA can reduce off-target cleavage (Cho et al. 2014).
5. Use of inducible Cas9 can prevent off-target effects (Nandy et al. 2019).
6. New Cas9 variants and anti-CRISPR proteins are natural sources for reducing off-target effects (Nakamura et al. 2019).

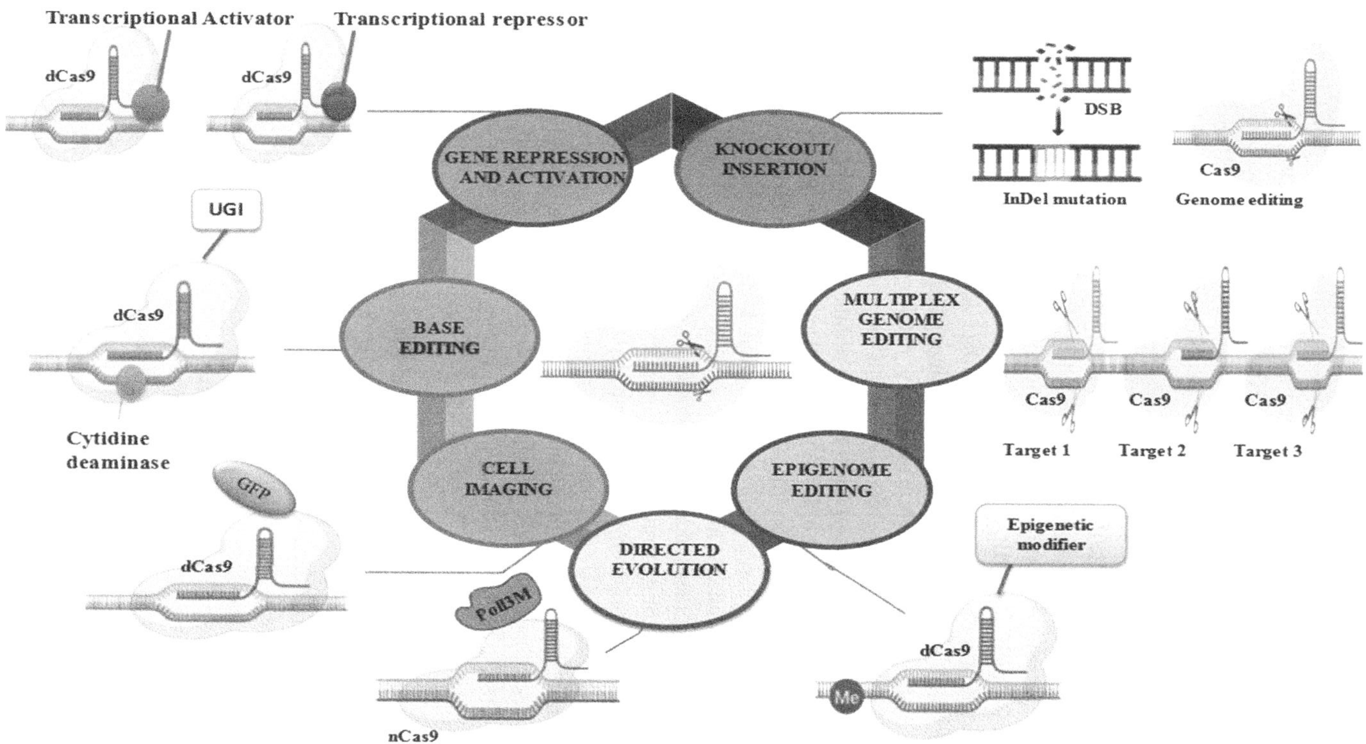

Fig. 1.8 Applications of the CRISPR/Cas system. Knockout/insertion models produced with CRISPR/Cas9 (shown in dark blue), multiplex genome editing using multiple sgRNA (light blue), epigenome modifiers can be recruited with dCas9 for epigenetic modifications (shown in purple), directed evolution with CRISPR/Cas (light pink), cell imaging with CRISPR/dCas (orange), base editing by recruiting base editors to target site (green), transcriptional regulation with CRISPRi and CRISPRa (dark pink)

1.7 Anti-CRISPR System and Its Applications

The development of immunity against pathogens is a natural process in microorganisms and is essential for the survival of all living beings. The CRISPR/Cas system is natural defense system adapted by microorganisms to protect themselves from viruses and transposons. However, viruses have adapted an anti-CRISPR system to protect their DNA from cleavage by the CRISPR/Cas9 defense system. Extensive structural studies have revealed that Cas9 requires a PAM recognition sequence to identify foreign DNA with high efficiency and specificity (Gleditzsch et al. 2019). Conversely, viruses have developed mechanisms to modify PAM sequences or anti-CRISPR proteins to target CRISPR-PAM interaction regions to block CRISPR activity. Several anti-CRISPR proteins (Acrs) with diverse inhibition strategies have been recognized. The inhibition of Cas9 from *Neisseria meningitidis* by AcrIIC1, AcrIIC2, and AcrIIC3 in bacterial and human cells was recently reported. Biochemical and structural data have revealed that AcrIIC2 inhibits the RNA binding ability of Cas9 by interacting with the bridge helix of Cas9 and preventing loading of DNA onto Cas9, resulting in inhibition of CRISPR activity. AcrIIC3 blocks Cas9 by interacting with the non-conserved surface of the HNH domain and the REC lobe of Cas9 to form a dimeric AcrIIC3-Cas9 complex; this complex prevent the loading of DNA onto Cas9 and results in deactivation of the CRISPR system (Zhu et al. 2019). Anti-CRISPR proteins are natural "off switches" of the Cas9 protein and have been recently used to regulate dCas9 activity in eukaryotic cells. Nakamura et al. (2019) have reported that cells pre-engineered with anti-CRISPR molecules are prone to gene editing (Nakamura et al. 2019). This indicates the possibility of generating "write-protected" cells that possess the ability to block gene editing. Moreover, they demonstrated the use of anti-CRISPR proteins to control CRISPR-based gene regulation. The application of CRISPR technology is highly compromised because of poor control of Cas9 expression and Cas9 activity. Therefore, the use of anti-CRISPR proteins to control Cas9 activity will allow more advanced dynamic control over gene regulation in the cell.

1.8 Biosafety Implications of CRISPR/Cas

CRISPR/Cas has emerged as the most popular tool for genome editing and has been adopted in every field of molecular biology and biotechnology in a very short period of time. CRISPR has been successfully used for a wide range of applications, from human gene therapy to transgene-free plants. It has been used to transform cells into molecular computers to exhibit conditional behavior under the control of molecular switches (Cribbs and Perera 2017). In addition, CRISPR has been used to produce disease models to study cancer, treat genetic diseases, and protect against human pathogens such as HIV (Sánchez-Rivera and Jacks 2015; Brokowski and Adli 2019). Furthermore, it has been used to prevent vector-borne diseases such as malaria and dengue fever (Adelman and Tu 2016). In agriculture, transgene-free plants have also been produced using CRISPR to address the regulatory issue in

GMO foods (Hartung and Schiemann 2014). These examples demonstrate the wide application of the CRISPR/Cas system both in humans and plants. Moreover, new CRISPR systems with more benefits are emerging. However, like any other technology, CRISPR too has its own challenges. Therefore, for the long-term application of CRISPR/Cas, it is important to address the regulatory, social, and ethical issues of this technology. Here, we have discussed the potential risks and ethical concerns related to CRISPR in human health and agriculture. We have also highlighted current regulations applicable to genome-edited animals and plants.

Legal and ethical considerations of CRISPR/Cas-mediated genome editing in humans are becoming increasingly important because of its potential application in gene editing of germline cells in humans. Although germline editing is helpful to understand human diseases, funding agencies and research regulatory bodies are reluctant to support this technology because of ethical and social concerns. Moreover, genomic changes in germline cells will be transferred to the next generation and consequently to future generations, thereby posing a potential threat to human lineage. Human germline editing is controversial as it may result in unexpected side effects in the long term. Nevertheless, a group of Chinese researchers has reported CRISPR editing in human embryos (Zhai et al. 2016) and germline mutations have also been inserted in animals for experimental purposes (Campbell and Eichler 2013). In 2018, a Chinese scientist claimed the first CRISPR-edited infant with mutations in the *CCR5* gene that conferred immunity against HIV (Li et al. 2019). This news invoked widespread criticism and highlighted the moral, ethical, and regulatory issues pertaining to CRISPR based genome editing (Li et al. 2019). Recently, China constituted a legal body to regulate human germline editing (Zhai et al. 2016). In the USA, the FDA has approved the use of CRISPR in genome editing therapies only in somatic cells due to prevailing ethical concerns. The US government does not fund projects related to germline gene therapy. Similarly, the European Union (EU) has imposed a ban on germline genome editing.

While human germline editing has become controversial due to unforeseen risks, CRISPR-edited agricultural products are approved for commercial use in the USA. For example, CRISPR-edited mushrooms and herbicide-tolerant soybean and corn are approved by regulatory authorities. Moreover, waxy corn by Dupont Pioneer and high oleic acid soybean are expected to obtain market approval from the FDA (Kuzma et al. 2016). Genetically modified crops such as Bt cotton and Bt maize were first introduced in the USA during 1996, and later adopted worldwide. However, the EU did not accept GMO due to the risks associated with transgene expression. However, CRISPR-edited crops differ from conventional GMOs because they do not require the insertion of a transgene in the plant genome. By using RNPs, in-vitro transcribed mRNA, or transiently expressed CRISPR/Cas, the edited plants circumvent the requirement of transgene insertion (Yin et al. 2015; Ilardi and Tavazza 2015). Transgene-free CRISPR-edited plants are entitled to bypass the conventional GMO regulations of the FDA, USDA, and APHIS in the USA (Friedrichs et al. 2019). There are three global regulatory systems that regulate the availability of agricultural GMO products: a product-based system, such as those in the USA, and process-based systems, such as those in the EU; both are novel trait-based regulatory systems. In the USA, the FDA, USDA, APHIS, and EPA regulate the availability of CRISPR-based agricultural products. However, these regulations also cover plant pests, and product can enter the regulatory process if a plant pest or viral DNA is inserted in plant DNA. Their regulations also cover edited insects that are detrimental to crop health. CRISPR-edited plant products that do not contain pest DNA can bypass these regulations. Currently, the EU does not approve any CRISPR-based agricultural product for commercialization (Sprink et al. 2016).

Another controversial application of the CRISPR/Cas system is gene drive. This tool enables researchers to force the inheritance of a specific trait and edit the genetics of entire population. Under natural circumstances, any specifically modified trait would have a 50% chance of inheritance to the offspring. Therefore, any release of GMOs in the environment would not pose a large risk for the population. In contrast, CRISPR-based gene drive ensures that mutations on one chromosome will also be copied on the partner chromosome. Thus, the modified trait will be inherited to all future generations. CRISPR drives have been proposed to eradicate diseases by eliminating disease-causing vectors, such as mosquitos and insects/pests in humans and agriculture, respectively. However, once released, it would not be easy to regulate CRISPR drives and prevent irreversible change in the entire population. Hence, any release of CRISPR drives may pose serious consequences to ecology, environment, and the food chain. Therefore, scientists have cautioned about release of edited insects pests in the environment and called for strict regulations on gene drive. In conclusion, although CRISPR has revolutionized the areas of agriculture and health sciences, CRISPR-edited organisms can pose serious risks to the environment and human health if released and commercialized without appropriate biosafety regulations. The products of CRISPR should be commercialized with caution to protect human health and future generations.

1.9 Future Prospects

The CRISPR technology has witnessed enormous progress during the last few years and has been recognized as one of the breakthrough discoveries of the twenty-first century. The development of the CRISPR/Cas system has transformed medical and agricultural sciences by providing novel benefits. For example, CRISPR has been demonstrated to act as molecular camera to record different events in DNA. This application may facilitate DNA-based data storage in the future and allow recording of cellular events. CRISPR technology has immense potential to eradicate vector-borne diseases such as malaria, dengue, COVID-19, as well as viral diseases in plants. In addition, it has been highlighted as miracle tool against antibiotic resistance. The use of sgRNA libraries to screen genome-wide mutant populations will accelerate new discoveries in functional genomics in plants and animals. It is our assumption that CRISPR-based directed evolution will allow the development of novel protein functions beyond natural evolution in the future. Moreover, base editing will empower the correction of genetic diseases in humans as well as develop new traits in plants. Genome-wide screens and multiplex genome editing will provide exciting new opportunities to study gene families and gene interaction studies in a very short time. CRISPR-based biotic and abiotic stress improvement, yield increase, and photosynthetically efficient crops will enhance food security and reduce nutritional challenges in developing countries. Studying the functions of *cis* regulatory elements and small RNAs will also help to improve agricultural output. Human gene therapy, disease modeling in animals, and the introduction of new traits in animals will further improve human health. CRISPR-based RNA editing will facilitate the characterization of genes whose disruption in DNA results in a lethal phenotype. Furthermore, spatiotemporal, inducible expression will further help characterize new genes in a tissue-specific and cell-specific manner. CRISPR will also facilitate development in chloroplast and mitochondrial gene editing.

Although CRISPR/Cas systems have become a revolutionary tool for genome editing and are adopted globally in a very short span of time, it still faces several limitations, such as large size, low specificity, low HDR efficiency, and inferior choice of delivery tools to wield its full potential. For example, the large size of Cas9 limits it packaging in viral vectors. Moreover, various Cas proteins with different PAM sequence requirements have been identified; however, PAM sequence specificity is a limitation. In plants, the efficiency of HDR is very low. Consequently, the scope of CRISPR/Cas to replace a gene or change a gene or promoter at a particular site can be very limited. The delivery of Cas as RNPs is challenging, but necessary to produce transgene-free crops to allay public perception about CRISPR crops. Moreover, process-based regulation in the EU is an obstacle in achieving economic benefits of CRISPR. Finally, transformation methods are not well established in some crops and are time- and labor-intensive. Nevertheless, CRISPR will continue to revolutionize agriculture, health, and food security.

References

Adelman ZN, Tu Z (2016) Control of mosquito-borne infectious diseases: sex and gene drive. Trends Parasitol 32(3):219–229

Amitai G, Sorek R (2016) CRISPR–Cas adaptation: insights into the mechanism of action. Nat Rev Microbiol 14(2):67

Argast GM, Stephens KM, Emond MJ, Monnat RJ Jr (1998) I-PpoI and I-CreI homing site sequence degeneracy determined by random mutagenesis and sequential in vitro enrichment. J Mol Biol 280(3):345–353

Arnould S, Perez C, Cabaniols J-P, Smith J, Gouble A, Grizot S, Epinat J-C, Duclert A, Duchateau P, Pâques F (2007) Engineered I-CreI derivatives cleaving sequences from the human XPC gene can induce highly efficient gene correction in mammalian cells. J Mol Biol 371(1):49–65

Barrangou R, Marraffini LA (2014) CRISPR-Cas systems: prokaryotes upgrade to adaptive immunity. Mol Cell 54(2):234–244

Barrangou R, Fremaux C, Deveau H, Richards M, Boyaval P, Moineau S, Romero DA, Horvath P (2007) CRISPR provides acquired resistance against viruses in prokaryotes. Science 315 (5819):1709–1712

Beumer KJ, Trautman JK, Bozas A, Liu J-L, Rutter J, Gall JG, Carroll D (2008) Efficient gene targeting in Drosophila by direct embryo injection with zinc-finger nucleases. PNAS 105 (50):19821–19826

Beumer KJ, Trautman JK, Mukherjee K, Carroll D (2013) Donor DNA utilization during gene targeting with zinc-finger nucleases. G3 (Bethesda) 3(4):657–664

Bibikova M, Golic M, Golic KG, Carroll D (2002) Targeted chromosomal cleavage and mutagenesis in Drosophila using zinc-finger nucleases. Genetics 161(3):1169–1175

Bikard D, Euler CW, Jiang W, Nussenzweig PM, Goldberg GW, Duportet X, Fischetti VA, Marraffini LA (2014) Exploiting CRISPR-Cas nucleases to produce sequence-specific antimicrobials. Nat Biotechnol 32(11):1146

Boch J, Scholze H, Schornack S, Landgraf A, Hahn S, Kay S, Lahaye T, Nickstadt A, Bonas U (2009) Breaking the code of DNA binding specificity of TAL-type III effectors. Science 326 (5959):1509–1512

Bolotin A, Quinquis B, Sorokin A, Ehrlich SD (2005) Clustered regularly interspaced short palindrome repeats (CRISPRs) have spacers of extrachromosomal origin. Microbiologia 151 (8):2551–2561

Bozas A, Beumer KJ, Trautman JK, Carroll D (2009) Genetic analysis of zinc-finger nuclease-induced gene targeting in Drosophila. Genetics 182(3):641–651

Brokowski C, Adli M (2019) CRISPR ethics: moral considerations for applications of a powerful tool. J Mol Biol 431(1):88–101

Brooks C, Nekrasov V, Lippman ZB, Van Eck J (2014) Efficient gene editing in tomato in the first generation using the clustered regularly interspaced short palindromic repeats/CRISPR-associated 9 system. Plant Physiol 166(3):1292–1297

Brouns SJ, Jore MM, Lundgren M, Westra ER, Slijkhuis RJ, Snijders AP, Dickman MJ, Makarova KS, Koonin EV, Van Der Oost J (2008) Small CRISPR RNAs guide antiviral defense in prokaryotes. Science 321(5891):960–964

Burstein D, Harrington LB, Strutt SC, Probst AJ, Anantharaman K, Thomas BC, Doudna JA, Banfield JF (2017) New CRISPR–Cas systems from uncultivated microbes. Nature 542 (7640):237

Campbell CD, Eichler EE (2013) Properties and rates of germline mutations in humans. Tigerpaper 29(10):575–584

Capecchi MR (1989) Altering the genome by homologous recombination. Science 244 (4910):1288–1292

Carte J, Wang R, Li H, Terns RM, Terns MP (2008) Cas6 is an endoribonuclease that generates guide RNAs for invader defense in prokaryotes. Genes Dev 22(24):3489–3496

Chandrasegaran S, Carroll D (2016) Origins of programmable nucleases for genome engineering. J Mol Biol 428(5):963–989

Charpentier E, Richter H, van der Oost J, White MF (2015) Biogenesis pathways of RNA guides in archaeal and bacterial CRISPR-Cas adaptive immunity. FEMS Microbiol Rev 39(3):428–441

Chavez A, Scheiman J, Vora S, Pruitt BW, Tuttle M, Iyer EP, Lin S, Kiani S, Guzman CD, Wiegand DJ (2015) Highly efficient Cas9-mediated transcriptional programming. Nat Methods 12(4):326

Chen B, Gilbert LA, Cimini BA, Schnitzbauer J, Zhang W, Li G-W, Park J, Blackburn EH, Weissman JS, Qi LS (2013) Dynamic imaging of genomic loci in living human cells by an optimized CRISPR/Cas system. Cell 155(7):1479–1491

Chevalier B, Turmel M, Lemieux C, Monnat RJ Jr, Stoddard BL (2003) Flexible DNA target site recognition by divergent homing endonuclease isoschizomers I-CreI and I-MsoI. J Mol Biol 329 (2):253–269

Cho SW, Kim S, Kim Y, Kweon J, Kim HS, Bae S, Kim J-S (2014) Analysis of off-target effects of CRISPR/Cas-derived RNA-guided endonucleases and nickases. Genome Res 24(1):132–141

Christian M, Cermak T, Doyle EL, Schmidt C, Zhang F, Hummel A, Bogdanove AJ, Voytas DF (2010) Targeting DNA double-strand breaks with TAL effector nucleases. Genetics 186 (2):757–761

Citorik RJ, Mimee M, Lu TK (2014) Sequence-specific antimicrobials using efficiently delivered RNA-guided nucleases. Nat Biotechnol 32(11):1141

Cong L, Ran FA, Cox D, Lin S, Barretto R, Habib N, Hsu PD, Wu X, Jiang W, Marraffini LA (2013) Multiplex genome engineering using CRISPR/Cas systems. Science 339(6121):819–823

Cribbs AP, Perera SM (2017) Focus: genome editing: science and bioethics of CRISPR-Cas9 gene editing: an analysis towards separating facts and fiction. Yale J Biol Med 90(4):625

Deltcheva E, Chylinski K, Sharma CM, Gonzales K, Chao Y, Pirzada ZA, Eckert MR, Vogel J, Charpentier E (2011) CRISPR RNA maturation by trans-encoded small RNA and host factor RNase III. Nature 471(7340):602

Deveau H, Barrangou R, Garneau JE, Labonté J, Fremaux C, Boyaval P, Romero DA, Horvath P, Moineau S (2008) Phage response to CRISPR-encoded resistance in Streptococcus thermophilus. J Bacteriol 190(4):1390–1400

Dixon JR, Jung I, Selvaraj S, Shen Y, Antosiewicz-Bourget JE, Lee AY, Ye Z, Kim A, Rajagopal N, Xie W (2015) Chromatin architecture reorganization during stem cell differentiation. Nature 518(7539):331

Dong D, Ren K, Qiu X, Zheng J, Guo M, Guan X, Liu H, Li N, Zhang B, Yang D (2016) The crystal structure of Cpf1 in complex with CRISPR RNA. Nature 532(7600):522

Feng Z, Zhang B, Ding W, Liu X, Yang D-L, Wei P, Cao F, Zhu S, Zhang F, Mao Y (2013) Efficient genome editing in plants using a CRISPR/Cas system. Cell Res 23(10):1229

Feng Z, Mao Y, Xu N, Zhang B, Wei P, Yang D-L, Wang Z, Zhang Z, Zheng R, Yang L (2014) Multigeneration analysis reveals the inheritance, specificity, and patterns of CRISPR/Cas-induced gene modifications in Arabidopsis. PNAS 111(12):4632–4637

Fineran PC, Gerritzen MJ, Suárez-Diez M, Künne T, Boekhorst J, van Hijum SA, Staals RH, Brouns SJ (2014) Degenerate target sites mediate rapid primed CRISPR adaptation. PNAS 111 (16):E1629–E1638

Fonfara I, Richter H, Bratovič M, Le Rhun A, Charpentier E (2016) The CRISPR-associated DNA-cleaving enzyme Cpf1 also processes precursor CRISPR RNA. Nature 532(7600):517

Friedrichs S, Takasu Y, Kearns P, Dagallier B, Oshima R, Schofield J, Moreddu C (2019) An overview of regulatory approaches to genome editing in agriculture. Biotechnol Res Innov 3 (2):208–220

Fu Y, Sander JD, Reyon D, Cascio VM, Joung JK (2014) Improving CRISPR-Cas nuclease specificity using truncated guide RNAs. Nat Biotechnol 32(3):279

Gantz VM, Jasinskiene N, Tatarenkova O, Fazekas A, Macias VM, Bier E, James AA (2015) Highly efficient Cas9-mediated gene drive for population modification of the malaria vector mosquito Anopheles stephensi. PNAS 112(49):E6736–E6743

Gao H, Smith J, Yang M, Jones S, Djukanovic V, Nicholson MG, West A, Bidney D, Falco SC, Jantz D (2010) Heritable targeted mutagenesis in maize using a designed endonuclease. Plant J 61(1):176–187

Gao L, Cox DB, Yan WX, Manteiga JC, Schneider MW, Yamano T, Nishimasu H, Nureki O, Crosetto N, Zhang F (2017) Engineered Cpf1 variants with altered PAM specificities. Nat Biotechnol 35(8):789

Garneau JE, Dupuis M-È, Villion M, Romero DA, Barrangou R, Boyaval P, Fremaux C, Horvath P, Magadán AH, Moineau S (2010) The CRISPR/Cas bacterial immune system cleaves bacteriophage and plasmid DNA. Nature 468(7320):67

Gilbert LA, Larson MH, Morsut L, Liu Z, Brar GA, Torres SE, Stern-Ginossar N, Brandman O, Whitehead EH, Doudna JA (2013) CRISPR-mediated modular RNA-guided regulation of transcription in eukaryotes. Cell 154(2):442–451

Gleditzsch D, Pausch P, Müller-Esparza H, Özcan A, Guo X, Bange G, Randau L (2019) PAM identification by CRISPR-Cas effector complexes: diversified mechanisms and structures. RNA Biol 16(4):504–517

Gomaa AA, Klumpe HE, Luo ML, Selle K, Barrangou R, Beisel CL (2014) Programmable removal of bacterial strains by use of genome-targeting CRISPR-Cas systems. MBio 5(1):e00928–e00913

Groenen PM, Bunschoten AE, Soolingen DV, Errtbden JDV (1993) Nature of DNA polymorphism in the direct repeat cluster of Mycobacterium tuberculosis; application for strain differentiation by a novel typing method. Mol Microbiol 10(5):1057–1065
Gudbergsdottir S, Deng L, Chen Z, Jensen JV, Jensen LR, She Q, Garrett RA (2011) Dynamic properties of the Sulfolobus CRISPR/Cas and CRISPR/Cmr systems when challenged with vector-borne viral and plasmid genes and protospacers. Mol Microbiol 79(1):35–49
Haft DH, Selengut J, Mongodin EF, Nelson KE (2005) A guild of 45 CRISPR-associated (Cas) protein families and multiple CRISPR/Cas subtypes exist in prokaryotic genomes. PLoS Comput Biol 1(6):e60
Hale CR, Majumdar S, Elmore J, Pfister N, Compton M, Olson S, Resch AM, Glover 3rd CVC, Graveley BR, Terns RM, Terns MP (2012) Essential features and rational design of CRISPR RNAs that function with the Cas RAMP module complex to cleave RNAs. Mol Cell 45 (3):292–302
Hale CR, Zhao P, Olson S, Duff MO, Graveley BR, Wells L, Terns RM, Terns MP (2009) RNA-guided RNA cleavage by a CRISPR RNa-Cas protein complex. Cell 139:945–956
Hammond A, Galizi R, Kyrou K, Simoni A, Siniscalchi C, Katsanos D, Gribble M, Baker D, Marois E, Russell S (2016) A CRISPR-Cas9 gene drive system targeting female reproduction in the malaria mosquito vector Anopheles gambiae. Nat Biotechnol 34(1):78
Hartung F, Schiemann J (2014) Precise plant breeding using new genome editing techniques: opportunities, safety and regulation in the EU. Plant J 78(5):742–752
Haurwitz RE, Jinek M, Wiedenheft B, Zhou K, Doudna JA (2010) Sequence-and structure-specific RNA processing by a CRISPR endonuclease. Science 329(5997):1355–1358
Hochstrasser ML, Taylor DW, Bhat P, Guegler CK, Sternberg SH, Nogales E, Doudna JA (2014) CasA mediates Cas3-catalyzed target degradation during CRISPR RNA-guided interference. PNAS 111(18):6618–6623
Hochstrasser ML, Taylor DW, Kornfeld JE, Nogales E, Doudna JA (2016) DNA targeting by a minimal CRISPR RNA-guided cascade. Mol Cell 63(5):840–851
Holkers M, Maggio I, Henriques SF, Janssen JM, Cathomen T, Gonçalves MA (2014) Adenoviral vector DNA for accurate genome editing with engineered nucleases. Nat Methods 11(10):1051
Hsu PD, Lander ES, Zhang F (2014) Development and applications of CRISPR-Cas9 for genome engineering. Cell 157(6):1262–1278
Hu X, Meng X, Liu Q, Li J, Wang K (2018) Increasing the efficiency of CRISPR-Cas9-VQR precise genome editing in rice. Plant Biotechnol J 16(1):292–297
Huang C-H, Lee K-C, Doudna JA (2018) Applications of CRISPR-Cas enzymes in cancer therapeutics and detection. Trends Cancer 4(7):499–512
Hwang WY, Fu Y, Reyon D, Maeder ML, Tsai SQ, Sander JD, Peterson RT, Yeh JJ, Joung JK (2013) Efficient genome editing in zebrafish using a CRISPR-Cas system. Nat Biotechnol 31 (3):227
Ilardi V, Tavazza M (2015) Biotechnological strategies and tools for Plum pox virus resistance: trans-, intra-, cis-genesis, and beyond. Front Plant Sci 6:379
Jackson RN, Wiedenheft B (2015) A conserved structural chassis for mounting versatile CRISPR RNA-guided immune responses. Mol Cell 58(5):722–728
Jackson RN, Golden SM, van Erp PB, Carter J, Westra ER, Brouns SJ, van der Oost J, Terwilliger TC, Read RJ, Wiedenheft B (2014) Crystal structure of the CRISPR RNA–guided surveillance complex from Escherichia coli. Science 345(6203):1473–1479
Jacobs TB, LaFayette PR, Schmitz RJ, Parrott WA (2015) Targeted genome modifications in soybean with CRISPR/Cas9. BMC Biotechnol 15(1):16
Jiang W, Marraffini LA (2015) CRISPR-Cas: new tools for genetic manipulations from bacterial immunity systems. Annu Rev Microbiol 69:209–228
Jiang W, Bikard D, Cox D, Zhang F, Marraffini LA (2013) RNA-guided editing of bacterial genomes using CRISPR-Cas systems. Nat Biotechnol 31(3):233

Jinek M, Chylinski K, Fonfara I, Hauer M, Doudna JA, Charpentier E (2012) A programmable dual-RNA–guided DNA endonuclease in adaptive bacterial immunity. Science 337 (6096):816–821
Kennedy EM, Kornepati AV, Goldstein M, Bogerd HP, Poling BC, Whisnant AW, Kastan MB, Cullen BR (2014) Inactivation of the human papillomavirus E6 or E7 gene in cervical carcinoma cells by using a bacterial CRISPR/Cas RNA-guided endonuclease. J Virol 88(20):11965–11972
Khan SH (2019) Genome-editing technologies: concept, pros, and cons of various genome-editing techniques and bioethical concerns for clinical application. Mol Ther Nucl Acids 16:326
Kim Y-G, Cha J, Chandrasegaran S (1996) Hybrid restriction enzymes: zinc finger fusions to Fok I cleavage domain. PNAS 93(3):1156–1160
Kim HJ, Lee HJ, Kim H, Cho SW, Kim J-S (2009) Targeted genome editing in human cells with zinc finger nucleases constructed via modular assembly. Genome Res 19(7):1279–1288
Kim Y, Kweon J, Kim A, Chon JK, Yoo JY, Kim HJ, Kim S, Lee C, Jeong E, Chung E (2013) A library of TAL effector nucleases spanning the human genome. Nat Biotechnol 31(3):251
Kim S, Kim D, Cho SW, Kim J, Kim J-S (2014) Highly efficient RNA-guided genome editing in human cells via delivery of purified Cas9 ribonucleoproteins. Genome Res 24(6):1012–1019
Konermann S, Brigham MD, Trevino AE, Hsu PD, Heidenreich M, Cong L, Platt RJ, Scott DA, Church GM, Zhang F (2013) Optical control of mammalian endogenous transcription and epigenetic states. Nature 500(7463):472
Koonin VE, Makarova SK, Zhang F (2017) Diversity, callsification and evolution of CRISPR-Cas systems. Curr Opin Microbiol 37:67–78
Kuzma J, Kokotovich A, Kuzhabekova A (2016) Attitudes towards governance of gene editing. Asian Biotechnol Dev Rev 18(1):69–92
Lanctôt C, Cheutin T, Cremer M, Cavalli G, Cremer T (2007) Dynamic genome architecture in the nuclear space: regulation of gene expression in three dimensions. Nat Rev Genet 8(2):104
Lawrenson T, Shorinola O, Stacey N, Li C, Østergaard L, Patron N, Uauy C, Harwood W (2015) Induction of targeted, heritable mutations in barley and Brassica oleracea using RNA-guided Cas9 nuclease. Genome Biol 16(1):258
Li J-F, Norville JE, Aach J, McCormack M, Zhang D, Bush J, Church GM, Sheen J (2013) Multiplex and homologous recombination–mediated genome editing in Arabidopsis and Nicotiana benthamiana using guide RNA and Cas9. Nat Biotechnol 31(8):688
Li Z, Liu Z-B, Xing A, Moon BP, Koellhoffer JP, Huang L, Ward RT, Clifton E, Falco SC, Cigan AM (2015) Cas9-guide RNA directed genome editing in soybean. Plant Physiol 169 (2):960–970
Li J-R, Walker S, J-b N, Zhang X-Q (2019) Experiments that led to the first gene-edited babies: the ethical failings and the urgent need for better governance. J Zhejiang Univ Sci B 20(1):32–38
Liang Z, Zhang K, Chen K, Gao C (2014) Targeted mutagenesis in Zea mays using TALENs and the CRISPR/Cas system. J Genet Genomics 41(2):63–68
Lillestøl RK, Shah SA, Brügger K, Redder P, Phan H, Christiansen J, Garrett RA (2009) CRISPR families of the crenarchaeal genus Sulfolobus: bidirectional transcription and dynamic properties. Mol Microbiol 72(1):259–272
Lucas P, Otis C, Mercier J-P, Turmel M, Lemieux C (2001) Rapid evolution of the DNA-binding site in LAGLIDADG homing endonucleases. Nucleic Acids Res 29(4):960–969
Ma H, Naseri A, Reyes-Gutierrez P, Wolfe SA, Zhang S, Pederson T (2015) Multicolor CRISPR labeling of chromosomal loci in human cells. PNAS 112(10):3002–3007
Maeder ML, Thibodeau-Beganny S, Osiak A, Wright DA, Anthony RM, Eichtinger M, Jiang T, Foley JE, Winfrey RJ, Townsend JA (2008) Rapid "open-source" engineering of customized zinc-finger nucleases for highly efficient gene modification. Mol Cell 31(2):294–301
Maggio I, Stefanucci L, Janssen JM, Liu J, Chen X, Mouly V, Gonçalves MA (2016) Selection-free gene repair after adenoviral vector transduction of designer nucleases: rescue of dystrophin synthesis in DMD muscle cell populations. Nucleic Acids Res 44(3):1449–1470
Makarova KS, Grishin NV, Shabalina SA, Wolf YI, Koonin EV (2006) A putative RNA-interference-based immune system in prokaryotes: computational analysis of the predicted

enzymatic machinery, functional analogies with eukaryotic RNAi, and hypothetical mechanisms of action. Biol Direct 1(1):7
Mali P, Yang L, Esvelt KM, Aach J, Guell M, DiCarlo JE, Norville JE, Church GM (2013) RNA-guided human genome engineering via Cas9. Science 339(6121):823–826
Mao Y, Zhang H, Xu N, Zhang B, Gou F, Zhu J-K (2013) Application of the CRISPR–Cas system for efficient genome engineering in plants. Mol Plant 6(6):2008–2011
Marraffini LA, Sontheimer EJ (2008) CRISPR interference limits horizontal gene transfer in staphylococci by targeting DNA. Science 322(5909):1843–1845
Marraffini LA, Sontheimer EJ (2010) Self versus non-self-discrimination during CRISPR RNA-directed immunity. Nature 463(7280):568
Mashimo T, Takizawa A, Voigt B, Yoshimi K, Hiai H, Kuramoto T, Serikawa T (2010) Generation of knockout rats with X-linked severe combined immunodeficiency (X-SCID) using zinc-finger nucleases. PLoS One 5(1):e8870
Meckler JF, Bhakta MS, Kim M-S, Ovadia R, Habrian CH, Zykovich A, Yu A, Lockwood SH, Morbitzer R, Elsäesser J (2013) Quantitative analysis of TALE–DNA interactions suggests polarity effects. Nucleic Acids Res 41(7):4118–4128
Miao J, Guo D, Zhang J, Huang Q, Qin G, Zhang X, Wan J, Gu H, Qu L-J (2013) Targeted mutagenesis in rice using CRISPR-Cas system. Cell Res 23(10):1233
Miller JC, Tan S, Qiao G, Barlow KA, Wang J, Xia DF, Meng X, Paschon DE, Leung E, Hinkley SJ (2011) A TALE nuclease architecture for efficient genome editing. Nat Biotechnol 29(2):143
Mladenov E, Iliakis G (2011) Induction and repair of DNA double strand breaks: the increasing spectrum of non-homologous end joining pathways. Mutat Res 711(1-2):61–72
Mojica FJ, Garrett RA (2013) Discovery and seminal developments in the CRISPR field. In: CRISPR-Cas systems. Springer, New York, pp 1–31
Mojica FJ, Díez-Villaseñor C, Soria E, Juez G (2000) Biological significance of a family of regularly spaced repeats in the genomes of Archaea, Bacteria and mitochondria. Mol Microbiol 36(1):244–246
Mojica FJ, García-Martínez J, Soria E (2005) Intervening sequences of regularly spaced prokaryotic repeats derive from foreign genetic elements. J Mol Evol 60(2):174–182
Moscou MJ, Bogdanove AJ (2009) A simple cipher governs DNA recognition by TAL effectors. Science 326(5959):1501–1501
Mussolino C, Alzubi J, Fine EJ, Morbitzer R, Cradick TJ, Lahaye T, Bao G, Cathomen T (2014) TALENs facilitate targeted genome editing in human cells with high specificity and low cytotoxicity. Nucleic Acids Res 42(10):6762–6773
Nakamura M, Srinivasan P, Chavez M, Carter MA, Dominguez AA, La Russa M, Lau MB, Abbott TR, Xu X, Zhao D (2019) Anti-CRISPR-mediated control of gene editing and synthetic circuits in eukaryotic cells. Nat Commun 10(1):194
Nam KH, Haitjema C, Liu X, Ding F, Wang H, DeLisa MP, Ke A (2012) Cas5d protein processes pre-crRNA and assembles into a cascade-like interference complex in subtype IC/Dvulg CRISPR-Cas system. Structure 20(9):1574–1584
Nandy S, Pathak B, Zhao S, Srivastava V (2019) Heat-shock-inducible CRISPR/Cas9 system generates heritable mutations in rice. Plant Direct 3(5):e00145
Nekrasov V, Staskawicz B, Weigel D, Jones JD, Kamoun S (2013) Targeted mutagenesis in the model plant Nicotiana benthamiana using Cas9 RNA-guided endonuclease. Nat Biotechnol 31 (8):691
Nelles DA, Fang MY, Aigner S, Yeo GW (2015) Applications of Cas9 as an RNA-programmed RNA-binding protein. BioEssays 37(7):732–739
Nemudryi A, Valetdinova K, Medvedev S, Zakian S (2014) TALEN and CRISPR/Cas genome editing systems: tools of discovery. Acta Nat 6(3):22
Niewoehner O, Jinek M (2016) Structural basis for the endoribonuclease activity of the type III-A CRISPR-associated protein Csm6. RNA 22(3):318–329

Orlowski J, Boniecki M, Bujnicki JM (2007) I-Ssp6803I: the first homing endonuclease from the PD-(D/E) XK superfamily exhibits an unusual mode of DNA recognition. Bioinformatics 23 (5):527–530

Pattanayak V, Lin S, Guilinger JP, Ma E, Doudna JA, Liu DR (2013) High-throughput profiling of off-target DNA cleavage reveals RNA-programmed Cas9 nuclease specificity. Nat Biotechnol 31(9):839

Peric-Hupkes D, Meuleman W, Pagie L, Bruggeman SW, Solovei I, Brugman W, Gräf S, Flicek P, Kerkhoven RM, van Lohuizen M (2010) Molecular maps of the reorganization of genome-nuclear lamina interactions during differentiation. Mol Cell 38(4):603–613

Piatek A, Ali Z, Baazim H, Li L, Abulfaraj A, Al-Shareef S, Aouida M, Mahfouz MM (2015) RNA-guided transcriptional regulation in planta via synthetic dC as9-based transcription factors. Plant Biotechnol J 13(4):578–589

Plagens A, Richter H, Charpentier E, Randau L (2015) DNA and RNA interference mechanisms by CRISPR-Cas surveillance complexes. FEMS Microbiol Rev 39(3):442–463

Price AA, Sampson TR, Ratner HK, Grakoui A, Weiss DS (2015) Cas9-mediated targeting of viral RNA in eukaryotic cells. PNAS 112(19):6164–6169

Puchta H (2005) The repair of double-strand breaks in plants: mechanisms and consequences for genome evolution. J Exp Bot 56(409):1–14

Qi LS, Larson MH, Gilbert LA, Doudna JA, Weissman JS, Arkin AP, Lim WA (2013) Repurposing CRISPR as an RNA-guided platform for sequence-specific control of gene expression. Cell 152 (5):1173–1183

Ramakrishna S, Dad A-BK, Beloor J, Gopalappa R, Lee S-K, Kim H (2014) Gene disruption by cell-penetrating peptide-mediated delivery of Cas9 protein and guide RNA. Genome Res 24 (6):1020–1027

Ramirez CL, Foley JE, Wright DA, Müller-Lerch F, Rahman SH, Cornu TI, Winfrey RJ, Sander JD, Fu F, Townsend JA (2008) Unexpected failure rates for modular assembly of engineered zinc fingers. Nat Methods 5(5):374

Ran FA, Hsu PD, Lin C-Y, Gootenberg JS, Konermann S, Trevino AE, Scott DA, Inoue A, Matoba S, Zhang Y (2013) Double nicking by RNA-guided CRISPR Cas9 for enhanced genome editing specificity. Cell 154(6):1380–1389

Rath D, Amlinger L, Rath A, Lundgren M (2015) The CRISPR-Cas immune system: biology, mechanisms and applications. Biochimie 117:119–128

Rollins MF, Schuman JT, Paulus K, Bukhari HS, Wiedenheft B (2015) Mechanism of foreign DNA recognition by a CRISPR RNA-guided surveillance complex from Pseudomonas aeruginosa. Nucleic Acids Res 43(4):2216–2222

Römer P, Hahn S, Jordan T, Strauß T, Bonas U, Lahaye T (2007) Plant pathogen recognition mediated by promoter activation of the pepper Bs3 resistance gene. Science 318(5850):645–648

Sampson TR, Weiss DS (2014) Exploiting CRISPR/C as systems for biotechnology. Bioessays 36 (1):34–38

Sampson TR, Saroj SD, Llewellyn AC, Tzeng Y-L, Weiss DS (2013) A CRISPR/Cas system mediates bacterial innate immune evasion and virulence. Nature 497(7448):254

Sánchez-Rivera FJ, Jacks T (2015) Applications of the CRISPR–Cas9 system in cancer biology. Nat Rev Cancer 15(7):387–395

Sander JD, Joung JK (2014) CRISPR-Cas systems for editing, regulating and targeting genomes. Nat Biotechnol 32(4):347

Sapranauskas R, Gasiunas G, Fremaux C, Barrangou R, Horvath P, Siksnys V (2011) The Streptococcus thermophilus CRISPR/Cas system provides immunity in Escherichia coli. Nucleic Acids Res 39(21):9275–9282

Schneider R, Grosschedl R (2007) Dynamics and interplay of nuclear architecture, genome organization, and gene expression. Genes Dev 21(23):3027–3043

Schornack S, Meyer A, Römer P, Jordan T, Lahaye T (2006) Gene-for-gene-mediated recognition of nuclear-targeted AvrBs3-like bacterial effector proteins. J Plant Physiol 163(3):256–272

Schumann K, Lin S, Boyer E, Simeonov DR, Subramaniam M, Gate RE, Haliburton GE, Chun JY, Bluestone JA, Doudna JA (2015) Generation of knock-in primary human T cells using Cas9 ribonucleoproteins. PNAS 112(33):10437–10442

Schwank G, Koo B-K, Sasselli V, Dekkers JF, Heo I, Demircan T, Sasaki N, Boymans S, Cuppen E, van der Ent CK (2013) Functional repair of CFTR by CRISPR/Cas9 in intestinal stem cell organoids of cystic fibrosis patients. Cell Stem Cell 13(6):653–658

Semenova E, Jore MM, Datsenko KA, Semenova A, Westra ER, Wanner B, Van Der Oost J, Brouns SJ, Severinov K (2011) Interference by clustered regularly interspaced short palindromic repeat (CRISPR) RNA is governed by a seed sequence. PNAS 108(25):10098–10103

Shah SA, Erdmann S, Mojica FJ, Garrett RA (2013) Protospacer recognition motifs: mixed identities and functional diversity. RNA Biol 10(5):891–899

Shalem O, Sanjana NE, Zhang F (2015) High-throughput functional genomics using CRISPR–Cas9. Nat Rev Genet 16(5):299–311

Shan Q, Wang Y, Li J, Zhang Y, Chen K, Liang Z, Zhang K, Liu J, Xi JJ, Qiu J-L (2013) Targeted genome modification of crop plants using a CRISPR-Cas system. Nat Biotechnol 31(8):686

Shukla VK, Doyon Y, Miller JC, DeKelver RC, Moehle EA, Worden SE, Mitchell JC, Arnold NL, Gopalan S, Meng X (2009) Precise genome modification in the crop species Zea mays using zinc-finger nucleases. Nature 459(7245):437

Smargon AA, Cox DB, Pyzocha NK, Zheng K, Slaymaker IM, Gootenberg JS, Abudayyeh OA, Essletzbichler P, Shmakov S, Makarova KS (2017) Cas13b is a type VI-B CRISPR-associated RNA-guided RNase differentially regulated by accessory proteins Csx27 and Csx28. Mol Cell 65(4):618–630

Smith J, Grizot S, Arnould S, Duclert A, Epinat J-C, Chames P, Prieto J, Redondo P, Blanco FJ, Bravo J (2006) A combinatorial approach to create artificial homing endonucleases cleaving chosen sequences. Nucleic Acids Res 34(22):e149–e149

Sorek R, Lawrence CM, Wiedenheft B (2013) CRISPR-mediated adaptive immune systems in bacteria and archaea. Annu Rev Biochem 82:237–266

Sprink T, Eriksson D, Schiemann J, Hartung F (2016) Regulatory hurdles for genome editing: process-vs. product-based approaches in different regulatory contexts. Plant Cell Rep 35 (7):1493–1506

Staals RH, Agari Y, Maki-Yonekura S, Zhu Y, Taylor DW, Van Duijn E, Barendregt A, Vlot M, Koehorst JJ, Sakamoto K (2013) Structure and activity of the RNA-targeting Type III-B CRISPR-Cas complex of Thermus thermophilus. Mol Cell 52(1):135–145

Staals RH, Zhu Y, Taylor DW, Kornfeld JE, Sharma K, Barendregt A, Koehorst JJ, Vlot M, Neupane N, Varossieau K (2014) RNA targeting by the type III-A CRISPR-Cas Csm complex of Thermus thermophilus. Mol Cell 56(4):518–530

Sternberg SH, Doudna JA (2015) Expanding the biologist's toolkit with CRISPR-Cas9. Mol Cell 58(4):568–574

Stoddard BL (2005) Homing endonuclease structure and function. Q Rev Biophys 38(1):49–95

Svitashev S, Young JK, Schwartz C, Gao H, Falco SC, Cigan AM (2015) Targeted mutagenesis, precise gene editing, and site-specific gene insertion in maize using Cas9 and guide RNA. Plant Physiol 169(2):931–945

Svitashev S, Schwartz C, Lenderts B, Young JK, Cigan AM (2016) Genome editing in maize directed by CRISPR–Cas9 ribonucleoprotein complexes. Nat Commun 7:13274

Terns RM, Terns MP (2014) CRISPR-based technologies: prokaryotic defense weapons repurposed. TIG 30(3):111–118

Townsend JA, Wright DA, Winfrey RJ, Fu F, Maeder ML, Joung JK, Voytas DF (2009) High-frequency modification of plant genes using engineered zinc-finger nucleases. Nature 459 (7245):442

Urnov FD, Rebar EJ, Holmes MC, Zhang HS, Gregory PD (2010) Genome editing with engineered zinc finger nucleases. Nat Rev Genet 11(9):636

Van der Oost J, Jore MM, Westra ER, Lundgren M, Brouns SJ (2009) CRISPR-based adaptive and heritable immunity in prokaryotes. Trends Biochem Sci 34(8):401–407

Van Der Oost J, Westra ER, Jackson RN, Wiedenheft B (2014) Unravelling the structural and mechanistic basis of CRISPR–Cas systems. Nat Rev Microbiol 12(7):479–492

Vercoe RB, Chang JT, Dy RL, Taylor C, Gristwood T, Clulow JS, Richter C, Przybilski R, Pitman AR, Fineran PC (2013) Cytotoxic chromosomal targeting by CRISPR/Cas systems can reshape bacterial genomes and expel or remodel pathogenicity islands. PLoS Genet 9(4):e1003454

Voytas DF (2013) Plant genome engineering with sequence-specific nucleases. Annu Rev Plant Biol 64:327–350

Wang Y, Cheng X, Shan Q, Zhang Y, Liu J, Gao C, Qiu J-L (2014) Simultaneous editing of three homoeoalleles in hexaploid bread wheat confers heritable resistance to powdery mildew. Nat Biotechnol 32(9):947

Westra ER, Semenova E, Datsenko KA, Jackson RN, Wiedenheft B, Severinov K, Brouns SJ (2013) Type IE CRISPR-cas systems discriminate target from non-target DNA through base pairing-independent PAM recognition. PLoS Genet 9(9):e1003742

Wright AV, Nuñez JK, Doudna JA (2016) Biology and applications of CRISPR systems: harnessing nature's toolbox for genome engineering. Cell 164(1-2):29–44

Xiao Y, Luo M, Hayes RP, Kim J, Ng S, Ding F, Liao M, Ke A (2017) Structure basis for directional R-loop formation and substrate handover mechanisms in type I CRISPR-Cas system. Cell 170(1):48–60

Xie K, Yang Y (2013) RNA-guided genome editing in plants using a CRISPR–Cas system. Mol Plant 6(6):1975–1983

Xie F, Ye L, Chang JC, Beyer AI, Wang J, Muench MO, Kan YW (2014) Seamless gene correction of β-thalassemia mutations in patient-specific iPSCs using CRISPR/Cas9 and piggyBac. Genome Res 24(9):1526–1533

Yamano T, Nishimasu H, Zetsche B, Hirano H, Slaymaker IM, Li Y, Fedorova I, Nakane T, Makarova KS, Koonin EV (2016) Crystal structure of Cpf1 in complex with guide RNA and target DNA. Cell 165(4):949–962

Yang H, Gao P, Rajashankar KR, Patel DJ (2016) PAM-dependent target DNA recognition and cleavage by C2c1 CRISPR-Cas endonuclease. Cell 167(7):1814–1828

Yin K, Han T, Liu G, Chen T, Wang Y, Yu AYL, Liu Y (2015) A geminivirus-based guide RNA delivery system for CRISPR/Cas9 mediated plant genome editing. Sci Rep 5:14926

Zegans ME, Wagner JC, Cady KC, Murphy DM, Hammond JH, O'Toole GA (2009) Interaction between bacteriophage DMS3 and host CRISPR region inhibits group behaviors of Pseudomonas aeruginosa. J Bacteriol 191(1):210–219

Zetsche B, Gootenberg JS, Abudayyeh OO, Slaymaker IM, Makarova KS, Essletzbichler P, Volz SE, Joung J, Van Der Oost J, Regev A (2015) Cpf1 is a single RNA-guided endonuclease of a class 2 CRISPR-Cas system. Cell 163(3):759–771

Zhai X, Ng V, Lie R (2016) No ethical divide between China and the West in human embryo research. Dev World Bioeth 16(2):116–120

Zhao L, Bonocora RP, Shub DA, Stoddard BL (2007) The restriction fold turns to the dark side: a bacterial homing endonuclease with a PD-(D/E)-XK motif. EMBO J 26(9):2432–2442

Zhao H, Sheng G, Wang J, Wang M, Bunkoczi G, Gong W, Wei Z, Wang Y (2014) Crystal structure of the RNA-guided immune surveillance Cascade complex in Escherichia coli. Nature 515(7525):147

Zhu Y, Gao A, Zhan Q, Wang Y, Feng H, Liu S, Gao G, Serganov A, Gao P (2019) Diverse mechanisms of CRISPR-Cas9 inhibition by type IIC anti-CRISPR proteins. Mol Cell 74 (2):296–309

CRISPR/Cas-Based Techniques in Plants

2

Zulqurnain Khan, Barkha Binyameen, Ummara Waheed, Muhammad Salman Mubarik, Muhammad Zubair Ghouri, Tahmina Sattar, and Asim Razzaq

Abstract

CRISPR techniques have been used in plants and animals with equal success. Ever since the emergence of CRISPR technology, scientists have been trying to find new CRISPR tools to address the concerns associated with the use of this technology. So, with the advent of time, new CRISPR tools have been discovered which are more precise and have broad spectrum with less limitations. The CRISPR toolbox has a number of tools for targeted modifications at epigenetic, genetic, transcriptional, and posttranscriptional level. For epigenetic modifications, dCas9 (catalytically dead Cas9) can be fused with Krüppel-associated box (KRAB), LSD1, acetyltransferase, DNA methyltransferase, DNA demethylase and some other effector domains. For targeted modifications at genomic level, Cas9 can be fused with different effector domains such as FokI, nickases, recombinases, transposases etc. For modifications at posttranscriptional level, Cas13 and CRISPRi approaches have been used successfully. CRISPR techniques are tremendous and have broad-spectrum applications compared to previous genome editing tools: ZFNs (zinc finger nucleases) and TALENs (Transcription activator like effector nucleases). Modularity of CRISPR technology has made it a technology of choice for the researchers and scientists all over

Z. Khan (✉) · B. Binyameen · U. Waheed · T. Sattar · A. Razzaq
Department of Biotechnology, Institute of Plant Breeding and Biotechnology (IPBB), MNS University of Agriculture, Multan, Pakistan
e-mail: zulqurnain.khan@mansuam.edu.pk; ummara.waheed@mnsuam.edu.pk

M. S. Mubarik
Centre of Agricultural Biochemistry and Biotechnology (CABB), University of Agriculture, Faisalabad, Pakistan

M. Z. Ghouri
Center for Advanced Studies in Agriculture and Food Security (CASAFS), University of Agriculture, Faisalabad, Pakistan

A. Ahmad et al. (eds.), *CRISPR Crops*,
https://doi.org/10.1007/978-981-15-7142-8_2

the world. In this chapter, an overview of the CRISPR techniques has been given with details of their applications for different purposes.

2.1 Introduction

In 1987, microbiologists observed a series of repetitive elements in the genome of *E. coli* which were interspersed with unique sequences described as clustered regularly interspaced short palindromic repeats (CRISPR) (Ishino et al. 1987). For a number of bacteria and archaea, different structures of CRISPR were discovered but for spacers, it almost took 20 years to derive from the invaded DNAs to be detected. The sequence of spacers matched with phage DNAs, reflecting a recollection of past infections (Bolotin et al. 2005). CRISPR loci are used for the transcription of RNAs and it further processed into CRISPR-derived RNAs (crRNAs) for the recognition and initiation of cutting of the foreign DNA. So far, five major types of CRISPR-based bacterial immune systems have been identified which will be expanded in future. Currently, there is a need to characterize other unclassified variants of CRISPR (Makarova et al. 2015). The system based on different CRISPR-associated (Cas) proteins helps in acquiring the new spacers, proliferating crRNA, and killing sequences of infectious nucleic acids. In order to introduce a targeted double-strand break (DSBs) for genome editing, a CRISPR system is required with Cas9 endonuclease along with crRNA and trans-activating RNA (tracrRNA) (Fig. 2.1). Both the RNAs have been combined to form a single-guided RNA (sgRNA), to edits a gene of interest, a target sequence and cloning procedure are required accordingly (Jinek et al. 2012). The CRISPR system, scientists are using today is based on only one gRNA and a Cas9 which is enough to target a sequence and further redesigning and cloning (Fig. 2.2). CRISPR/Cas system efficiently creates DSBs at the target site which is further utilized for targeted genome editing based on cellular repair mechanisms: NHEJ and HDR (Fig. 2.3). But there is a limitation in selecting a target sequence, it needs an adjacent protospacer motif (PAM), defined by the species origin of the Cas9 protein used. In standard applications "NGG" PAM at the $3'$ end of the target is required where the Cas9 is taken from *Streptococcus pyogenes*. It is not a big problem because NGG sequences are common in any coding sequence. Thus, genome editing by using CRISPR/Cas9 can now be implemented to almost all the species. Other ways like stable plant translation and transient expression can be achieved by the direct transformation of genes into the plant cells. It can be accomplished by absorption of DNA by polyethylene glycol (PEG), DNA microinjection, particle bombardment, and electroporation into the protoplasts (Altpeter et al. 2005). All these methods used in plants are not based on *Agrobacterium*-mediated transformation method only. Therefore, targeted editing needs only the genes for sgRNA and Cas9 without the need of various vectors. The very first reports on plant genome editing based on the CRISPR/Cas9 were published in 2013 (Nekrasov et al. 2013; Shan et al. 2013; Feng et al. 2013; Li et al. 2013; Xie and Yang 2013). As a proof of the concept, a *phytoene*

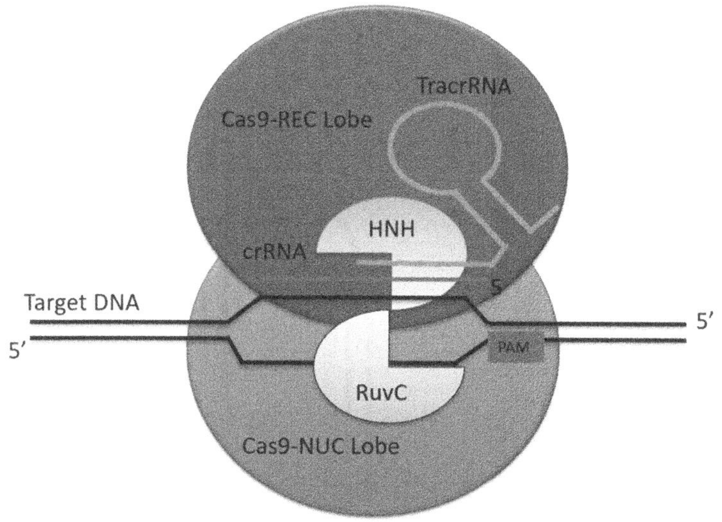

Fig. 2.1 crRNA, TracrRNA, and Cas9 complex shown in the figure is responsible to target the attacking viruses in the bacteria. Cas9 proteins have two lobes each having a nuclease domain (HNH and RuvC) responsible to cut the double strand of the DNA. TracrRNA makes a duplex with crRNA which scans the target site for complementary bases to recruit the Cas9 nuclease for creating DSB at the target site. PAM region is a protospacer adjacent motif which is needed to target the specific site in the DNA

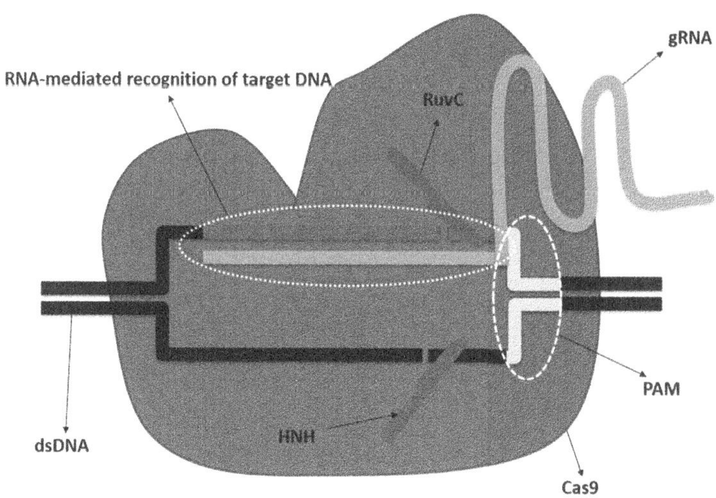

Fig. 2.2 CRISPR/Cas9-gRNA complex targeting a double-stranded DNA

desaturase synthase (*PDS*) gene was targeted by *Agrobacterium*-mediated transient expression of sgRNA and Cas9 in *Nicotiana benthamiana* leaves and *Arabidopsis*. For editing of PDS gene in tobacco, *Arabidopsis* and rice, a direct DNA transfer to

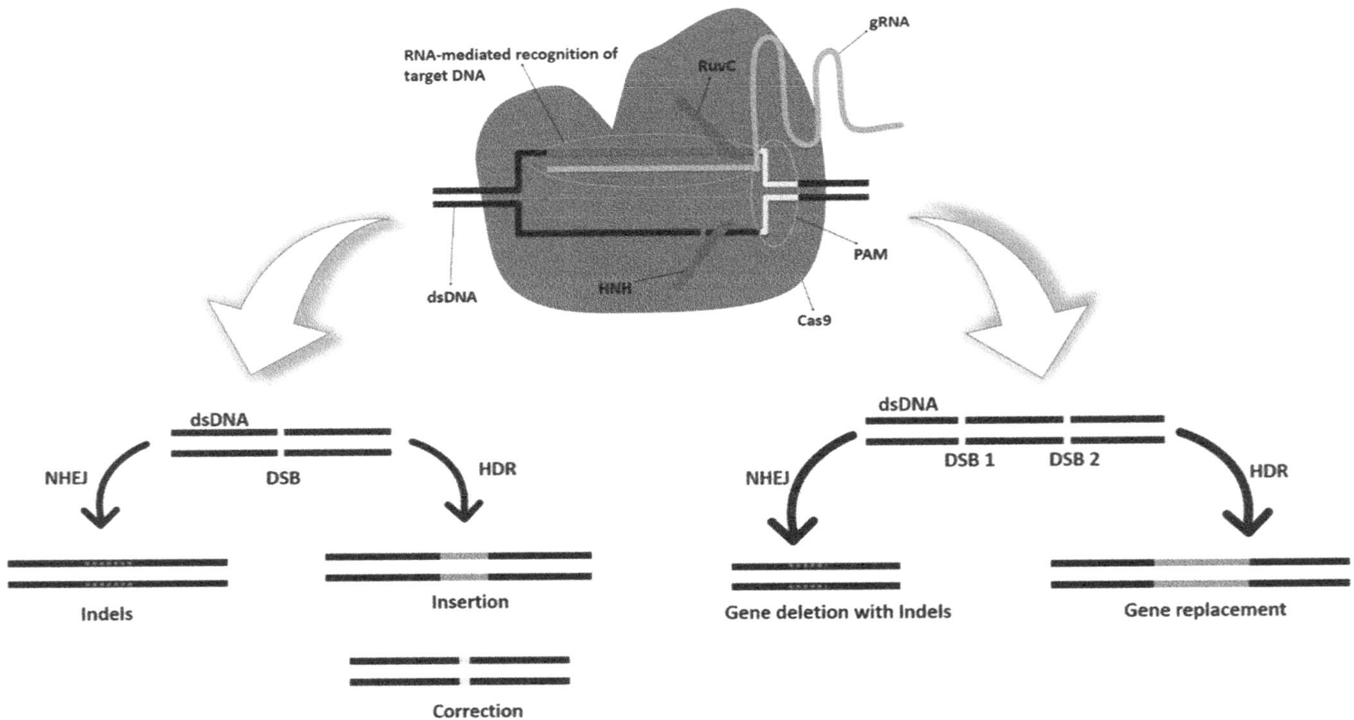

Fig. 2.3 CRISPR/Cas9-mediated DSBs and its repair in plant cell. A DSB mediated by CRISPR/Cas9 is repaired in eukaryotic cell using two repair mechanisms. The figure is showing the creation and repair of DSB in the cell. DSB is repaired through NHEJ or HDR. Based on the objective of the researcher, indels, gene deletion, gene insertion, gene correction, or gene replacement can be done.

protoplasts for transient expression of Cas9 and sgRNA was also used. In all mutations, InDels (Insertion and/or Deletions) mutations could be identified, with various mutation frequencies (Nekrasov et al. 2013; Li et al. 2013; Shan et al. 2013). For genome modification, the transformation remains a barrier in many crop plants for the efficient genome editing in plant breeding. So, there is a need to develop strategies in the future for genome editing in crop plants with more effective and high-performance technologies (Altpeter et al. 2016). More recently, for DNA-free gene targeting, use of in vitro pre-assembled Cas9-sgRNA ribonucleoproteins (RNPs) has been successfully demonstrated. RNPs along with particles were bombarded to immature embryos and regenerated plants were picked with editing events (Woo et al. 2015). CRISPR/Cas9 system with Cas9's deactivated or dead nuclease domain (known as dCas9) may also be used to target specific DNA sequences as a DNA-binding protein. Compared to zinc fingers (ZFs) and transcription activator like effectors (TALEs), dCas9 can also be combined with other effector domains for various purposes such as gene activation, gene suppression, epigenetic mark alteration, etc. (Khan et al. 2019). Genome editing is a tool which is used for identifying the gene function and also for improving crops. The most recent development in CRISPR/Cas9 method is using multiple single-guide RNA (sgRNAs) molecules to direct DSBs in the genome (Xing et al. 2014), having a potential to revolutionize farming. Unfortunately, the effectiveness of all sgRNAs is not equal, and it is hard for bioinformatics to predict their effectiveness. In cotton (*Gossypium hirsutum* L.), the transformation process is very lengthy and labor-intensive, so, there is a need to minimize the risk of using an unsuccessful sgRNA that could result in a transgenic plant without having a desired mutation induced by CRISPR system (Gao et al. 2017). CRISPR/Cas9 system has shown a great promise to resolve agricultural problems very quickly. This system can be used to alter genome sequence of any organism to achieve the desired trait including plants.

Compared to other contemporary genome editing tools like ZFNs and TALENs, CRISPR/Cas9 is cheaper, quicker, reliable, and highly effective even at multiplex stage. This technology is evolving very fast in the field of plant genome editing. This technology is becoming user-friendly for developing the crop plants of non-transgenic genome edited to counteract the harmful effects of climate change and ensuring food security for future increasing population (Haque et al. 2018).

2.2 CRISPR and Its Types

It is notable that CRISPR content and distribution vary in the closely related strains considerably. CRISPR, for example, occurs in the genus *Mycobacterium Tuberculosis*, but not in the *M. leprae*. On the other hand, some bacteria contain the same sequence of CRISPR like *E. coli*, *Mycobacterium avium*, *Archaeoglobus fulgidus*, and *Methanothermobacter thermoautotrophicus* (Jansen et al. 2002). In one genome, the number of CRISPR arrays varies from 1 to 18, and the number of repeat units varies from 2 to 374 in one CRISPR array (Marraffini and Sontheimer 2010). Based on the CRISPR database (https://crispr.i2bc.paris-saclay.fr), as of May

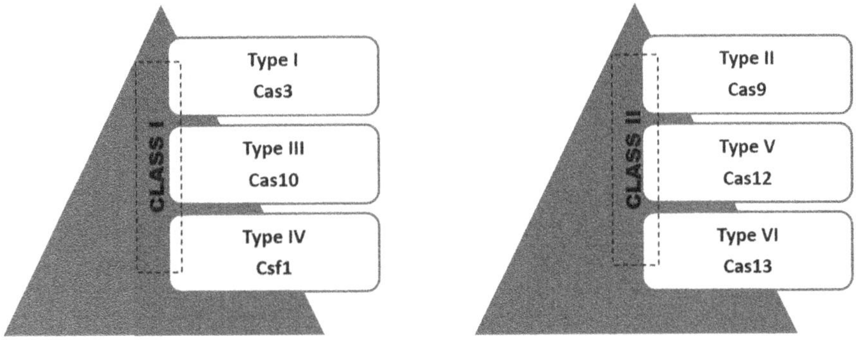

Fig. 2.4 CRISPR/Cas classes and types with examples

2017, 202 (87%) of the 232 archaeal species and 3059 (45%) of the 6782 bacterial species were identified as CRISPRs. Interestingly, a survey of 1724 genome designs suggested that CRISPR-Cas systems are much less common in microbial environmental communities about 10.4% in bacteria and 10.1% in archaea. Lack of CRISPR-Cas systems among major bacterial lineages having no cultivated representatives is the main reason behind this difference. The current classification of CRISPR-Cas system has two categories, which is based on the encoded effector proteins (Shmakov et al. 2017). A general description of classes and types with examples is given in Fig. 2.4. Class 1 of CRISPR-Cas systems consists of 4–7 Cas proteins and based on multi-subunit effector complexes in uneven stoichiometry. In archaea and bacteria, this system is widespread, includes all hyperthermophiles, which comprises about 90% of all CRISPR-Cas loci identified. While Class 2 consists on the remaining 10%, uses a single multidomain effector protein, and is found in bacteria exclusively (Burstein et al. 2017).

Each class of CRISPR is further divided into three types, Class 1 based on I, III, and IV while Class 2 has II, V, and VI types. With the existence of unique signature protein, it is easy to distinguish the types: Cas3 for type I, Cas9 for type II, and Cas10 for type III. The Type I and Type III based on Csm/Cmr complexes and multimeric effector complexes which is known as the CRISPR-associated antiviral defense complex (Cascade), are evolutionarily related and architecturally identical (Venclovas 2016; Rouillon et al. 2013). Unlike all other CRISPR-Cas systems, the Type IV systems are functionally uncharacterized and do not include the Cas1 and Cas2 nuclease adaptation module (Makarova et al. 2015; Rouillon et al. 2013).

Notably, it is known that the subtype III-B systems, effector modules use spacers provided by type I systems and testifying the modularity of CRISPR-Cas systems (Garrett et al. 2011).

While many genomes which encode type IV systems do not carry identifiable CRISPR loci, it is hypothesized that type IV systems are similar to subtype III-B system, So, they have the ability to use crRNAs from various CRISPR arrays (Makarova et al. 2015). Finally, each group is categorized further into multiple subtypes (I-A to F and U and III-A to D in class 1; II-A to C, V-A to E and U, and VI-A to C in class 2) on the basis of characteristic gene arrangements and additional signature genes (Rouillon et al. 2013; Koonin et al. 2017).

2.2.1 Class 2 System Is Suitable for Genome Editing

Class 2 CRISPR-Cas system is a more attractive choice for genome editing because it is based on effector complexes with simple architecture. Several different effectors of Class 2 were registered, include Type II Cas9, Type (V) Cas12a (formerly Cpf1), Type (V) Cas12b (C2c1), and the Type (VI) Cas13a (C2c2), Cas13b (C2c3). The Cas 9 is the best studied and more common multidomain effector protein; the two unrelated domains of crRNA-dependent endonucleases, RuvC and HNH, are responsible for the cleavage of targeted and non-targeted DNA strands in crRNA-target DNA complex. A trans-activating crRNA (tracrRNA) encodes by a Type II CRISPR-Cas loci that evolved from appropriate CRISPR. It is important for pre-crRNA processing and also for target recognition in Type II systems. The Cas9-crRNA complex is a molecular mechanism for cleavage of targeted DNA, as shown in Fig. 2.5. DNA-Cas9-crRNA complex crystal structure (Nishimasu et al. 2014) was studied at the atomic level. In the genome of several bacteria and archaea, a gene known as *Cpf1* is present at a place where it is adjacent to Cas1, Cas2, and a

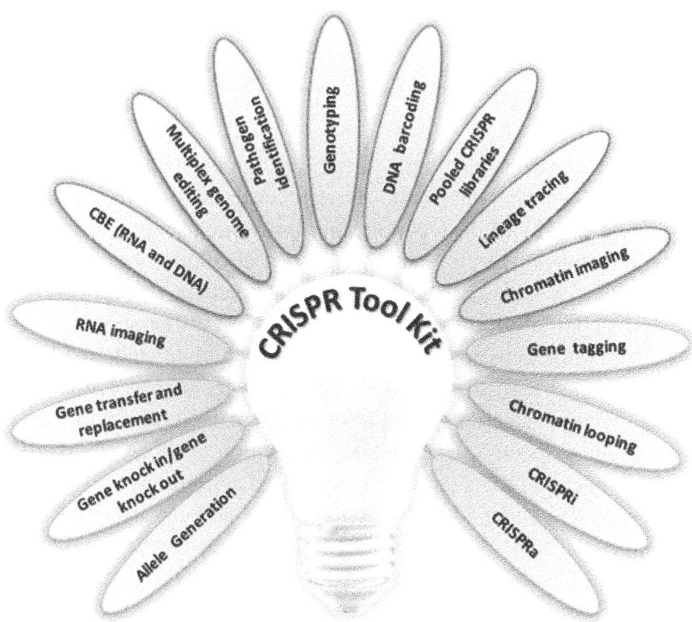

Fig. 2.5 CRISPR tool kit. A number of features have been given in the CRISPR toolkit showing diverse application for multiple purposes

CRISPR array (Shmakov et al. 2017). The type V effector prototype Cas12a (*Cpf1*) contains nuclease-like two domains, but one domain is missing that is HNH. However, the Cas12a-crRNA-target DNA complex is structurally analyzed which revealed a unique fold of second nuclease domain that is functionally equivalent to Cas9's HNH domain (Yamano et al. 2016). A single RNA-guided nuclease Cas12a does not require a Cas9 activity which is essential for tracrRNA (Zetsche et al. 2015, 2017) in identification of neighboring PAM and in its cleavage pattern that specifies the targeted strand and the protein is also different from Cas9.

The discovery of Class 2, two distantly related effector proteins; Cas9 and Cas12a, has proved that a big fraction of diversity exits among different variants of CRISPR/Cas system. A direct bioinformatic search was carried out with Cas12b (Type V) and Cas13a and Cas13b (Type VI), distinguished from Cas9 and Cas12a (Shmakov et al. 2015), and it was found that Type V effectors also require a tracrRNA for the targeted activity like Cas9. Although, most of the reported CRISPR-Cas systems have been found to target DNA only but recently dicovered multi-component systems, Type III-A and Type III-B, have the ability to target RNA as well (Jiang et al. 2016). In comparison, type VI effectors, Cas13a and Cas13b, also target RNAs and mediate RNA interference. Instead of RuvC-like nuclear domains, Type V (Cas13a to Cas13d) contains two Higher Eukaryotes and Prokaryotes Nucleotide-binding (HEPN) domains (Abudayyeh et al. 2016). The discovery of Class 2 CRISPR system is presenting more exciting opportunities for genome engineering at different levels in a variety of organisms (Murugan et al. 2017).

2.3 CRISPR Toolkit

There have been numerous attempts to modify genes in plants in the past, but CRIPSR/Cas9's efficiency was found most precise and specific (Nekrasov et al. 2013; Shan et al. 2013). With technology improvements, several highly efficient CRISPR/Cas9 vector delivery systems have been developed for plant GE, such as viral infection supersession, cis-element gene destruction, genomic deletion, gene knockout, and genome editing multiplexes. Continuous progress in this editing toolkit has enabled the delivery of the Cas9 system to plant cells in a more precise, accurate, and targeted manner, including the discovery of new Cas9 variants. The CRISPR-Cas immune bacterial system has rapidly evolved into an effective genome editing tool in many species. This system is based on a gRNA (guided RNA) that guides the Cas9 nuclease to its target site. Scientific community is widely exploiting the application of RNA-guided Cas9 technology in cell cultures (Ran et al. 2013), animals (Yang et al. 2014; Wang et al. 2013), and plants (Bortesi and Fischer 2015; Belhaj et al. 2015). On the plant side, RNA-guided genome engineering via Cas9 has been used in a variety of strategies, ranging from single and/or multiple gene knockouts (Shan et al. 2013; Fauser et al. 2014) to targeted donor sequence insertions (Schiml et al. 2014) or even targeted transcriptional regulation by combining transcriptional activation or repressor domains into an inactivated Cas9

(Piatek et al. 2015). A striking feature of gRNA-Cas9 is that it enables the simultaneous targeting of several sequences. While similar technologies such as the ZFNs (Beerli and Barbas III 2002) or the TAL effectors require the recoding of a new protein for each target sequence. A 20-nt change in the RNA guide with the gRNA-Cas9 is sufficient, paving the way for multiplex editing and design of complex regulatory circuits among other engineering options (Bogdanove et al. 2010). Direct transfection of Cas9 and guide RNAs into plant protoplasts followed by single-cell plant regeneration has been shown to be successful for genome editing in rice and tobacco, but the efficiency remained relatively poor, and besides, for many crop species, full plant regeneration from protoplasts is not currently feasible (Nielsen and Voigt 2014). The use of agrobacterium-mediated T-DNA transformation, accompanied by callus induction and organogenic plant regeneration (or floral dip transformation for Arabidopsis) is a promising alternative for plants. In this case, T-DNA-delivered gRNA-Cas9 can also integrate into the genome and continue its function in somatic tissues in addition to acting transiently during callus formation (Eeckhaut et al. 2013). It is important to extend the ability to combine different gRNAs with Cas9 within a single T-DNA in order to exploit the full potential of the T-DNA strategy, as it has been shown that all-in-one plasmid approaches significantly increase editing efficiency (Barrangou et al. 2007). Over the past seven years, the CRISPR-Cas toolbox has been used frequently because it consists on simpler RNA-guided DNA recognition system for genetic manipulations. This toolbox also provides many beneficial scientific tools used for designing desired genetic traits, also curing the genetic diseases and gives new approaches like live cell imaging, high-performance functional genomics screens, and also the point of care diagnostics.

CRISPR provides RNA-guided adaptive immunity against foreign genetic elements by directing nuclease to bind and cut different sequences of nucleic acids (Marraffini and Sontheimer 2008). Bacteria capture snippets of foreign genetic elements through a process called adaptation and incorporate them into their CRISPR genomic array. CRISPR array transcription creates CRISPR RNAs (crRNAs) that bind to Cas nucleases and provides specificity by matching target nucleic acids with base (Brouns et al. 2008). CRISPR/Cas has emerged as a new tool for targeting DNA using single-guide RNA (sgRNA) to enable genetic editing of any genome region (Jasin and Rothstein 2013). This system consists of two elements, a chimeric sgRNA and a CRISPR-associated protein (Cas9), which precisely unwinds and cleaves the target DNA, the cleavage site being determined by complementarity with the sgRNA (Cong et al. 2013). The only restriction to target a DNA sequence in this process is the presence of the PAM region. It has been shown that the CRISPR system is extremely useful for site-specific genome engineering. Recently, the deactivated version of Cas9 protein called dCas9 for programmable RNA-dependent DNA-binding protein was developed in bacterial and human cells (Wang et al. 2013). Targeting nuclease-inactive Cas9 protein (dCas9) to a gene's coding region can block RNA polymerase binding and elongation, resulting in dramatic transcription suppression. In addition, it was also reported that dCas9 can be modulated to recruit different protein effectors (activators or repressors) to DNA in a highly specific manner (Jinek et al. 2012) for activation (CRISPRa) or suppression/interference (CRISPRi) a gene. More recently, the fusion of dCas9 with the repressor domain of the Krüppel-associated box (KRAB) resulted in an effective transcriptional interference (Parsi et al. 2017; Gilbert et al. 2014). Furthermore, CRISPRi was also used for multiplexed endogenous gene control (Kearns et al. 2014; Gilbert et al. 2013) and stable gene repression with silencing efficiency typically achieved by RNAi while minimally impacting non-targeted gene transcription.

2.4 Variants of CRISPR Used in Genome Editing in Plants

CRISPR/Cas9 system is very fast, highly specific, and powerful which results in a few off-target events. Therefore, this system is most promising for the modification of plant genome. It is anticipated that CRISPR/Cas9 system has a great impact on research of applied and basic plant biology. It must have an important impact on crop breeding. Genome editing facilitates best genome engineering technique because it avoids time-consuming backcrossing in conventional breeding schemes and consistent changes directly in best cultivars. Simultaneously alteration of many traits is a more effective and successful approach to the pyramid breeding given by CRISPR/Cas9 (Bolotin et al. 2005). NHEJ-mediated gene knockout is the most direct application of CRISPR/Cas9. For the improvement of crop yield and to provide the host with resistance against target pathogens, negative regulators of disease resistance and grain growth can be modified. Certain forms of gene modification can also be used for agricultural purposes, like controlling epigenetic modulation and gene expression. The most significant step in CRISPR/Cas9-mediated gene editing is creation of a special sgRNA expression cassette, and it serves as a guide system for the Cas9/sgRNA complex consisting of 98 nucleotides with a targeted sequence based on 20 nucleotides (Nishimasu et al. 2014). RNA polymerase III is used in plants to transcribe sgRNA and its expression is primarily regulated by promoters of U3 or U6 (Li et al. 2013). Since the expression cassettes of the promoters sgRNA: U3/U6 are very small in size, it is possible to use approximately ~300–600 bp, overlapping PCR or adaptor ligation may be used to create such expression cassettes (Xie and Yang 2013). In 2015, based on the PCR technique, Ma and colleagues developed a robust cloning-free approach for developing sgRNA expression cassette. The CRISPR/Cas9 system has been constructed using the Gibson assembly or cloning strategy to directly clone the sgRNA expression cassette into binary vectors (Wang et al. 2013). Gao and Zhao used a ribozyme mechanism in another approach to generate sgRNA by transcribing pre-RNA via RNA polymerase II, whereby inducible or constitutive promoters can be connected to obtain the desired role of sgRNA (Ma et al. 2015). Cas9's construction consists of a coding sequence of 4107 bp. The Cas9 coding sequence must be fused with the nuclear localization signal (NLS) for Cas9 nuclear position in eukaryotes. Plant–bias codons have been used to design highly efficient and optimized Cas9 expression cassettes in plants (Soybean and *Medicago truncatula*) for improved GE (Michno et al. 2015).

For example, the use of codon-optimized Cas9p in rice and the Gramineae family has been enhanced by enhancing GC contents (Baltes et al. 2014), which imitate the Gramineae family genes (Ali et al. 2018). The expression of Cas9 in dicots and monocots for highly targeted gene editing can be obtained in *A. thaliana*, rice, and maize using callus-based transformation approaches. For CRISPR/Cas9-mediated GE, cargo-vector containing both sgRNA and Cas9 expression cassettes must be transported to targeted sites in plant cells. Floral dip and biolistic approaches are generally performed for cargo-vector transformation. For plant conversion, advanced approaches such as complex ribonucleoprotein, plasmid delivery, and virus-mediated delivery systems are now being implemented. There are some drawbacks in using the virus-mediated delivery system, but there have been several experiments conducted in plants using the virus delivery system (Jiang et al. 2013). Researchers typically use a transient expression system to move the vector to protoplast to analyze the CRISPR/Cas9 toolkit's feasibility. For the direct delivery of Cas9 gene expression cassettes (Wolter and Puchta 2017), biolistic and PEG-mediated transformation techniques can be used. However, due to heritable targeted mutations, regenerating plants from protoplast represents a major drawback associated with this approach in many plant species. Agrobacterium-mediated transformation is a highly scientific approach to stable transformation, in dicot and monocots, of the CRISPR/Cas9 system (Char et al. 2017; Zhang et al. 2018). The CRISPR/Cas9 GE tool has gained remarkable importance in agriculture; however, in plant GE there are some disadvantages that limit its use. The major concerns associated with this technology are non-specific targets, HR inefficiency, constriction of PAM sequences, frequency of cargo-vectors, and many others. Several specific creative moves are routinely integrated into the GE network to overcome these issues with the advancement of plant biotechnology.

Different CRISPR variants have been developed so far, which eliminate the need of DSB to induce targeted variations i.e. Base Editing technique allows targeted single base replacement in the particular region of the DNA. In this technique, effector domain(s) associated with CRISPR is/are altered or replaced to achieve single base replacement in the DNA. The first DNA base editors perform deamination of exocyclic amines of the targeted cytosine to generate uracil. This type of base editor is called Cytosine base editor as it converts a C•G base pair to a T•A base pair (Komor et al. 2016a, b). The purpose of cytosine deamination can be achieved by using an APOBEC1 cytidine deaminase that was first used by Liu and coworkers. APOBEC1 cytidine deaminase is capable of recognizing ssDNA as a substrate, but is incapable of acting on dsDNA (Harris et al. 2002). Fusion of APOBEC1 to dead Cas9 from *Streptococcus pyogenes* (dCas9, a mutant of Cas9 containing D10A and H840A) resulted in base editor 1 (BE1) (Komor et al. 2016a, b).When bound to its cognate DNA, dCas9 performs local denaturation of the DNA duplex to generate an R-loop in which the DNA strand not paired with the guide RNA exists as a disordered single-stranded bubble (Nishimasu et al. 2014). Fusion to dCas9 presents the target site to APOBEC1 in high effective molarity and enables cytosine base editor to deaminate cytosines located in a variety of different sequence motifs, albeit with differing efficacies (Komor et al. 2016a, b).

The second type of base editor is adenine base editor that converts an A•T base pair into a G•C base pair. This base editor is of particular importance as it is a promising tool to reverse pathogenic point mutations. Adenine contains an exocyclic amine that can be deaminated to alter its base-pairing preferences. This base editor actually performs the deamination of adenosine that yields inosine. Although, inosine in the third position of a tRNA anticodon is well known to pair with A, U, or C in mRNA during translation, in the context of a polymerase active site, inosine exhibits the base-pairing preference of guanosine (Komor et al. 2016a, b; Gaudelli et al. 2017a, b; Yan et al. 2018).

2.5 CRISPR as Precision Breeding Technique

By using sequence specific nucleases, a targeted plant genome editing has been achieved and it has a vast potential for the crop improvement to provide sustainable productive agricultural system and to resolve the emerging demands of global food (Liu et al. 2013). Traditionally, the mutation and conventional plant breeding techniques were being used for crop improvement but now the existing plant variation is declining, so that these techniques are going to be constrained and it hinders the production for future feeding (Chen and Gao 2014). Like CRISPR/Cas9 system, there is an insistent need for the strategies of crop improvement and having a novel genome editing techniques, which are able to enhance the crop significant functions and also make advanced valuable products (Zhang and Zhou 2014). In few years, CRISPR/Cas system has been developed so vast with numerous examples of the targeted mutagenesis and transcriptional control targeted regulation in crop and as well as in some model plants which clearly shows the evolutionary important feature of this new system (Khatodia and Khurana 2014). To develop the designer crop, a Cas9 nuclease mediated targeted genome editing is used and it is also extensively used for gene replacement, gene knockouts, interrogating gene function, multiplex editing and transcription modulation in plants and animals (Khatodia et al. 2016).

By the CRISPR/Cas9 system, a routine targeted mutagenesis in the plants will definitely open new dimensions in the research of plant biology. For inactivation study, the approaches of conventional mutagenesis were unable to access every gene, because these approaches were based on the random nature of gene integration. For inaccessible genes, the mutants can be generated by the technology of CRISPR/Cas9, multiple loci can be mutated by it and it creates large deletions, which can help in plant breeding without the introduction of transgene. However, genetically modified (GM) crops could have been a possible way out for crop improvement, if the disagreements concerning the environmental and health implications of GM crops could be evaded (Hilbeck et al. 2011). For genome manipulation, a DNA construct can be inserted randomly into the several chromosomes directly by using genetic engineering technique but this results in an undesirable effect because of random insertion. So, for large intensive changes, this technique is not much favorable, like in case of changing a whole metabolic pathway in a plant (Lau et al. 2014). Through plant breeding techniques which have been commonly used, new traits can be introduced into cultivated varieties by means of existing a natural genetic variation and random mutagenesis.

In a plant genome, a precise modification can be introduced by using CRISPR/Cas system which firmly inherits. To make free transgenic plants, a transgene region could be removed after targeted gene editing during crop variety improvement (Xu et al. 2015; Mahfouz et al. 2014; Kanchiswamy et al. 2015). The new and more faster techniques of plant breeding are ZFNs, TALENs, and Cas9/sgRNA system. All these techniques perform better than traditional breeding methods, it can produce a null segregant line which lacks the transgenic insert (Woo et al. 2015; Podevin et al. 2012; Schaart et al. 2016; Araki and Ishii 2015). Both the techniques, new breeding technique and classical breeding technique, produce identical plants. According to the end product, these plants should be evaluated (Woo et al. 2015; Hartung and Schiemann 2014; Giddings et al. 2012). Thus, the plants in this case which are edited by CRISPR have chances to be out of the present GMO regulations, but still this technology creates various new challenges for the social acceptance and regulation of genetically edited crops (Jones 2015; Voytas and Gao 2014; Kanchiswamy et al. 2015).

Currently, the new plant breeding techniques are in debate for legislation on genetically modified organisms (GMO). Regulatory authorities in relation with advisory authorities are considering to classify of GE crop varieties which are produced by method of genome editing as non-GM for increasing social acceptance (Voytas and Gao 2014; Lusser et al. 2012; Araki and Ishii 2015). They suggested a concept of GE crops regulation which is based on the regulatory responses from the authorities present in world by categorizing the (genome edited organisms) GEO under the process-based GMO and product-based GMO regulations. They recommend that the crops should not be considered as GMO which have been engineered at gene or genome level by using the CRISPR/Cas9 technology under the product-based GMO regulation, because the end product by using this technology is not really a GMO. So, there is a need for CRISPR/Cas technology to be optimized in plants, as this technology of editing has the potential to produce the valuable agronomic traits in varieties of plants (Quétier 2016). For understanding the complex quantitative traits, the classical breeding techniques can better supplement by CRISPR/Cas system, whose combinations of genes are selected by using traditional breeding method (Fichtner et al. 2014). The GE crops has proven to be more acceptable which have the appropriate regulatory structure than those plants which contains the foreign DNA (Osakabe and Osakabe 2015; Lozano-Juste and Cutler 2014). Plants which do not contain any foreign DNA in their genome will not be a GMO, indicated by Agricultural Department of the USA. However, in the near future, it is accepted by the European Commission to publish the regulatory uncertainty of genome editing (Jones 2015). The extent of CRISPR technology will depend on the public perception and performance of the varieties of GE crop in applied crop breeding and applied plant research which benefits the world's food security (Wolt et al. 2016; Belhaj et al. 2015).

2.6 Power of CRISPR/Cas Targeting Multiple Sites

Improving crops through genome editing, compared to traditional breeding approaches, requires a more advanced level of bioinformatics and genomics, as it is based on the genes which are functionally characterized (Nogué et al. 2016). In addition to genomic knowledge, it depends on efficient processes of transformation and regeneration to improve crops through genome editing. It is therefore necessary to advance plant transformation to make further effort to harness the genome editing potential completely (Altpeter et al. 2016). The extremely helpful property of CRISPR is multiplex editing because it significantly accelerates the process of breeding and could be blend with the production of double-haploid (DH) (Dwivedi et al. 2015) and speed breeding (Watson et al. 2018). DH lines are developed by crossing an inducer line, in which the zygote lost the haploid chromosome set, followed by genome doubling which finally results in a single generation of homozygous crop. The method provides basis for speed breeding by not only shortening the breeding process but also increasing genetic gain. By simple knock-out mutations, many features present in the gene coding sequence can be modified, but to generate new alleles by modification of gene expression, editing in regulatory sequences is required. The Lippman laboratory developed a genetic method (Rodríguez-Leal et al. 2017) which seems very promising, having the potential to unlock the changes in regulatory parts of genome for dosage effects. Traditionally, a cumbersome process is adapting preferred allelic variants to a range of breeding germplasms. Now, the transcriptional alleles can be generated directly which are most desirable and can be selected in the context of specific genetic background by this new genetic method. It has more wide applicability further than the new regulatory variants generation. For the generation of a set of new variants, the genetic method can be combined with a genome editing approach that is appropriate for a specific location. It might be combining with multiplex nickases, base editing system, or the new EvolvR system in addition to the multiplex Cas9 approach (Halperin et al. 2018). The Cas9 nickase fusion with adenine deaminase or cytidine (Komor et al. 2016a, b; Gaudelli et al. 2017a, b) enables precise A-to-G or C-to-T base editing within a specified range of sequence. Human APOBEC3A is used as a deaminase or with additional minor modifications (Zong et al. 2018); the software of base editing has been recently further adapted for the plants. By this, the deamination window has improved from the position of protospacer 3–9 to the positions of protospacer 1–17 and it enhances the effectiveness of deamination in rich GC sequence contexts. Additionally, the utility of base editing to generate new transcriptional alleles was demonstrated by Zong et al. (2018). They have targeted the three regulatory elements in wheat protoplasts promoter (TaVRN1-A1) which is involved in the regulation of vernalization, by using enhanced base editor. They detected a number of mutations through deep-sequencing, in these three targeted regulatory elements. To delete the different splicing isoforms, base editing can also be used by inducing the G to A transformations in respective 5′ splice site. Thus, it is possible to exclude corresponding mature mRNA forms and the specific splicing events (Xue et al. 2018). Another way to create the site-specific genetic diversity is given by the

EvolvR system (Halperin et al. 2018). It relies on union of a Cas9 nickase between the error-prone polymerase domain. This allows all nucleotides to be diversified at a particular location, within a tunable window size of 350 bp. Therefore, if needed a large collection of random mutations at a specific location, EvolvR has a benefit over the simple editing in terms of the greater editing window and greater mutation diversity. Finally, the de novo domestication which is mediated by CRISPR offers another exciting new possibility. This allows the exploitation of wild crop relatives as an important source of the allele mining, which might expand the crop germplasm pool extensively. In view of the resistance of a wild plant to wide range of stresses and genetic impoverishment of many crops (Østerberg et al. 2017), It will be proved very useful. Unfortunately, new crop varieties developed though genome editing technique are hindered in numerous areas by strict regulation on GMO, especially the areas which adhere to a product-based regulatory framework like European Union. The European Court of Justice's recent decision about establishing the targeted mutagenesis by using genome editing tools is subject to strict GMO legislation, even the product is totally free of transgenes (ECJ 2018). This is an important obstacle to the progress and innovation in these fields. In order to derive the benefits and restore the innovations from the technique of genome editing, a switch to a product-based regulatory framework is immediately required in Europe. Luckily, most of the countries do not face such an innovation impediment, no doubt that CRISPR/Cas9 will continue to revolutionize plant breeding on a global scale.

2.7 DNA-Free Genome Editing via CRISPR/Cas

More than fifty-two plant species have been targeted using CRISPR technology raning from model plants to cereals and cash crops and vegetable to economically less important plant species (Metje-Sprink et al. 2019). Such methods have been used as proof of concept for basic research and in most of its applications to investigate gene functions. Nevertheless, some market-oriented features have been discussed which include improved agronomic characteristics, enhanced quality of food, increasing tolerance to biotic and abiotic stress and herbicides. CRISPR systems which is originated from bacterial species are used for selective gene editing and existing CRISPR systems already have a lot of modifications such as (1) enhancing their precision, (2) altering their adjacent protospacer motif, (3) altering the ability to cut DNA, and (4) fusing them with additional proteins. In addition, techniques have become available for plant transformation using *Agrobacterium tumefaciens* or *Rhizobium rhizogenes* and new approaches. Some of them are used for transforming only the protein-RNA complexes or proteins; without using recombinant DNA it makes possible to modify the genome. As a consequence, the host genome did not incorporate with foreign DNA (Metje-Sprink et al. 2019). Currently, CRISPR/Cas9 is used to perform DNA-free genome editing either through direct delivery of isolated RNPs or through transient expression of mRNAs encoding for guided RNA and Cas nuclease. The Cas-gRNA complex is already functional and pre-assembled upon delivery when using the RNPs, when using the mRNA, it must be confirmed that the editors are transcribed and the complex must be assembled which causes a short delay in operation. There are two problems in DNA-free transformation: (1) delivery through plant cell wall; (2) regeneration of plant from cell or tissue to wall-free cells. The first tissue which was used for DNA-free genome editing were protoplasts, as polyethylene glycol (PEG) mediated fusion can easily target them. In the vesicles of PEG, the RNA or complex of RNP were found and fused with protoplasts. The system allows about 10% level of an average editing which is lower than the systems of DNA-based (Andersson et al. 2018; Svitashev et al. 2016). The process is well organized in potato from transfection to crops regeneration; it is easy for a single gene to change all of its four copies in 3–4% of shoots which are regenerated (Andersson et al. 2018). In grapevine and apple, the system of RNPs transformation works well, but there is a need to work on optimization of regeneration of edited protoplast to develop DNA free genome edited crop plants (Malnoy et al. 2016). A single-cell alga called *Chlamydomonas reinhardtii* was effectively transformed by using the method of electroporation along with RNPs. Although, a number of functional protocols have been reported for regeneration of potatoes, petunia, tobacco, lettuce and soy but the regeneration rate is very low (Kim et al. 2017; Woo et al. 2015; Subburaj et al. 2016). Recently, the immature embryos have been used for the DNA-free transformation. The mRNA and RNPs biolistic delivery can target immature embryos (Liang et al. 2017; Svitashev et al. 2016; Zhang et al. 2016). Furthermore, it is possible for transmitting the naked DNA into the plants, but it needs to be modified to convert DNA-free Genome Editing tools. Researchers have been using protoplast microinjection for transformation since 1983 (Griesbach and Lawson 1983). In plants, it has been demonstrated for the delivery of DNA into oil palm protoplast (Masani et al. 2014) but yet it has not been tested for the RNP delivery. The biolistic delivery of RNPs into plant cells has been optimized. For this purpose, gold-activated mesoporous silica nanoparticles are used to load the proteins (Martin-Ortigosa and Wang 2014). Biolistic delivery of RNPs is not yet well established in plants. In planta particle bombardment (iPB) has been use in wheat targeting shoot apical meristem (SAM) of imbibed seeds through gold particles coated with CRISPR/Cas9 plasmids (Hamada et al. 2017). Moreover, it was claimed that the method may be used to produce DNA-free genome edited crops. The transformation of immature plant tissue is based on a new method. The new method of transforming mature plant tissue is cell penetrating peptides (CPPs). CPPs are a class of small, positively charged peptides with the ability of penetrate cell membranes. Recently, they have been proven to be able to bind site-specific nucleases (Rádis-Baptista et al. 2017). Demonstration of DNA-free genome editing is still needed in plants to take advantage of its potential. New methods of transformation and cell transmission of single cells, like microfluidic systems or sonicative cell disruption, and intense light beams have been investigated for delivery of macromolecules into the cells but have not yet been demonstrated in plants (Han et al. 2015; Schlicher et al. 2006). Delivery of nucleic acids or mRNA or proteins (RNPs) of the GE tools have been tested for creating DNA-free plants. So, the future of GE in plants is very bright addressing regualtory issues as well.

2.8 CRISPR/Cas Driven Gene Drive

The technologies of gene drive are able to disperse the engineered genes throughout the targeted population more quickly than by simple genetic inheritance (Burt 2003). In nature, selfish genetic elements generate multiple copies across the whole genome and enhance their own transmission without any harmful effects on the individual (Werren 2011). CRISPR/Cas9 endonuclease system has been derived from bacteria having the ability to cut the target DNA sequence based on 20bp complementarity of guide RNA (gRNA) (Sander and Joung 2014). When CRISPR/Cas9 is used as a part of gene drive, it generates mutation anywhere in the genome sequence which is complementary to gRNA and it will make copies of itself (Sander and Joung 2014); as a consequence, the individual heterozygotes for mutation will convert into homozygotes (Esvelt et al. 2014). Compared to previous technologies, CRISPR/Cas9 provides flexibility, simplicity, and precision in targeting gene, so that, with little training and some special equipment, any laboratory of standard molecular biology could adopt this technology (Sander and Joung 2014). Most importantly, as compared to all previously used technologies of gene drive, CRISPR/Cas9 is considered inexpensive and efficient (Esvelt et al. 2014). With the help of CRISPR-based gene drive technology, the rapid spread of DNA cassette can be achieved in the targeted species. There are three elements in the cassette: gene which encodes the Cas9 protein, a gene that codes the gRNA which targets specific site in the genome, and the last one is flanking sequence which helps the cassette to insert at a specific given target site (Gantz et al. 2015). At designed position in the genome, the construct is able to copy and paste itself, as a result it propagates through a population. There is a 50% chance of normal allele to be inherited by each offspring according to mendelian inheritance. In comparison, gene drive cassette has 90% chance to be transmitted to next generation due to its ability to copy itself through whole genome (Gantz et al. 2015). Gene drive can detour vagaries of evolution by manipulating the mutations and heredity and there transmission to the next generation. Exact mutations directed by CRISPR/Cas-based gene drive are based on cut and paste of desired DNA sequence. A few individuals, having genes with the ability to modify host DNA sequences, released within a small fraction of the population may lead to complete gene drive within the entire population (Burt 2003). The cassette of gene drive can be designed for the transportation of a payload gene into targeted genome to manipulate the metabolism and its physiology; it also can be designed to integrate at specific position in the genome to knock out the gene. If gene drive displaces the male and female specific genes which are essential for the reproduction, it leads to the eradication of the species (Burt 2003). So, it is easy to identify the genes which are essential for the reproduction than those genes which alter species characteristics. Research and applications of gene drive are growing very rapidly. The technology of CRISPR/Cas9 for a gene editing is 7–8 years old. In 2015, the very first report of gene drive organism (mosquitoes and fruit flies) was published (Gantz et al. 2015). In the past 2 years, the Indian TATA group and Gates foundation invested approximately US $140 million on gene drive for the improvement in crop productivity and for controlling the vectors of disease. Recently,

Dupont, Bayer, and Monsanto signed the license agreement with the companies of biotech to use technology of CRISPR/Cas9 for developing plant varieties (Begley 2016).

2.9 Add-Ons in the CRISPR Technology

Base Editing has now become a prominent tool for engineering different crops' genomes. Fusion of CRISPR-Cas9 and activation-induced cytidine deaminase (Target-AID) for point mutagenesis at genomic regions specified by single-guide RNAs (sgRNAs) has been used in rice to target multiple genes. Multiple herbicide-resistance point mutations were induced by multiplexed base editing (Shimatani et al. 2017). In another study, a CRISPR–Cas9 nickase-cytidine deaminase fusion is done to achieve targeted conversion of cytosine to thymine from position 3–9 within the protospacer in both protoplasts and regenerated rice, wheat, and maize plants at frequencies of up to 43.48% (Zong et al. 2017).

One of the most recent advancements in CRISPR is Prime editing technique that also induces mutation without making double-strand break. Prime editing is a versatile and precise genome editing tool with the ability to directly write new genetic information into a specified DNA site using a catalytically impaired Cas9 endonuclease fused to an engineered reverse transcriptase (RT). RT is programmed with a prime editing guide RNA (pegRNA) that both specifies the target site and encodes the desired edit. More than 175 edits in human cells are performed, including targeted insertions, deletions, and all 12 types of point mutations, without requiring double-strand breaks or donor DNA templates (Anzalone et al. 2019).

2.10 Future Prospects

Since the launch of the first CRISPR/Cas-based genome editing method, several new CRISPR/Cas9 versions have been developed and used to revolutionize biotechnology. CRISPR/Cas-based technologies provide an accessible and adaptable way to change, regulate, and view genomes in a wide range of fields from biological research to biotechnology applications. CRISPR/Cas tools have been considerably accelerating the pace of research from the genetics to discovery of genes which are contributing to the improvement of crops. CRISPR/Cas-orientated biotechnology is increasing rapidly and is helping the world's masses to achieve food security and food safety. CRISPR/Cas9 applications in agriculture are already producing products for various markets outside the clinic. In this ever-expanding scientific era, the CRISPR/Cas toolkit is firmly positioned on the cutting edge of genome editing and genetic techniques in general. Contemporary CRISPR-based technologies, particulary for gain and loss of gene function, open new horizons in disease management, insect resistance, and functional genomics. In spite of the fact that many new, traditional, disease-resistant, and productive plant variant elites are manufactured, the pace of such techniques is not sufficient to meet today's needs. In the changing climatic conditions, plant breeders have big challenges in developing biotic and abiotic stress resistant plant varieties with high yield and growth rate. In crop development, nitrogen is a critical limiting factor. The majority of the genes of the nitrogen fixation (*nif*) and their relative levels of expression have been characterized (Temme et al. 2012). In order to decreased our dependence on inorganic fertilizers, transfer of genetic elements (Nod factor signaling pathways) from legumes into cereals such as wheat could be a way to increase nitrogen of the cereals by fixation of atmospheric nitrogen. The main aim of synthetic biology is also to build regulatory systems to manipulate the behavior of plants and create new features that improve crop productivity (Jusiak et al. 2016). For multiplex gene activation, repression, and editing of epigenome, dCas9-mediated gene regulations provide unparallel opportunities for designing synthetic transcription factors to build ever more complex, programmable, and efficient gene circuits. So, CRISPR-based genome editing has completely revolutionized our conventional ways of targeting genome. The approach of CRISPR targeting to any part of the DNA is something worth noticing. By using this, we can alter or replace any gene to harvest the desired product. Intronic region in crops is entailed with some vital genes that can be activated using CRISPR. If these genes are activated, it is likely that we will get to see some novel characters in plants. Using marvelous CRISPR technology for plant improvement will be very helpful for the plant breeders to acheive their longstanding goal of developing climate-smart plant varieties.

References

Abudayyeh OO, Gootenberg JS, Konermann S, Joung J, Slaymaker IM, Cox DB, Shmakov S, Makarova KS, Semenova E, Minakhin L (2016) C2c2 is a single-component programmable RNA-guided RNA-targeting CRISPR effector. Science 353(6299):5573

Ali Z, Eid A, Ali S, Mahfouz MM (2018) Pea early-browning virus-mediated genome editing via the CRISPR/Cas9 system in Nicotiana benthamiana and Arabidopsis. Virus Res 244:333–337

Altpeter F, Baisakh N, Beachy R, Bock R, Capell T, Christou P, Daniell H, Datta K, Datta S, Dix PJ (2005) Particle bombardment and the genetic enhancement of crops: myths and realities. Mol Breed 15(3):305–327

Altpeter F, Springer NM, Bartley LE, Blechl AE, Brutnell TP, Citovsky V, Conrad LJ, Gelvin SB, Jackson DP, Kausch AP (2016) Advancing crop transformation in the era of genome editing. Plant Cell 28(7):1510–1520

Andersson M, Turesson H, Olsson N, Fält AS, Ohlsson P, Gonzalez MN, Samuelsson M, Hofvander P (2018) Genome editing in potato via CRISPR-Cas9 ribonucleoprotein delivery. Physiol Plant 164(4):378–384

Anzalone AV, Randolph PB, Davis JR, Sousa AA, Koblan LW, Levy JM, Chen PJ, Wilson C, Newby GA, Raguram A (2019) Search-and-replace genome editing without double-strand breaks or donor DNA. Nature 576(7785):149–157

Araki M, Ishii T (2015) Towards social acceptance of plant breeding by genome editing. Trends Plant Sci 20(3):145–149

Baltes NJ, Gil-Humanes J, Cermak T, Atkins PA, Voytas DF (2014) DNA replicons for plant genome engineering. Plant Cell 26(1):151–163

Barrangou R, Fremaux C, Deveau H, Richards M, Boyaval P, Moineau S, Romero DA, Horvath P (2007) CRISPR provides acquired resistance against viruses in prokaryotes. Science 315 (5819):1709–1712

Beerli RR, Barbas CF III (2002) Engineering polydactyl zinc-finger transcription factors. Nat Biotechnol 20(2):135
Begley S (2016) Monsanto licenses CRISPR technology to modify crops—with key restrictions. https://www.statnews.com/2016/09/22/monsanto-licenses-crispr/. Accessed 22 Sep 2016
Belhaj K, Chaparro-Garcia A, Kamoun S, Patron NJ, Nekrasov V (2015) Editing plant genomes with CRISPR/Cas9. Curr Opin Biotechnol 32:76–84
Bogdanove AJ, Schornack S, Lahaye T (2010) TAL effectors: finding plant genes for disease and defense. Curr Opin Plant Biol 13(4):394–401
Bolotin A, Quinquis B, Sorokin A, Ehrlich SD (2005) Clustered regularly interspaced short palindrome repeats (CRISPRs) have spacers of extrachromosomal origin. Microbiology 151 (8):2551–2561
Bortesi L, Fischer R (2015) The CRISPR/Cas9 system for plant genome editing and beyond. Biotechnol Adv 33(1):41–52
Brouns SJ, Jore MM, Lundgren M, Westra ER, Slijkhuis RJ, Snijders AP, Dickman MJ, Makarova KS, Koonin EV, Van Der Oost J (2008) Small CRISPR RNAs guide antiviral defense in prokaryotes. Science 321(5891):960–964
Burstein D, Harrington LB, Strutt SC, Probst AJ, Anantharaman K, Thomas BC, Doudna JA, Banfield JF (2017) New CRISPR–Cas systems from uncultivated microbes. Nature 542 (7640):237
Burt A (2003) Site-specific selfish genes as tools for the control and genetic engineering of natural populations. Proc R Soc Lond Ser B 270(1518):921–928
Char SN, Neelakandan AK, Nahampun H, Frame B, Main M, Spalding MH, Becraft PW, Meyers BC, Walbot V, Wang K (2017) An agrobacterium-delivered CRISPR/Cas9 system for high-frequency targeted mutagenesis in maize. Plant Biotechnol J 15(2):257–268
Chen K, Gao C (2014) Targeted genome modification technologies and their applications in crop improvements. Plant Cell Rep 33(4):575–583
Cong L, Ran FA, Cox D, Lin S, Barretto R, Habib N, Hsu PD, Wu X, Jiang W, Marraffini LA (2013) Multiplex genome engineering using CRISPR/Cas systems. Science 339(6121):819–823
Dwivedi SL, Britt AB, Tripathi L, Sharma S, Upadhyaya HD, Ortiz R (2015) Haploids: constraints and opportunities in plant breeding. Biotechnol Adv 33(6):812–829
ECJ (2018) Judgment of 25 July 2018, Confédération Paysanne a.o., C-528/16, ECLI:EU: C:2018:583. https://eur-lex.europa.eu/legal-content/DE/TXT/?uri=CELEX:62016CJ0528. Accessed 23 Sep 2018.
Eeckhaut T, Lakshmanan PS, Deryckere D, Van Bockstaele E, Van Huylenbroeck J (2013) Progress in plant protoplast research. Planta 238(6):991–1003
Esvelt KM, Smidler AL, Catteruccia F, Church GM (2014) Emerging technology: concerning RNA-guided gene drives for the alteration of wild populations. elife 3:e03401
Fauser F, Schiml S, Puchta H (2014) Both CRISPR/C as-based nucleases and nickases can be used efficiently for genome engineering in A rabidopsis thaliana. Plant J 79(2):348–359
Feng Z, Zhang B, Ding W, Liu X, Yang D-L, Wei P, Cao F, Zhu S, Zhang F, Mao Y (2013) Efficient genome editing in plants using a CRISPR/Cas system. Cell Res 23(10):1229–1232
Fichtner F, Castellanos RU, Ülker B (2014) Precision genetic modifications: a new era in molecular biology and crop improvement. Planta 239(4):921–939
Gantz VM, Jasinskiene N, Tatarenkova O, Fazekas A, Macias VM, Bier E, James AA (2015) Highly efficient Cas9-mediated gene drive for population modification of the malaria vector mosquito *Anopheles stephensi*. Proc Natl Acad Sci 112(49):E6736–E6743
Gao W, Long L, Tian X, Xu F, Liu J, Singh PK, Botella JR, Song C (2017) Genome editing in cotton with the CRISPR/Cas9 system. Front Plant Sci 8:1364
Garrett RA, Vestergaard G, Shah SA (2011) Archaeal CRISPR-based immune systems: exchangeable functional modules. Trends Microbiol 19(11):549–556
Gaudelli NM, Komor AC, Rees HA, Packer MS, Badran AH, Bryson DI, Liu DR (2017a) Programmable base editing of A• T to G• C in genomic DNA without DNA cleavage. Nature 551(7681):464

Gaudelli NM, Komor AC, Rees HA, Packer MS, Badran AH, Bryson DI, Liu DR (2017b) Programmable base editing of A• T to G• C in genomic DNA without DNA cleavage. Nature 551(7681):464–471
Giddings LV, Potrykus I, Ammann K, Fedoroff NV (2012) Confronting the Gordian knot. Nat Biotechnol 30:208–209
Gilbert LA, Larson MH, Morsut L, Liu Z, Brar GA, Torres SE, Stern-Ginossar N, Brandman O, Whitehead EH, Doudna JA (2013) CRISPR-mediated modular RNA-guided regulation of transcription in eukaryotes. Cell 154(2):442–451
Gilbert LA, Horlbeck MA, Adamson B, Villalta JE, Chen Y, Whitehead EH, Guimaraes C, Panning B, Ploegh HL, Bassik MC (2014) Genome-scale CRISPR-mediated control of gene repression and activation. Cell 159(3):647–661
Griesbach R, Lawson R (1983) Improving plants through genetic engineering. Florists' Rev 9:2
Halperin SO, Tou CJ, Wong EB, Modavi C, Schaffer DV, Dueber JE (2018) CRISPR-guided DNA polymerases enable diversification of all nucleotides in a tunable window. Nature 560 (7717):248
Hamada H, Linghu Q, Nagira Y, Miki R, Taoka N, Imai R (2017) An in planta biolistic method for stable wheat transformation. Sci Rep 7(1):11443
Han X, Liu Z, Chan Jo M, Zhang K, Li Y, Zeng Z, Li N, Zu Y, Qin L (2015) CRISPR-Cas9 delivery to hard-to-transfect cells via membrane deformation. Sci Adv 1(7):e1500454
Haque E, Taniguchi H, Hassan M, Bhowmik P, Karim MR, Śmiech M, Zhao K, Rahman M, Islam T (2018) Application of CRISPR/Cas9 genome editing technology for the improvement of crops cultivated in tropical climates: recent progress, prospects, and challenges. Front Plant Sci 9:617
Harris RS, Petersen-Mahrt SK, Neuberger MS (2002) RNA editing enzyme APOBEC1 and some of its homologs can act as DNA mutators. Mol Cell 10(5):1247–1253
Hartung F, Schiemann J (2014) Precise plant breeding using new genome editing techniques: opportunities, safety and regulation in the EU. Plant J 78(5):742–752
Ishino Y, Shinagawa H, Makino K, Amemura M, Nakata A (1987) Nucleotide sequence of the iap gene, responsible for alkaline phosphatase isozyme conversion in Escherichia coli, and identification of the gene product. J Bacteriol 169 (12):5429-5433
Hilbeck A, Meier M, Römbke J, Jänsch S, Teichmann H, Tappeser B (2011) Environmental risk assessment of genetically modified plants-concepts and controversies. Environ Sci Eur 23(1):13
Jansen R, Embden JD, Gaastra W, Schouls LM (2002) Identification of genes that are associated with DNA repeats in prokaryotes. Mol Microbiol 43(6):1565–1575
Jasin M, Rothstein R (2013) Repair of strand breaks by homologous recombination. Cold Spring Harb Perspect Biol 5(11):a012740
Jiang W, Zhou H, Bi H, Fromm M, Yang B, Weeks DP (2013) Demonstration of CRISPR/Cas9/ sgRNA-mediated targeted gene modification in Arabidopsis, tobacco, sorghum and rice. Nucleic Acids Res 41(20):e188
Jiang W, Samai P, Marraffini LA (2016) Degradation of phage transcripts by CRISPR-associated RNases enables type III CRISPR-Cas immunity. Cell 164(4):710–721
Jinek M, Chylinski K, Fonfara I, Hauer M, Doudna JA, Charpentier E (2012) A programmable dual-RNA–guided DNA endonuclease in adaptive bacterial immunity. Science 337 (6096):816–821
Jones HD (2015) Regulatory uncertainty over genome editing. Nat Plants 1(14011):10
Jusiak B, Cleto S, Perez-Piñera P, Lu TK (2016) Engineering synthetic gene circuits in living cells with CRISPR technology. Trends Biotechnol 34(7):535–547
Kanchiswamy CN, Sargent DJ, Velasco R, Maffei ME, Malnoy M (2015) Looking forward to genetically edited fruit crops. Trends Biotechnol 33(2):62–64
Kearns NA, Genga RM, Enuameh MS, Garber M, Wolfe SA, Maehr R (2014) Cas9 effector-mediated regulation of transcription and differentiation in human pluripotent stem cells. Development 141(1):219–223

Khan Z, Khan SH, Ahmad A, Aslam S, Mubarik MS, Khan S (2019) CRISPR/dCas9-Mediated Inhibition of Replication of Begomoviruses. Int J Agric Biol 21(4):711–718

Khatodia S, Khurana SP (2014) Trending: the Cas nuclease mediated genome editing technique. Biotechnol Today 4(1):46–49

Khatodia S, Bhatotia K, Passricha N, Khurana S, Tuteja N (2016) The CRISPR/Cas genome-editing tool: application in improvement of crops. Front Plant Sci 7:506

Kim H, Kim S-T, Ryu J, Kang B-C, Kim J-S, Kim S-G (2017) CRISPR/Cpf1-mediated DNA-free plant genome editing. Nat Commun 8:14406

Komor AC, Kim YB, Packer MS, Zuris JA, Liu DR (2016a) Programmable editing of a target base in genomic DNA without double-stranded DNA cleavage. Nature 533(7603):420

Komor AC, Kim YB, Packer MS, Zuris JA, Liu DR (2016b) Programmable editing of a target base in genomic DNA without double-stranded DNA cleavage. Nature 533(7603):420–424

Koonin EV, Makarova KS, Zhang F (2017) Diversity, classification and evolution of CRISPR-Cas systems. Curr Opin Microbiol 37:67–78

Lau W, Fischbach MA, Osbourn A, Sattely ES (2014) Key applications of plant metabolic engineering. PLoS Biol 12(6):e1001879

Li J-F, Norville JE, Aach J, McCormack M, Zhang D, Bush J, Church GM, Sheen J (2013) Multiplex and homologous recombination–mediated genome editing in Arabidopsis and Nicotiana benthamiana using guide RNA and Cas9. Nat Biotechnol 31(8):688

Liang Z, Chen K, Li T, Zhang Y, Wang Y, Zhao Q, Liu J, Zhang H, Liu C, Ran Y (2017) Efficient DNA-free genome editing of bread wheat using CRISPR/Cas9 ribonucleoprotein complexes. Nat Commun 8:14261

Liu W, Yuan JS, Stewart CN Jr (2013) Advanced genetic tools for plant biotechnology. Nat Rev Genet 14(11):781–793

Lozano-Juste J, Cutler SR (2014) Plant genome engineering in full bloom. Trends Plant Sci 19 (5):284–287

Lusser M, Parisi C, Plan D, Rodriguez-Cerezo E (2012) Deployment of new biotechnologies in plant breeding. Nat Biotechnol 30(3):231

Ma X, Zhang Q, Zhu Q, Liu W, Chen Y, Qiu R, Wang B, Yang Z, Li H, Lin Y (2015) A robust CRISPR/Cas9 system for convenient, high-efficiency multiplex genome editing in monocot and dicot plants. Mol Plant 8(8):1274–1284

Mahfouz MM, Piatek A, Stewart CN Jr (2014) Genome engineering via TALENs and CRISPR/Cas9 systems: challenges and perspectives. Plant Biotechnol J 12(8):1006–1014

Makarova KS, Wolf YI, Alkhnbashi OS, Costa F, Shah SA, Saunders SJ, Barrangou R, Brouns SJ, Charpentier E, Haft DH (2015) An updated evolutionary classification of CRISPR–Cas systems. Nat Rev Microbiol 13(11):722

Malnoy M, Viola R, Jung M-H, Koo O-J, Kim S, Kim J-S, Velasco R, Nagamangala Kanchiswamy C (2016) DNA-free genetically edited grapevine and apple protoplast using CRISPR/Cas9 ribonucleoproteins. Front Plant Sci 7:1904

Marraffini LA, Sontheimer EJ (2008) CRISPR interference limits horizontal gene transfer in staphylococci by targeting DNA. Science 322(5909):1843–1845

Marraffini LA, Sontheimer EJ (2010) CRISPR interference: RNA-directed adaptive immunity in bacteria and archaea. Nat Rev Genet 11(3):181

Martin-Ortigosa S, Wang K (2014) Proteolistics: a biolistic method for intracellular delivery of proteins. Transgenic Res 23(5):743–756

Masani MYA, Noll GA, Parveez GKA, Sambanthamurthi R, Prüfer D (2014) Efficient transformation of oil palm protoplasts by PEG-mediated transfection and DNA microinjection. PLoS One 9(5):e96831

Metje-Sprink J, Menz J, Modrzejewski D, Sprink T (2019) DNA-free genome editing: past, present and future. Front Plant Sci 14(9):1957

Michno JM, Wang X, Liu J, Curtin SJ, Kono TJ, Stupar RM (2015) CRISPR/Cas mutagenesis of soybean and Medicago truncatula using a new web-tool and a modified Cas9 enzyme. GM Crops Food 6(4):243-252.

Murugan K, Babu K, Sundaresan R, Rajan R, Sashital DG (2017) The revolution continues: newly discovered systems expand the CRISPR-Cas toolkit. Mol Cell 68(1):15–25
Nekrasov V, Staskawicz B, Weigel D, Jones JD, Kamoun S (2013) Targeted mutagenesis in the model plant Nicotiana benthamiana using Cas9 RNA-guided endonuclease. Nat Biotechnol 31 (8):691
Nielsen AA, Voigt CA (2014) Multi-input CRISPR/Cas genetic circuits that interface host regulatory networks. Mol Syst Biol 10:11
Nishimasu H, Ran FA, Hsu PD, Konermann S, Shehata SI, Dohmae N, Ishitani R, Zhang F, Nureki O (2014) Crystal structure of Cas9 in complex with guide RNA and target DNA. Cell 156 (5):935–949
Nogué F, Mara K, Collonnier C, Casacuberta JM (2016) Genome engineering and plant breeding: impact on trait discovery and development. Plant Cell Rep 35(7):1475–1486
Osakabe Y, Osakabe K (2015) Genome editing with engineered nucleases in plants. Plant Cell Physiol 56(3):389–400
Østerberg JT, Xiang W, Olsen LI, Edenbrandt AK, Vedel SE, Christiansen A, Landes X, Andersen MM, Pagh P, Sandøe P (2017) Accelerating the domestication of new crops: feasibility and approaches. Trends Plant Sci 22(5):373–384
Parsi KM, Hennessy E, Kearns N, Maehr R (2017) Using an inducible CRISPR-dCas9-KRAB effector system to dissect transcriptional regulation in human embryonic stem cells. In: Eukaryotic transcriptional and post-transcriptional gene expression regulation. Springer, Cham, pp 221–233
Piatek A, Ali Z, Baazim H, Li L, Abulfaraj A, Al-Shareef S, Aouida M, Mahfouz MM (2015) RNA-guided transcriptional regulation in planta via synthetic dC as9-based transcription factors. Plant Biotechnol J 13(4):578–589
Podevin N, Devos Y, Davies HV, Nielsen KM (2012) Transgenic or not? No simple answer! EMBO Rep 13(12):1057–1061
Quétier F (2016) The CRISPR-Cas9 technology: closer to the ultimate toolkit for targeted genome editing. Plant Sci 242:65–76
Rádis-Baptista G, Campelo IS, Morlighem J-ÉR, Melo LM, Freitas VJ (2017) Cell-penetrating peptides (CPPs): From delivery of nucleic acids and antigens to transduction of engineered nucleases for application in transgenesis. J Biotechnol 252:15–26
Ran FA, Hsu PD, Wright J, Agarwala V, Scott DA, Zhang F (2013) Genome engineering using the CRISPR-Cas9 system. Nat Protoc 8(11):2281
Rodríguez-Leal D, Lemmon ZH, Man J, Bartlett ME, Lippman ZB (2017) Engineering quantitative trait variation for crop improvement by genome editing. Cell 171(2):470–480
Rouillon C, Zhou M, Zhang J, Politis A, Beilsten-Edmands V, Cannone G, Graham S, Robinson CV, Spagnolo L, White MF (2013) Structure of the CRISPR interference complex CSM reveals key similarities with cascade. Mol Cell 52(1):124–134
Sander JD, Joung JK (2014) CRISPR-Cas systems for editing, regulating and targeting genomes. Nat Biotechnol 32(4):347
Schaart JG, van de Wiel CC, Lotz LA, Smulders MJ (2016) Opportunities for products of new plant breeding techniques. Trends Plant Sci 21(5):438–449
Schiml S, Fauser F, Puchta H (2014) The CRISPR/C as system can be used as nuclease for in planta gene targeting and as paired nickases for directed mutagenesis in Arabidopsis resulting in heritable progeny. Plant J 80(6):1139–1150
Schlicher RK, Radhakrishna H, Tolentino TP, Apkarian RP, Zarnitsyn V, Prausnitz MR (2006) Mechanism of intracellular delivery by acoustic cavitation. Ultrasound Med Biol 32(6):915–924
Shan Q, Wang Y, Li J, Zhang Y, Chen K, Liang Z, Zhang K, Liu J, Xi JJ, Qiu J-L (2013) Targeted genome modification of crop plants using a CRISPR-Cas system. Nat Biotechnol 31(8):686
Shimatani Z, Kashojiya S, Takayama M, Terada R, Arazoe T, Ishii H, Teramura H, Yamamoto T, Komatsu H, Miura K (2017) Targeted base editing in rice and tomato using a CRISPR-Cas9 cytidine deaminase fusion. Nat Biotechnol 35(5):441–443

Shmakov S, Abudayyeh OO, Makarova KS, Wolf YI, Gootenberg JS, Semenova E, Minakhin L, Joung J, Konermann S, Severinov K (2015) Discovery and functional characterization of diverse class 2 CRISPR-Cas systems. Mol Cell 60(3):385–397

Shmakov S, Smargon A, Scott D, Cox D, Pyzocha N, Yan W, Abudayyeh OO, Gootenberg JS, Makarova KS, Wolf YI (2017) Diversity and evolution of class 2 CRISPR–Cas systems. Nat Rev Microbiol 15(3):169

Subburaj S, Chung SJ, Lee C, Ryu S-M, Kim DH, Kim J-S, Bae S, Lee G-J (2016) Site-directed mutagenesis in Petunia× hybrida protoplast system using direct delivery of purified recombinant Cas9 ribonucleoproteins. Plant Cell Rep 35(7):1535–1544

Svitashev S, Schwartz C, Lenderts B, Young JK, Cigan AM (2016) Genome editing in maize directed by CRISPR–Cas9 ribonucleoprotein complexes. Nat Commun 7:13274

Temme K, Zhao D, Voigt CA (2012) Refactoring the nitrogen fixation gene cluster from *Klebsiella oxytoca*. Proc. Natl Acad Sci 109(18):7085–7090

Venclovas Č (2016) Structure of Csm2 elucidates the relationship between small subunits of CRISPR-Cas effector complexes. FEBS Lett 590(10):1521–1529

Voytas DF, Gao C (2014) Precision genome engineering and agriculture: opportunities and regulatory challenges. PLoS Biol 12:6

Wang H, Yang H, Shivalila CS, Dawlaty MM, Cheng AW, Zhang F, Jaenisch R (2013) One-step generation of mice carrying mutations in multiple genes by CRISPR/Cas-mediated genome engineering. Cell 153(4):910–918

Watson A, Ghosh S, Williams MJ, Cuddy WS, Simmonds J, Rey M-D, Hatta MAM, Hinchliffe A, Steed A, Reynolds D (2018) Speed breeding is a powerful tool to accelerate crop research and breeding. Nat Plants 4(1):23

Werren JH (2011) Selfish genetic elements, genetic conflict, and evolutionary innovation. Proc Natl Acad Sci 108:10863–10870

Wolt JD, Wang K, Sashital D, Lawrence-Dill CJ (2016) Achieving plant CRISPR targeting that limits off-target effects. Plant Genome 9:3

Wolter F, Puchta H (2017) Knocking out consumer concerns and regulator's rules: efficient use of CRISPR/Cas ribonucleoprotein complexes for genome editing in cereals. Genome Biol 18 (1):43

Woo JW, Kim J, Kwon SI, Corvalán C, Cho SW, Kim H, Kim S-G, Kim S-T, Choe S, Kim J-S (2015) DNA-free genome editing in plants with preassembled CRISPR-Cas9 ribonucleoproteins. Nat Biotechnol 33(11):1162

Xie K, Yang Y (2013) RNA-guided genome editing in plants using a CRISPR–Cas system. Mol Plant 6(6):1975–1983

Xing H-L, Dong L, Wang Z-P, Zhang H-Y, Han C-Y, Liu B, Wang X-C, Chen Q-J (2014) A CRISPR/Cas9 toolkit for multiplex genome editing in plants. BMC Plant Biol 14(1):327

Xu R-F, Li H, Qin R-Y, Li J, Qiu C-H, Yang Y-C, Ma H, Li L, Wei P-C, Yang J-B (2015) Generation of inheritable and "transgene clean" targeted genome-modified rice in later generations using the CRISPR/Cas9 system. Sci Rep 5:11491

Xue C, Zhang H, Lin Q, Fan R, Gao C (2018) Manipulating mRNA splicing by base editing in plants. Sci China Life Sci 61(11):1293–1300

Yamano T, Nishimasu H, Zetsche B, Hirano H, Slaymaker IM, Li Y, Fedorova I, Nakane T, Makarova KS, Koonin EV (2016) Crystal structure of Cpf1 in complex with guide RNA and target DNA. Cell 165(4):949–962

Yan F, Kuang Y, Ren B, Wang J, Zhang D, Lin H, Yang B, Zhou X, Zhou H (2018) Highly efficient A· T to G· C base editing by Cas9n-guided tRNA adenosine deaminase in rice. Mol Plant 11 (4):631–634

Yang H, Wang H, Jaenisch R (2014) Generating genetically modified mice using CRISPR/Cas-mediated genome engineering. Nat Protoc 9(8):1956

Zetsche B, Gootenberg JS, Abudayyeh OO, Slaymaker IM, Makarova KS, Essletzbichler P, Volz SE, Joung J, Van Der Oost J, Regev A (2015) Cpf1 is a single RNA-guided endonuclease of a class 2 CRISPR-Cas system. Cell 163(3):759–771

Zetsche B, Heidenreich M, Mohanraju P, Fedorova I, Kneppers J, DeGennaro EM, Winblad N, Choudhury SR, Abudayyeh OO, Gootenberg JS (2017) Multiplex gene editing by CRISPR–Cpf1 using a single crRNA array. Nat Biotechnol 35(1):31

Zhang L, Zhou Q (2014) CRISPR/Cas technology: a revolutionary approach for genome engineering. Sci China Life Sci 57(6):639–640

Zhang Y, Liang Z, Zong Y, Wang Y, Liu J, Chen K, Qiu J-L, Gao C (2016) Efficient and transgene-free genome editing in wheat through transient expression of CRISPR/Cas9 DNA or RNA. Nat Commun 7:12617

Zhang S, Zhang R, Song G, Gao J, Li W, Han X, Chen M, Li Y, Li G (2018) Targeted mutagenesis using the Agrobacterium tumefaciens-mediated CRISPR-Cas9 system in common wheat. BMC Plant Biol 18(1):302

Zong Y, Wang Y, Li C, Zhang R, Chen K, Ran Y, Qiu J-L, Wang D, Gao C (2017) Precise base editing in rice, wheat and maize with a Cas9-cytidine deaminase fusion. Nat Biotechnol 35 (5):438

Zong Y, Song Q, Li C, Jin S, Zhang D, Wang Y, Qiu J-L, Gao C (2018) Efficient C-to-T base editing in plants using a fusion of nCas9 and human APOBEC3A. Nat Biotechnol 36 (10):950–953

Delivery Methods, Resources and Design Tools in CRISPR/Cas

3

Muhammad Aamir Aslam, Masooma Hammad, Aftab Ahmad, Josef Altenbuchner, and Hazrat Ali

Abstract

CRISPR (clustered regularly interspaced short palindromic repeats) and CRISPR associated system Cas9 (CRISPR/Cas9) system was discovered as the bacterial immune system against invading viruses. Its applications not only include genome editing but it is also being used for gene regulation, epigenome editing, chromatin engineering, and imaging. The most widely used CRISPR/Cas9 system which have been used for experimental genome editing consists of two components, i.e. Cas9 nuclease protein and sgRNA (a 20 nucleotide (nt) long RNA molecule). The programmability and specificity of this system is conferred by the gRNA molecule. These two components can either be delivered in the form of ribonucleoprotein complex (RNPs), in vitro transcribed mRNA or in the form of DNA which is transcribed in vivo to produce Cas9 and sgRNA. The efficiency of this system depends upon the proper delivery, efficient transcription, and optimal activity of sgRNA and Cas9 on intended target site and minimum unwanted genome perturbations, known as off-target effects. In this chapter, we will present a brief introduction and history of CRISPR/Cas followed by description of various tools used for its delivery and overview of gRNA design tools.

M. A. Aslam
Institute of Microbiology, University of Agriculture, Faisalabad, Pakistan
e-mail: aamir.aslam@uaf.edu.pk

M. Hammad · H. Ali (✉)
National Institute for Biotechnology and Genetic Engineering (NIBGE), Faisalabad, Pakistan

A. Ahmad
Center for Advanced Studies in Agriculture and Food Security (CASAFS), University of Agriculture, Faisalabad, Pakistan
e-mail: aftab.ahmad@uaf.edu.pk

J. Altenbuchner
Institute of Industrial Genetics, University of Stuttgart, Stuttgart, Germany
e-mail: josef.altenbuchner@iig.uni-stuttgart.de

A. Ahmad et al. (eds.), *CRISPR Crops*,
https://doi.org/10.1007/978-981-15-7142-8_3

Different resources are described which are used to validate the intended genome editing. Finally, we will conclude with future directions and the broader impact of CRISPR technologies.

3.1 CRISPR/Cas: An Introduction

In the microbial world full of threatening nucleic acids from invaders, bacteria have to protect themselves against phages and other bacteria. Thus, bacteria and archaea have evolved such a system in which they collect specific DNA fragments from invaders' genome called protospacer and incorporate them into repetitive region of their own genome (called Clustered Regularly Interspaced Short Palindromic Repeats or CRISPR) thus creating a "memory" of attack. Next time, when the same invader attacks this mechanism gets into action and cleaves foreign DNA (McGinn and Marraffini 2018). So basically CRISPR/Cas based immunity goes through three stages; adaptation phase, expression phase, and interference phase. Acquisition or adaptation phase starts when a short sequence of alien DNA is cleaved out of its genome and incorporated into CRISPR arrays in host genome (Roointan and Morowvat 2016). CRISPR arrays are usually stretches of 21–48 bp which are repetitively (up to 250 times) present at CRISPR loci and these repeats are interspaced by variable stretches or spaces of approximately 20–58 bp (Horvath et al. 2008). In the expression stage, CRISPR locus gets transcribed and a precursor CRISPR-RNA (crRNA) or pre-crRNA is synthesized. This pre-crRNA is processed to produce mature crRNA. The crRNA forms complex with CRISPR-associated proteins or Cas proteins, which is called ribonucleo protein (RNP) complex. In the interference phase, the complex guided by crRNA, based on complementarity, targets complementary sequence in attacker's genome and Cas proteins which are endonucleases, cleave foreign DNA thus protects the host from foreign attack (Roointan and Morowvat 2016). Approximately 45% bacteria and 90% archea have adapted this mode of immunity (Crawley et al. 2018). Based on *cas* gene content, CRISPR/Cas systems can be categorized into two groups (Wiedenheft et al. 2011) and on the basis of different protein complexes targeting DNA or RNA, there are six types along with 20 subtypes. Out of these, types I, II, and V destroy DNA, type III targets both DNA and RNA, type VI attacks RNA, while type IV is uncharacterized yet. There is a basic conserved mechanism of acquiring spacer in all types which is determined by genes *cas* I and *cas* 2 (McGinn and Marraffini 2018) whereas different protein complexes tackle invader nucleic acids differently. Type I CRISPR/Cas system utilizes a cascade made up of crRNA and multiple proteins for recognizing target DNA. After recognition, target is chopped down by a nuclease called Cas3. In the case of type III CRISPR/Cas system, nuclease Cas10 gets into the complex like that of type I cascade, targets foreign DNA and eventually degrades it. Unlike multiprotein complexes involved in type I and III, type II CRISPR/Cas system utilizes only one multidomain nuclease, Cas9, for its function of binding and cleavage (Gasiunas et al. 2014).

3.2 History of CRISPR

History of CRISPR/Cas system started with the discovery of palindromic repeats with intervening nonrepetitive sequences found in *E. coli* (Ishino et al. 1987) and later on in other bacteria (Van Soolingen et al. 1993) and archaea (Mojica et al. 1993) but their exact function was unknown. The term CRISPR was coined for the presence of these sequences in 2002. With the discovery of genes found adjacent to CRISPR loci in genomes which were termed as CRISPR-associated system or *cas*, a connection between CRISPR loci and *cas* genes was established (Jansen et al. 2002) but the exact function was still to be explored. The research on CRISPR/Cas system took a leap in 2005, when three groups Mojica et al. (2005), Pourcel et al. (2005), and Bolotin et al. (2005) independently outlined foreign chromosomal origin of CRISPR's nonrepeated sequences derived from other bacteria and bacteriophages. An interesting observation was made that bacteria containing foreign viral DNA sequences showed resistance to that particular virus which suggested the possibility of the involvement of this system in adaptive immunity of bacteria. In 2007, it was experimentally proved to be the fact by Barrangou et al. (2007). When status of CRISPR/Cas system as adaptive immune system had fully established, its potential use in gene editing procedures was suggested, which has previously been done by zinc fingers nucleases (ZFNs) and transcription activator-like effector nucleases (TALENs). ZFPs and TALENS belong to the class of programmable chimeric nucleases, which have DNA binding domain specific to the target, linked with DNA cleaving domain which is non-specific. In this way these procedures induce double stranded breaks in target DNA. But the problem with these methods is that they have to be designed every single time a DNA modification is required, that limits their frequent applicability. Unlike them, CRISPR/Cas system offers great flexibility for use and only new single guide RNA (sgRNA) is required for each targeting while Cas protein remains the same, enhancing their potential for routine applications (Gaj et al. 2013).

3.3 Type II CRISPR/Cas9 for Gene Editing

Due to its simplicity and ease to use, CRISPR/Cas system type II is the most studied and most adapted system for genome engineering. This system only contains two RNA components and a single nuclease called Cas9 for its function. The system starts with the transcription of CRISPR loci to generate a long RNA molecule called pre-crRNA which contains homologous sequences of previous invaders between repeated regions. Another RNA is transcribed from upstream of CRISPR locus which is called trans-activating CRISPR RNA or tracrRNA. It has complementarity with CRISPR repeat region so it gets attached to newly transcribed pre-crRNA thus creating double stranded RNA which is recognized and cleaved by an RNase called RNase III. In the result of this, a complex of crRNA and tracrRNA is formed which has now only one spacer sequence. This single spacer crRNA:tracrRNA complex is recruited by Cas9, a nuclease, thus an active ribonucleoprotein (RNP) complex is generated, which is ready to attack the invader nucleic acid which it had been previously exposed to (Thurtle-Schmidt and Lo 2018). The crRNA gets attached to its complementary sequence in the genome of incoming invader after Cas9 binds to a specific sequence called PAM (Protospacer *A*djacent *M*otif). There are different PAM sequences which are recognized by Cas9 of different bacterial or archeal origin. To date Cas9 of *Streptococcus pyogenes* (SpCas9) is the most studied and adapted one for genome editing. It recognizes PAM sequence 5′ NGG 3′. Two arginines at position 1333 and 1335 in Cas9 help in its positioning at 5′ site of PAM sequence by creating an interaction with two guanines of PAM, thus unwind invaders DNA for cleavage (Anders et al. 2014). When crRNA finds complementarity with foreign DNA sequence upstream of PAM, its forms hybrid of RNA and DNA in an R loop, which Cas9 cleaves. For nucleic acid cleavage, Cas9 has two domains, an HNH domain which nicks DNA strand complementary to crRNA while RuvC domain creates a nick in non-complementary strand (Gasiunas et al. 2012). Cas9 performs blunt end cleavage at three base pairs upstream to PAM sequence. This has lethal effect on invader and thus host gets protected by the foreign attack (Thurtle-Schmidt and Lo 2018).

Due to their ease of use, type II CRISPR/Cas9 system was quickly adopted by scientists for making deliberate changes in genomes. For gene editing purposes, CRISPR/Cas system relies on creating double stranded breaks (DSBs) in the DNA. Cells utilize two main mechanisms for the repair of these DSBs; either by Non-Homologous End Joining (NHEJ) or by Homology Directed Repair (HDR). NHEJ does not require a template sequence for repair and simply joins the broken ends thus there is great possibility of causing INDELs in DNA while HDR follows a homology template for repairing DSBs in DNA and therefore it is error free mechanism of repair. While using CRISPR/Cas system for gene/genome editing, both these mechanisms can be employed for different applications; by using error prone repair by NHEJ, targeted genes can be rendered inactive through frameshift mutations or INDELs while for precise genome editing or a homology template can be applied so mutations can be accurately created or repaired or even genes can be knocked in or out accurately (Ma et al. 2016). Doudna and Charpentier were the pioneers to use this system in 2012 for introducing double stranded breaks in DNA (Jinek et al. 2012). They also demonstrated that tracrRNA and crRNA could be fused into a single truncated version of RNA which still possessed cleaving activity (Jinek et al. 2012). Thus CRISPR/Cas9 system get condensed to just two components; a single guide RNA (sgRNA) for targeting and Cas9 for cleavage which was a breakthrough for its success in coming times. See Fig. 3.1a. for pictorial representation of mechanism of action CRISPR system for its use to experimentally perform gene editing after cutting of DNA on-target site and Fig. 3.1b for two different DNA repair pathways (HDR or NHEJ) to get desired genetic changes (gene insertion, deletion or random or specified mutation). To further reduce off-target effects of CRISPR/Cas9, the use of Cas9 as "nickase" (nCas9) instead of nuclease was suggested (Cong et al. 2013). It opened another paradigm of its utility, which was experimentally proved to be successful for more specific genome editing (Ran et al. 2013). Furthermore, it was demonstrated that by removing DNA cleavage activity of

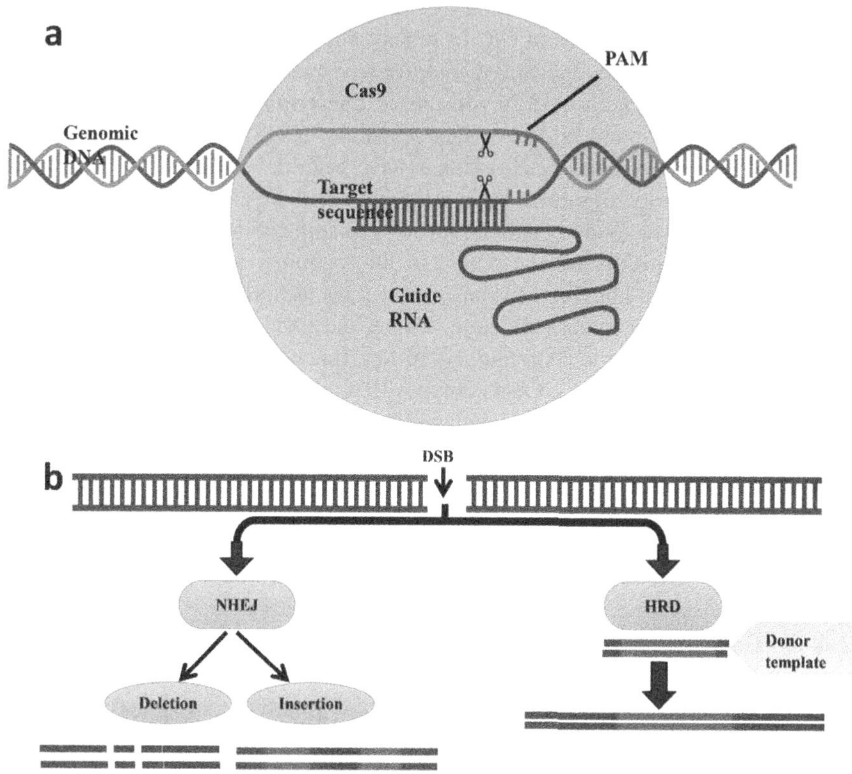

Fig. 3.1 Mechanism of action CRISPR/Cas system for its use to experimentally perform gene editing. (**a**) Cas9 protein makes a double stranded cut 3′ upstream of PAM region and on place where sgRNA binds to DNA template through complementarity between base pairs. (**b**) repair of this double stranded DNA cut can happen through two different repair pathways (HDR or NHEJ). If a foreign DNA template bearing homology to the region of cut is present, the foreign DNA can be inserted at the point of DNA cut through HDR pathway. In the absence of foreign DNA template cell can also repair the cut DNA through error prone NHEJ pathways and it can result in the deletion of base pair thus resulting in frame shift mutations or other types of mutations leading to protein dysfunctions and gene ablation

Cas9 while retaining its DNA binding capability, deactivated Cas9 (dCas9) can be used to disrupt transcription factors/and RNA polymerase; this way they can suppress expression of multiple genes as well (Qi et al. 2013). Moreover, Cas9 and sgRNA can be repurposed as transcription activator for enhanced gene expression (Mali et al. 2013a). Another recent breakthrough in type II CRISPR/Cas9 system was the use of catalytically inactive Cas9 with tRNA (transfer RNA) adenosine deaminase, in order to target and specifically convert base pairs from A/T to G/C and found 50% efficiency with fewer unintended effects. This technique added an invaluable tool to existing CRISPR/Cas toolbox especially to mend single point mutations involved in most diseases (Gaudelli et al. 2017). This technique has been adopted by scientist for precise gene editing in mouse (Liang et al. 2018; Yang et al.

2018) and plants, i.e. rice (Yan et al. 2018; Hua et al. 2018) even some of them induced reverse transitions from G/C to A/T as well (Komor et al. 2017). CRISPR/Cas systems have found applications in epigenetics as well to study and modulate relationships of epigenetic factors with genes transcription, i.e. by acetylation or methylation (Hilton et al. 2015). Another domain of CRISPR/Cas9 usage is inducible CRISPR/Cas9 systems. Such systems have been developed which have their DNA editing activity triggered by light (Nihongaki et al. 2015) or by certain chemicals which was found beneficial for in-vivo application in mice (Polstein and Gersbach 2015). sgRNA can be bound to fluorophores in order to locate genes within nuclei without bursting cells and nuclei. This technique termed as CASFISH has been employed in living (Ma et al. 2015a) and fixed human cells (Deng et al. 2015) although with some shortcomings of non-functional fusion (Ledford 2016). Quite recently, in some cases, Cas9 subtypes II-A and II-C were found to recognize target RNA based on its interaction with sgRNA without the need of PAM sequence. This mechanism has shown to be effective against DNA as well as RNA viruses thus further expanding the horizon of applicability of highly versatile CRISPR/Cas9 system (Strutt et al. 2018).

3.4 Genome Editing Reagents

The two prerequisites of CRISP/Cas system for genome editing are gRNA for targeting and Cas9 for cleavage. These reagents may be provided in different formats; as DNA, RNA, or as ribonucleoprotein complex (RNP). To be expressed from DNA, two plasmids are required, one containing DNA sequence of sgRNA and the other one for DNA sequence of Cas9 protein. Cas9 proteins have been isolated from different bacterial sources so both the plasmids should be carefully designed, must be codon optimized according to the codon used in the host genome (Shen et al. 2013). A suitable promoter is added immediately upstream to transcription start site (TSS) for Cas9 gene transcription to mRNA. Moreover, for efficient expression of Cas9 mRNA, untranslated regions (UTRs) are added. sgRNA has to be expressed only to mRNA so it is transcribed by RNA polymerase III (Thurtle-Schmidt and Lo 2018). In addition, homology templates for the repair of DSBs are added to the plasmid, in case when precise editing via homologous recombination is required (Altenbuchner 2016). DNA plasmids are the most frequently used reagents for genome editing but burden on host cells can be reduced by in-vitro transcription of Cas9 into mRNA, to which 5′ cap and 3′ poly-A tail are added and then, along with sgRNA, delivered to the host cell (Lo et al. 2013). In other instances, Cas9 can be expressed in-vitro as protein and then loaded with sgRNA and delivered to the cell in the form ribonucleoprotein complex (RNP) (Liang et al. 2015). In the following sections, different methods are described for the delivery of reagents meant for genome editing through CRISPR-Cas system in animals and plants.

3.5 Delivery Methods for CRISPR-Cas in Animals

3.5.1 Viral-Mediated Delivery

Using virus as delivery system for any exogenous DNA, there are two stages of host genome editing. During first stage, virus recognizes its specific host, enter into it and releases its genome; DNA into cell nucleus or RNA into cytoplasm for its replication. After the host cellular machinery replicates viral genome, virus leaves the host cell and can infect neighboring host cells. For deliberate editing of target genome, required material is packed into the virus and is allowed to infect the target cell where it enters and make desired changes (Vannucci et al. 2013). A wide range of virus based vectors are available for targeted genome editing; Adeno-associated viral vectors (AAVs), adeno viral vectors (AdVs), and lentiviral vectors (LVs), to name a few (Chandrasekaran et al. 2018). For an overview of different methods used for delivery of CRISPR-Cas system in animals and animal cells, see Fig. 3.2.

3.5.1.1 Adeno-Associated Viral Vector Mediated Delivery

Adeno-associated virus (AAV) is a single stranded DNA virus which belongs to the genus Dependovirus of family Parvoviridae. Its high demand for gene therapy is owing to its inability to cause any disease in humans. On its first attack, it causes negligible or no immunogenicity in its host adding to its profile as safe delivery method (Samulski and Muzyczka 2014). Moreover, it is capable of infecting a broad

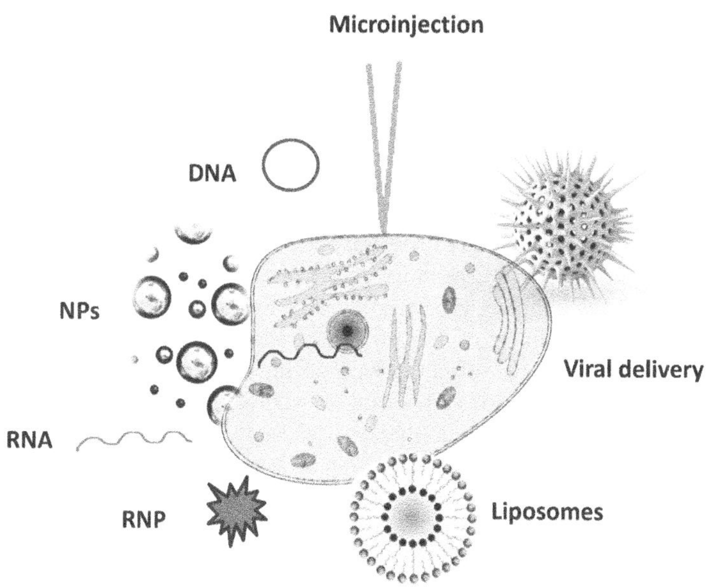

Fig. 3.2 Delivery methods commonly used to deliver CRISPR-Cas system in animal cells. Microinjection, Liposomes, Nanoparticles, and viruses are commonly used delivery methods to deliver CRISPR-Cas system inside cells

range of cells which helps it escape immune reaction of the host. A major advantage of using this system is that DNA delivered by this method can exist in host cell for indefinite time period; it can remain independently in cell or can be incorporated into the host genome making it suitable for long-term expression (Deyle and Russell 2009). Their high versatility have made AAV vectors serve various applications in all research areas; in vitro, ex vivo, and in vivo (Lino et al. 2018). Despite its benefits, the use of AAV faces many practical challenges. First of all, the average size of an AAV is approximately 20 nm and it is capable of accommodating 4.5–5 kB of genome in it. But alone the size sgRNA and SpCas9 is about 4.2 kB which leaves practically no space for additional genetic elements, i.e. repair templates in case of HDR, multiple gRNA, reporters or tags (Wu et al. 2010). This problem can be overcome by using minimal promoters/enhancers with Cas9 gene. As in case of correcting mutations involved in Duchenne muscular dystrophy (DMD) in mouse model, Long et al. (2016) used a minimized version of cytomegalovirus promoter along with SpCas9 which resulted in restoration of dystrophin protein and improved muscular function in mice. Another possibility to overcome packaging size restraint is to engineer host cell for the expression of Cas9 protein and to deliver only sgRNA into the cells using adeno-associated virus. This was employed by Carroll et al. (2016) for creating a mouse model for cardiovascular research. They transfected heart cells of mice by microinjection for the expression of Cas9 protein and only delivered sgRNA exogenously for *Myh6* gene deletion (Carroll et al. 2016). Many studies show that Cas9 and sgRNA can be separated into two different AAVs and then co-infecting the host with them thus allowing in vivo genome engineering in brain cells (Swiech et al. 2015) and retinal cells (Hung et al. 2016). This approach increases available construct size but it comes with the cost of screening those cells which have both virus particles infecting. Cas9 gene can be splitted into two halves; each half being packed in different viral particles and combines in host cell. The reconstituted Cas9 was found equally competent as intact protein (Truong et al. 2015; Chew et al. 2016).

3.5.1.2 Lentiviral Vector (LV) Mediated Delivery

LVs are derived from HIV-1, resulting in the production of a very efficient vector owing to its greater carrying capacity (9.7 kb), tropism, and the trait of infecting post-mitotic cells (Naldini et al. 1996b) and found useful for simultaneous delivery of sgRNA and Cas9. However, a Cas9 producing cell line is not required for this co-delivery (Shalem et al. 2014). LVs are considered the most widely utilized viral vectors for applications like gene therapy, especially when a stable and extended expression is required (Naldini et al. 1996a). Actually, LVs are also chosen for ex vivo gene therapy because of these properties, especially when hematopoietic stem cells (HSCs) and T cells (T lymphocytes) are being targeted. For instance, a recent study utilized LVs for T-lymphocyte based immunotherapy and for treating Wiskott–Aldrich syndrome and β-thalassemia (Naldini 2015). Integrase-deficient LVs carrying integrase mutations and long terminal repeats (LTRs) are used to increase the probability of insertional mutagenesis, however, complete elimination of the risk is not ensured (Philippe et al. 2006; Wanisch and Yáñez-Muñoz 2009).

Presently, a number of research groups are exploiting LV-based co-delivery to produce knockout libraries (Sanjana et al. 2014; Shalem et al. 2014). Apart from ex vivo gene therapy, LVs also have a role to play in in vivo gene therapy; development of CRISPR/Cas9 and LV-Cas9 are such examples (Choi et al. 2016). A dual vector system consisting of both CRISPR and LV Cas9 is being used to induce novel mutations in tumor suppressor as well as tumor accelerator genes (Aubrey et al. 2015). LVs have been successfully used for genetic manipulations using CRISPR system in a number of previous studies (Wang et al. 2014; Koike-Yusa et al. 2014).

3.5.1.3 Adenovirus-Mediated Delivery

Adenoviruses belong to the family Adenoviridae and are named so because of their isolation from adenoid tissues of humans in 1953. Due to its larger size, adenovirus can accommodate larger genomic elements. Cas9 gene, multiple sgRNAs or even homology-based repair templates can be easily packaged in a single viral particle. It allows the expression of sgRNA and Cas9 at equal rate maintaining their uninterrupted combined effect. Its non-integrating nature makes it suitable for transient expression of Cas9 in dividing cells. It is extensively applied in in vitro and in vivo delivery systems and has successfully been used in different cell lines. It has been used for in vivo genome engineering of mice for inducing Nonalcoholic steatohepatitis (NASH), a liver disease, by editing *Pten* gene involved in negative regulation of PI3-AKT pathway of cell proliferation and found successful even if it had provoked immune response in hepatocytes (Wang et al. 2015). A wide array of different human cell lines including osteosarcoma U2OS cells, hMSCs, cervix carcinoma HeLa cells, and myoblasts were successfully transduced with second generation adenovirus expressing SpCas9 or sgRNA and resulted in transduction efficiencies comparable to that of adenovirus with TALENS (high-titers up to 20×10^{10} transducing units/ml) and opens the avenues for editing human somatic cells in vivo (Maggio et al. 2014).

3.5.2 Non-viral Vectors

3.5.2.1 Cationic Vectors

Cationic vectors are positively charged vectors and are suitable for the uptake and delivery of anionic DNA of sgRNA or Cas9 to the cells. In this method nucleic acids are coupled with positively charged compounds. These are broadly classified into two categories based on either they have cationic polymer or cationic lipid. In case of polymer-based vectors for CRISPR/Cas components delivery, structural and physical properties of lipid compound are decisive for its efficiency. Polyethyleneimine (PEI) is the most commonly used cationic polymer due to its high molecular weight, more positive charges, branching pattern, and presence of secondary amine. All these properties make it suitable candidate to pass through and retain into the cell although it shows some toxicity to the cell. Linear PEI found to have more transfection efficiency than branched PEI (Wightman et al. 2001). Lipid based cationic vectors have also been utilized for delivering sgRNA and Cas9. Liposomes protected nucleic acid can evade endosomal targeting because liposomes get fused with endosomal membrane during its maturation phase. Liposome based delivery of sgRNA and Cas9 RNP complex to target genes in mouse hair cells got 80% success in gene editing (Zuris et al. 2015).

3.5.2.2 Cell-Penetrating Peptides

The presence of cellular barriers in the form of cell wall and cell membrane and cell membrane's low permeability hinders the intracellular delivery of cargos (Guidotti et al. 2017). Moreover, negative charge, high molecular weight, and vulnerability of DNA adds to the complexity (de Figueiredo et al. 2014). To deliver cargos intracellular, it was observed that certain proteins are capable of translocation effectively which were called as cell-penetrating peptides. Cell-penetrating peptides (CPPs) are small peptides of about 5–30 amino acids which have strong translocating capabilities and can cross cellular barriers without interacting with receptors through energy dependent, i.e. through endocytosis or energy independent processes, i.e. through direct penetration into cell membrane by creating transient pores (Guidotti et al. 2017). Based on their physical properties these peptides may possess cationic, hydrophobic/non-polar or amphipathic nature (Lino et al. 2018). At physiological pH, cationic peptides have net positive charge because these are rich in positively charged residues like arginines and lysines which contribute to their internalization into the cells. Hydrophobic peptides possess very low net charge at physiological pH due to the presence of hydrophobic amino acids, but show more affinity with non-polar components of cell membrane and can spontaneously cross cell membrane in energy independent fashion. Amphipathic class of CPPs contain hydrophobic amino acids, i.e. arginines and lysines along with hydrophobic amino acids, i.e. alanine, valine, leucine, isoleucine, etc. (Zaro and Shen 2015). These peptides are either naturally existing or chimeric peptides and are prepared by covalently linking hydrophobic stretch for cellular uptake to hydrophilic signal containing prolines, arginines, and lysines for translocation into nucleus (Morris et al. 2008). Different classes of CPPs have been used either individually or in combinations for in-vitro and ex-vivo purposes but there in vivo applications are still limited (Morris et al. 2008). By coupling CPP with Cas9 in thioester bond and sgRNA complexed with CPP, genome editing was made in many cell lines including human embryonic stem cells (hESCs), embryonic carcinoma cells, HeLa cells, HEK293T cells, and dermal fibroblasts and showed less off-target effects (Ramakrishna et al. 2014). But their efficiency relies on the peptide type and cell type and have variable efficiency in case of CRISPR/Cas components delivery (Lino et al. 2018).

3.5.3 Other Non-viral Methods

Lipid based nanoparticles are commonly used for delivering various therapeutic agents to cells for quite a long time. They have also been used to deliver nucleic acids into cells either DNA or RNA. Naked nucleic acids always face resistance by cell in the form of enzymatic degradation or chopping down by lysozymes. To make foreign DNA enter safely into the cell or nucleus, it has to be encapsulated in such an entity which can dodge cellular checkpoints. Lipid nanoparticles presents an easy solution to this problem as they do not provoke immunogenic response like viral delivery systems and can be used for in vivo, ex vivo, and in vitro applications (Wang et al. 2016). Earlier, lipid nanoparticles were used with RNA interference but later they were used for CRISPR/Cas as well (Thi et al. 2017). CRISPR/Cas components can be delivered in the form of nucleic acids, i.e. as DNA or mRNA of Cas9 with sgRNA or as Cas9 protein with sgRNA in complex but nucleic acids showed better results (Yin et al. 2016; Wang et al. 2016). Lipofectamine is positively charged commercially available lipofection reagent that can bind anionic nucleic acids. This reagent helps reduce electrostatic repulsion of anionic nucleic acids by negatively charged membranes. It can transfect mitotic as well as non-mitotic cells. For an effective transfection, positively charged lipid has to be formulated with a helper lipid which is neutral in nature. In this way, cationic lipid binds and protect anionic nucleic acids while neutral lipid bound with cationic lipid helps fuse this complex with endosomal membranes (Dalby et al. 2004). Lipofectamine has been used in different humans and animal cells to successfully deliver sgRNA/Cas9 cargos (Schwank et al. 2013; Sakuma et al. 2014; Liang et al. 2015; Zuris et al. 2015). There are fewer success stories of CRISPR/Cas delivery through this method (Guidotti et al. 2017). Efforts are being made to enhance its activity by screening more compounds, to add more functional groups and to protect these particle from immune response (Lino et al. 2018). For an overview of different methods used for delivery of CRISPR-Cas reagents in animals and animal cells, see Fig. 3.2.

3.5.4 Physical Methods

In order to avoid problems regarding viral vectors, many physical methods for delivery of Cas9 have been developed and utilized. These methods are simpler, target specific, highly reproducible, and therefore utilized extensively for in vivo and ex vivo applications.

3.5.4.1 Microinjection

Microinjection is considered "benchmark" with 100% efficiency of delivering CRISPR components into targeted cells (Wang et al. 2013b; Horii et al. 2014). Microinjection is relatively simpler technique and by using a microscope, cargos are delivered through micro needle (0.5–5.0 m diameter) into the cell. Injecting cargos may be in the form of DNA sequences of Cas9 and sgRNA, mRNA of both of them by in vitro transcription, or sgRNA with Cas9 protein. Microinjection offers several advantages, i.e. it allows precise quantities of cargos to be delivered in order to avoid off-target cleavage, provision of large quantities of cargos unlike restricted size of packaging in viral vectors and most importantly it can circumvent cellular barriers. All these properties make it a viable option for ex vivo and in vitro delivery application, but in vivo applications are still limited. There are various formats for introducing cargos into the cell based on particular application required. Some scientists prefer delivering DNA both for sgRNA and Cas9 directly into cellular nucleus, in order to avoid in vitro mRNA synthesis. But single stranded DNA or plasmid DNA can easily get incorporated randomly into the genome, may lead to unintended gene disruption and can affect cellular functions. Moreover, nuclear integration of delivery cargos may constitutively express Cas9 protein adding to the complexity of the situation (Nakagawa et al. 2015; Chuang et al. 2017). To solve these problems, some scientists suggested direct delivery of mRNA for Cas9 and sgRNA into cytoplasm of the cell so that Cas9 mRNA gets translated and Cas9 protein binds sgRNA and carry it to nucleus for DNA modifications. So far, it has been found to be a better approach to minimize off-target effects (Horii et al. 2014). For an overview of different methods used for delivery of CRISPR-Cas system in animals and animal cells, see Fig. 3.2.

3.5.4.2 Electroporation

Electroporation is the method of introducing cargos into the buffer suspended cells by using high voltage electrical pulses to temporarily create very small holes (of nanosize) in the cellular membrane. Owing to its flexibility, higher transformation efficiencies and high throughput nature, it is one of the most established techniques of genetic manipulations especially in vitro and ex vivo as it does not rely on cell type. But it is less preferred for in vivo applications as cells have to be exposed to higher voltages. Mammalian cells are more sensitive to longer duration of high voltage pulses than bacterial cells, and cell lines are comparatively more tolerant than zygotes. Different formats of delivery cargos have been attempted by electroporation. Use of RNP in various human cell lines showed less off-target but this method has also been found effective with DNA plasmids, even for multiplexing (Schumann et al. 2015; Matano et al. 2015).

3.5.4.3 Delivery of RNPs Using Gesicles

The term "microvesicle" encompasses a heterogenous group of extracellular membranous organelles having 20–1000 nm diameter which are known to be involved in normal cellular functions; immunity, cell proliferation, and communication, etc. (Breakefield et al. 2011). Mangeot et al. (2011) induced the formation of gesicles from human cells by spiking them with glycoprotein of vesicular stomatitis virus (VSV-G). They named these vesicles as "gesicles" with reference to glycoprotein. The exact mechanism for their biosynthesis is not well understood but they are assumed to be of endosomal origin (Roy and Noad 2008).

Gesicles present a new delivery vehicle for therapeutic agents. Being native elements, they do not pose threat of provoking immune responses or toxicity in host cells. This tool is efficient in simultaneous delivery of Cas9 and sgRNA. For cargos delivery, first of all expression cassette for protein or mRNA must be transfected to the producer cell ex vivo, which is then induced by glycoprotein to produce gesicles loaded with delivery cargos. Finally, these loaded gesicles are isolated and used for delivery (Breakefield et al. 2011). Quinn et al. (2016) used ligand dependent dimerization method named "iDimerize technology" for CRISPR/Cas RNP complex packaging into gesicles. They delivered these gesicles in various human iPS (Induced pluripotent stem) cells and found them more efficient than plasmid based delivery with no off-target effects along with stable and tightly controlled release of CRISPR/Cas components (Quinn et al. 2016).

3.6 Delivery Methods for CRISPR-Cas in Plants

In plants, first reports of application of CRISPR/Cas came in 2013 by various groups for the genome editing of model plants *Arabidopsis thaliana* and *Nicotiana benthamiana* (Li et al. 2013; Nekrasov et al. 2013; Shan et al. 2013b), after which a new horizon of genome editing opened, since then many crop and non-crop plants genomes have been edited by CRISPR/Cas technique. Genome editing efficiency of CRISPR/Cas system is dependent on various factors including nature of target DNA, plant species, expression of sgRNA, GC contents and Cas9 and structure of sgRNA and nature of host genome (Bortesi and Fischer 2015). Plant genome editing CRISPR/Cas based vectors usually contain a strong promoter such as that of cauliflower mosaic virus (CMV) derived 35S RNA polymerase II promoter for the expression of Cas9 and U3 or U6 RNA polymerase III promoter for sgRNA transcription (Ma et al. 2016). For multiplexing, sgRNA may be provided independently or in the form of polycistronic transcript which is further processed by endogenous or exogenous nucleases into individual sgRNAs (Xing et al. 2014; Ma et al. 2015b).

3.6.1 Agrobacterium-Mediated DNA Delivery

Nature has an excellent genetic engineer in the form of *Agrobacterium*, which has the potential to infect a broad category of plants and integrate the part of its DNA called T-DNA into the host plant genome. Certain virulence genes encoded by Ti plasmid aid this interkingdom transmission of DNA and T-DNA transfer is mediated by type IV secretion system. T-DNA accompanying some bacterial proteins enters the plant cell nucleus and gets randomly incorporated to the plant genome (Nester 2015). This potential of *Agrobacterium* has been recruited by scientists to integrate DNA fragments of their interest in to the plant genome. *Agrobacterium*-mediated DNA delivery is considered a reliable approach for stable transformation of CRISPR/Cas cargos in dicotyledonous and monocotyledonous plants (Char et al. 2017; Zhang et al. 2018a). Most of the genome editing events in plants involve the use of *Agrobacterium*-mediated DNA delivery and this method has been utilized for a broad range of plants; from model plants, e.g. Arabidopsis, tobacco, *Nicotiana benthamiana*, petunia to important crop plants, e.g. rice, barley, soybean, Brassica, and maize (Hiei and Komari 2008). *Agrobacterium*-mediated floral dip method is generally used for genome editing in *Arabidopsis* and for Cas9 transformed tobacco, TRV carrying sgRNA has been introduced by agro-infiltration method. In case of dicots and monocots, i.e. rice, maize, tomato, potato and poplar, stable transformation by *Agrobacterium*-mediated delivery into callus, tissues or immature embryos is used (Ma et al. 2016).

This method has been deployed for single gene targeting as well as for multiplexing. Multiplexing has been done in various food and non-food crops in order to manipulate their intrinsic traits. In plants, NHEJ is more prominent method of DSB repair (Puchta 2005) which has been harnessed for the production of mutant alleles. This method has been utilized to develop resistance/tolerance against various abiotic stresses, e.g. resistance against herbicides (Li et al. 2016) and biotic stresses, e.g. tolerance against viral (Macovei et al. 2018; Tashkandi et al. 2018; Chandrasekaran et al. 2016), bacterial (Zhou et al. 2015), and fungal diseases (Wang et al. 2018; Zhang et al. 2018b). Nutritional properties of plants have been manipulated, i.e. amylose and anthocyanin content in rice, lycopene in tomato (Li et al. 2018), steroidal glycoalkaloids (SGA) in potato (Nakayasu et al. 2018). Moreover, plant growth hormone, i.e. gibberellin levels have been modulated in barley and *Brassica* (Lawrenson et al. 2015; Yang et al. 2017). For an overview of different methods used for delivery of CRISPR-Cas system in plants, see Fig. 3.2.

3.6.2 Viral Vectors Mediated Delivery

RNA and DNA based viral vectors can serve as an efficient delivery method for plant cells. RNA viruses (Tobacco rattle virus) are preferable than DNA viruses, such as Bean yellow dwarf virus (BYDV), Cabbage leaf curl virus (CLCV), Wheat dwarf virus (WDV) in a way that they do not get integrated into plant genome (Ran et al. 2017). One of the most widely used viral vector is Tobacco rattle virus (TRV), which has bipartite genome comprised of two positive sense strands; TRV(RNA 1) is involved in virus movement and replication and TRV (RNA 2) being non-essential for virus' survival can be replaced with multiple cloning sites for the insertion of target sequences (Senthil-Kumar and Mysore 2014). Recombinant TRV vector is transformed into *Agrobacterium*, which then infects and transfers it to the plant cells, where viral RNAs are replicated and transferred to other cells through meristematic tissues. They can even reach germline cells which obviates the need of tissue culturing although genome editing can also be done by tissue culture method (Mahas et al. 2019). TRV is highly efficient delivery method especially for multiplexing because of its natural tendency to infect majority of plant species, its small size offers greater opportunity for cloning and library construction and its easy introduction via *Agrobacterium*-mediated delivery (Zaidi and Mansoor 2017). A unique TRV mediated delivery approach has been developed by Ali et al. (2015) which included generation of Cas9 overexpressing transgenic lines of *N. benthamiana* and then delivering sgRNA packaged in TRV by *Agrobacterium*. This method facilitates multiplexing without the need of laborious tissue culture procedures. For an overview of different methods used for delivery of CRISPR-Cas system in plants, see Fig. 3.3.

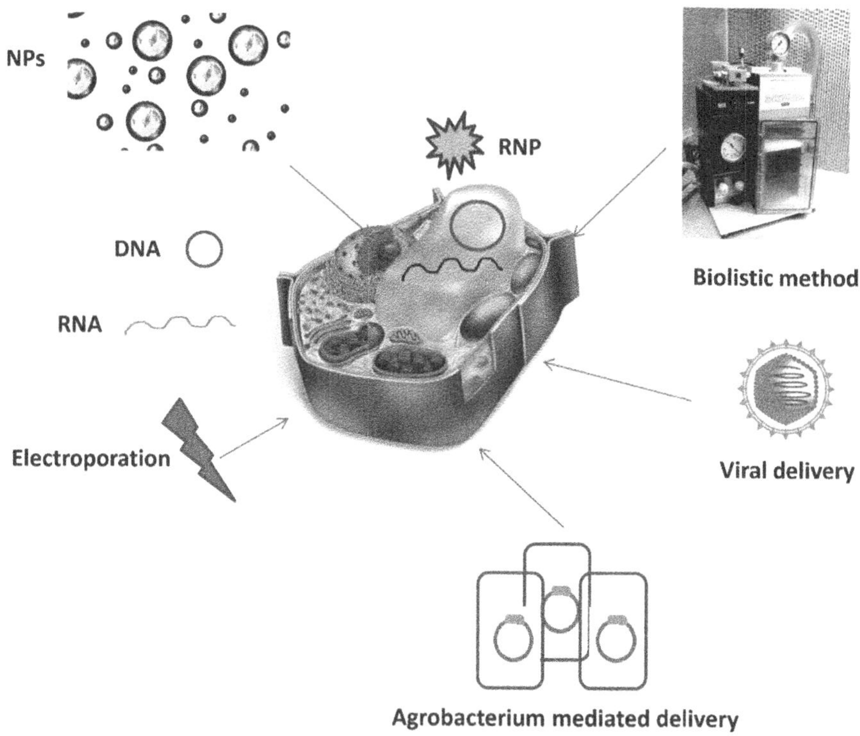

Fig. 3.3 Delivery methods commonly used to deliver CRISPR-Cas system in plant cells. Agrobacterium, electroporation, biolistic method, nanoparticles, and viruses are commonly used delivery methods to deliver CRISPR-Cas system inside cells in plants

Geminiviruses are single stranded circular DNA (SSD) viruses with genome size 2.5–3.0 kb. BYDV and CLCV are widely used DNA viruses. Within host cells, SSD gets converted to double stranded intermediate with the help of host DNA polymerase and initiates expression of viral proteins; replication initiation protein (Rep) being the only protein essential for viral replication as it initiates rolling circle amplification RCA. Then, single stranded DNA get encapsulated to form virions, which may infect neighboring cells through plasmodesmata (Ran et al. 2017). Geminiviruses can be exploited as vector for genome engineering due to its broad host range, high replication rate, requirement of only single protein, Rep, for replication which can be expressed under endogenous and exogenous promoters (Baltes et al. 2014), its ability to replicate through homologous recombination dependent mechanism may serve as ideal tool for precise editing through homologous recombination (Richter et al. 2016). Baltes et al. (2014) initiated the use of geminivirus vectors for transient expression of ZFN, TALENs, and CRISPR/Cas9 and delivery of templates for repair. This method has been employed in tobacco for precise gene editing by homologous recombination and showed better efficiency than *Agrobacterium* mediated delivery (Yin et al. 2015; Čermák et al. 2015). In wheat, an optimized wheat dwarf virus (WDV) system was used for gene targeting (Gil-Humanes et al. 2017).

3.6.3 Biolistic Delivery

Biolistic delivery method employs the use of gene gun to deliver DNA coated microprojectiles into the plant cells and DNA gets randomly incorporated into recipient's genome. This method is routinely used for plant transformation owing to its reasonable efficiency, provision of delivering multiple constructs simultaneously and availability to almost every plant species. The downside of this method is the delivery of multiple copies of foreign gene into the same cell which may lead to genes altered or no expression in the transgenic plants. This technique has been extensively used with TALEN or CRISPR/Cas in wheat, rice and *Brachypodium* for creating accurate mutations in genes (Shan et al. 2013a, b). In soybean (*Glycine max*), it has been used with homology templates along with CRISPR/Cas system for precise editing of ALS1 gene (Li et al. 2015). In maize, all the reagents along with homology templates for repair (ALS2 gene) were delivered through biolistic method and found some insertion events which *Agrobacterium* method was unable to do. Resulting plants were chlorsulfuron-resistant (Svitashev et al. 2015). In rice, two sites in ALS1 (acetolactate synthase) gene were targeted by two gRNAs and amino acid substitution was done at W548L and S627I by particle bombardment method for delivering plasmid and homology templates. This method was found successful in developing tolerance against BS (bispyribac sodium) herbicide (Sun et al. 2016). Biolistic method has also been used for delivery of non-DNA cargos for genome editing in order to avoid regulations for GM plants. RNP delivery has been employed in wheat and maize (Liang et al. 2017; Svitashev et al. 2016). For the delivery of protein-based cargos, Martin-Ortigosa and Wang have developed another version of biolistic method which they named as "proteolistics." In this method proteins to be delivered were mixed with microprojectile suspension containing gold particles, dehydrated and bombarded. This unique version of biolistic is best suited for DNA free delivery (Martin-Ortigosa and Wang 2014).

One of the limitations of this method is the requirement of expensive equipment and its accessories, laborious procedure for the preparation of particles coated with DNA/RNA/protein and preparation of plant tissue for bombardment. To circumvent these problems, another inexpensive method called Whisker mediated delivery method was developed. In this method, suspension of culture cells, DNA, and whiskers of silicon carbide fiber (SCF) were mixed and shaken vigorously so that SCF generate small holes in the plant cells which would allow DNA to enter intact cell. This method has been conveniently exploited in culture cells of Black Mexican Sweet (BMS), maize (*Zea mays*), and tobacco (*Nicotiana tabacum*) (Kaeppler et al. 1990). It has been utilized for stable transformation in maize, wheat, and rice (Kaeppler et al. 1992; Rakoczy-Trojanowska 2002). For an overview of different methods used for delivery of CRISPR-CAS system in plants, see Fig. 3.3.

3.6.4 PEG Mediated Delivery

Poly Ethylene Glycol (PEG) mediated delivery is used for protoplast transfection. Protoplasts are plant cells which have their cell wall removed with cell wall degrading (cellulases and pectinases) enzymes by using an osmoticum. Protoplasts may be formed from different plant tissues such as leaves, roots, stems pollen, anthers. PEG mediated protoplast transfection is quite useful in the study of transient gene expression, cellular pathways including protein interactions and their subcellular localization, etc. (Sahab et al. 2019). It has been employed with homology templates for precise genome editing in wheat, rice, soybean, and maize (Shan et al. 2014; Sun et al. 2015; Feng et al. 2016). PEG mediated transformation has been used for transient transfection of potato protoplasts for targeting three different sites of granule-bound starch synthase (*GBSS*) gene (Andersson et al. 2017). Protoplast of grapes and apple cultivars and wheat have been targeted using CRISPR/Cas system with RNP preassembled complex (Malnoy et al. 2016; Liang et al. 2017). Similarly, protoplasts of tobacco, *Arabidopsis thaliana*, rice, and lettuce were successfully transformed with RNP complex and up to 46% frequencies were achieved and mutation were even transmissible to germline as well which showed the potential of this method for DNA free genome editing (Woo et al. 2015). Along with transient assays, in some plants, PEG mediated method has shown to be capable of generation of stable lines as well. Canola (*Brassica napus*) an industrially important oil seed plant was demonstrated to generate transient as well as stable lines from leaf mesophyll cells using this method (Sahab et al. 2019).

3.6.5 Nanoparticles Mediated Delivery

Nanoparticles (NPs) have been extensively used as delivery tool for drugs and biomolecules inside the cells. They offer a great potential for intracellular delivery of editing cargos, due to their flexibility to be tuned by chemical and physical means, so that they can penetrate cell membranes. Although quite extensively used as delivery method for mammalian cells, the use of nanoparticles for plant transformation is scarce due to the presence of thick cell wall which presents a hurdle to the successful entry of editing cargos into the plant cell which has to be chemically or mechanically forced. Passive delivery of biomolecules into plants is only possible in somatic tissues. In case of plants, first attempt of NP use for delivering DNA and chemicals was done in 2007 in *Nicotiana tabacum* by using 100–200-nm gold-capped MSNs (mesoporous silica nanoparticles) loaded with GFP plasmid were delivered to plant cotyledons through biolistic method where GFP was expressed (Larue et al. 2014). This study demonstrated the feasibility for use of this method for DNA delivery into plants as well. Gold MSNs have also demonstrated protein delivery capabilities which can be harnessed for DNA free delivery of genome editors in plants. Cre recombinase enzyme was loaded on gold MSNs and delivered by biolistic method to maize plants and 20% of embryos were regenerated to callus with desired modification (Martin-Ortigosa et al. 2014). Although various delivery strategies need some chemical or mechanical force for delivery of NPs, still there smaller size and larger surface area bear the advantage that less DNA is required as compared to other commonly exploited delivery methods such as in case of PEG-protoplast transfection (Torney et al. 2007). Recently, some plant species were transformed in vitro and in vivo with passive delivery of NPs including model plants such as *Arabidopsis thaliana* roots (Chang et al. 2013; Jiang et al. 2014), *N. tabacum* protoplasts and crop plants such as *Triticum aestivum* (wheat) and *Eruca sativa* with different biomolecules (siRNA, double stranded DNA, plasmid DNA) (Demirer et al. 2018). With the application of nuclease based editing methods, NP based delivery holds great promise to be used for target genome editing in plants (Cunningham et al. 2018). For an overview of different methods used for delivery of CRISPR-CAS system in plants, see Fig. 3.3.

3.7 Emerging Delivery Techniques

3.7.1 Streptolysin O

Streptolysin O (SLO) is a toxin produced by various strains of beta-hemolytic group A of Streptococci for causing infection. It is activated by some reducing agent like thiol for its activation and targets cholesterols in eukaryotic cell membrane and damages membrane integrity by creating arch like structures (Bhakdi et al. 1985). Although believed to be damaging for cells, considering its interactive nature for cellular membranes, SLO can be tuned to be used as delivery method. This toxin has been used for increasing membrane permeability reversibly and found effective in intracellular delivery of proteins up to 100 kDa with keeping cells viable (Walev et al. 2001). It has been utilized for siRNA delivery (Brito et al. 2008) and protein labeling for imaging live cells (Teng et al. 2017), but its potential for CRISPR/Cas delivery has not observed yet. Naturally concerns are there for its in vivo usage but it may serve as a viable option for in vitro studies.

3.7.2 Multifunctional Envelope-Type Nanodevices (MENDs)

MENDS are non-viral based delivery systems in which DNA is condensed and complexed with poly-L-lysine (PLL). A lipid film is developed over this complex. This technique was originated at Hokkaido University by Harashima group and they called it "programmed packaging" for stepwise delivery into the targeted organelles (Yamada et al. 2012). Lipid enveloped DNA/PLL was found to be tenfold more effective than naked DNA/PLL and 1000-fold by adding stearyl octaarginine which adds to the cell penetration of the complex (Kogure et al. 2004). Nakamura et al. (2012) added tetra-lamellar MEND for targeted delivery to nucleus and mitochondria and PEG-peptide-DOPE-conjugated MEND for bladder cancer cells specifically. The lipid layer can further be modified functionally by adding different decorations, for example, PEG, beneficial for enhanced vascular circulation, lipids for more efficient endosomal evasion, CCPs for improved cellular delivery and nuclear tags for guidance to nucleus (Kogure et al. 2004). After seen their multifaceted applications in delivering cargos of short interfering RNA (siRNA), plasmid DNA, and therapeutic agents of Bacillus Calmette-Guerin (BCG), work must be done for their in vivo and in vitro usage (Lino et al. 2018).

3.7.3 Lipid-Coated Mesoporous Silica Particles

Lipid-coated mesoporous silica particles are prepared by covering mesoporous nanoparticles with lipid bilayer shell. This technique was originally developed at the University of New Mexico and Sandia National Laboratories by Liu et al. (2009). Although they have not been utilized for CRISPR/Cas system, their improved surface area capable of supporting large cargos and provision of modification in particle and pore size for intended needs make it an attractive option for future use effective gene delivery (Du et al. 2014; Durfee et al. 2016). Until now, these particles have been served in vitro and in vivo as delivery system for chemical based cancer therapeutics (Wang et al. 2013a; Su et al. 2017), imaging agents, and phototherapy agents (Mackowiak et al. 2013). But whether they can carry heavy weight RNP complex or DNA is still a challenge and needs further investigation (Lino et al. 2018).

3.7.4 Inorganic Nanoparticles

There are versatile options for usage of inorganic nanoparticles as potential delivery agent based on their history of their usage in similar fields. There is a whole range of available inorganic nanoparticles from mesoporous silica nanoparticles (Luo et al. 2014), dense silica nanoparticles (Luo and Saltzman 2000) to carbon nanotubes (Bates and Kostarelos 2013). But their usage for CRISPR/Cas delivery is not reported yet. Gold nanoparticles have shown to be potent delivery agent for CRISPR/Cas9 (Mout et al. 2017). Although they possess greater potential for their future usage for their additional advantages over conventional viral/non-viral based vector, i.e. higher simplicity, reproducibility, size distribution, stability, and chemical functionality, but their full utility is still to come (Lino et al. 2018).

3.8 gRNA Design and Design Tools

The basic working principle of CRISPR-Cas system relies on gRNA to guide Cas endonuclease to intended site of genomic modification in the genome. There are many factors which can influence the outcome of the CRISPR-based experiments. These factors include delivery methods of Cas9 and sgRNAs, activity of Cas9 and target site selection through designing sgRNAs. In the above section of this chapter, various tools for the delivery of sgRNA have been described, in this section of this chapter, the current criteria for designing sgRNAs and different tools for designing sgRNAs are described. Those factors are described which affect the specificity of Cas9/sgRNA at target site and also factors which determine the activity at off-target sites. The websites, software, and tools are described which help in the optimum design of sgRNA to get high specificity at target site and minimum activity at off-target site.

3.8.1 sgRNA Design Criteria and Target Efficiency

3.8.1.1 PAM: A Basic Requirement for Designing gRNA

While employing CRISPR-Cas based system, a target site is selected for intended genetic modifications. The target site consists of two parts (1) protospacer, which have exact complementarity to 20 nucleotides on 5′ end of designed sgRNA and (2) protospacer adjacent motif (PAM), which is present immediately next to the 3′ end of the target sequence and this is required for efficient binding of Cas9 protein and carrying out its nuclease activity. CRISPR/Cas system works by identifying specific target sites in genome, which comprises of protospacer sequence in DNA and adjacent PAM. For designing of gRNA, to be used in a genetic modification experiment, the basic rule is very simple, i.e. one can scan for the presence of all PAM sites in the intended genomic region and select adjacent 20 nucleotides (protospacer) for designing gRNA (Cui et al. 2018). The number of sequences adjacent to PAM depend upon the type of PAM sequence and this number is 20 for the most widely used SpCas9, NGG PAM. For the most widely used SpCas9, NGG is the optimal PAM sequence but NAG and NGA could also act as PAM sequence and in this case the activity of Cas9 protein is much limited. Because NGG PAM sequence occurs on approximately every 8 bp in the human genome, this sequence is relatively common in many other genomes as well, therefore the SpCas9 can target many sites occurring frequently in a given genome (Cong et al. 2013). For providing more flexibility to researcher regarding target site selection, variants of SpCas9 have been developed which display altered PAM specificities. One such variant is SpCas9 (VRER) which recognizes NGCG PAM site. This was also reported to exhibit greater on-target specificity than wild-type SpCas9 (Kleinstiver et al. 2015). Other different PAM sequences can be used in combination with other different Cas9 proteins, which have different PAM site requirements, derived from other bacterial species. For example, *Staphylococcus aureus* Cas9 recognizes NNGRR PAM site. Additionally, this Cas9 protein was demonstrated to exhibit greater on-target specificity compared with SpCas9 (Ran et al. 2015). Although such new versions of Cas9 are emerging, most CRISPR design tools are modeled for SpCas9 and utilize NGG or NAG PAM consensus sites for sgRNA design by default. Because of the simple NGG PAM sequence requirements, which occurs very frequently in genomes, *S. pyogenes*' Cas9 (spCas9) is used in many different applications. Different naturally occurring Cas endonucleases have been described with different PAM sequence requirements. Researchers can choose for different PAM sequences, associated with different Cas enzymes, while designing gRNA. See

Table 3.1 Naturally occurring major Cas endonucleases along with their characteristics

Nuclease	PAM sequence	sgRNA length (bp)	Cutting site	Reference
spCas9	NGG	20	~3 bp 5′of PAM	Jinek et al. (2012)
FnCas9	NGG	20	~3 bp 5′of PAM	Hirano et al. (2016)
St1Cas9	NNAGA AW	20	~3 bp 5′of PAM	Gasiunas et al. (2012)
St3Cas9	NGGNG	20	~3 bp 5′of PAM	Gasiunas et al. (2012)
SaCas9	NNGR RT	21	~3 bp 5′of PAM	Mojica et al. (2009)
NmCas9	NNNNG ATT	24	~3 bp 5′of PAM	Hou et al. (2013)
CjCas9	NNNNACAC	22	~3 bp 5′of PAM	Kim et al. (2017)
sCpf1	TTTV	24	19/24 bp 3′of PAM	Yamano et al. (2016)
LbCpf1	TTTV	24	19/24 bp 3′of PAM	Yamano et al. (2016)

Table 3.1 for different naturally occurring Cas nucleases and their characteristics including associated PAM sequences.

3.8.1.2 On-Target Efficiency Prediction

As described earlier, the most basic requirement for sgRNA design is presence of PAM and adjacent protospacer region on the target DNA which have exact complementary to the 20 nucleotides on the 5′ end of designed sgRNA. In theory, when there is exact complementarity of these 20 nucleotides on-target DNA to sgRNA, the sgRNA/Cas9 complex should make a cut to DNA, but previous studies suggests that cutting efficiency varies significantly among sgRNAs and that some target sites were inherently more effective than others (Jinek et al. 2012; Cong et al. 2013; Mali et al. 2013b; Yang et al. 2013; Wang et al. 2014). This suggests that the intrinsic factors related to sequence and position of this sequence in the target DNA could influence the efficiency of cutting. Therefore, predictive models are required for identifying target sites with high efficiency. The earlier versions of sgRNA design tools were just sequence pattern recognition programs, which recognized the occurrence of PAM sequences (Upadhyay and Sharma 2014). Additionally, some information on the location of target site within gene was also incorporated to help researchers conclude about the likelihood of the functional effects of CRISPR/Cas9 on-target site. With increasing knowledge, more factors have been discovered which are known to affect the cutting efficiencies of gRNA. After quantitatively modeling these factors, sequence-based prediction models have been designed, which can successfully predict the efficacy of sgRNA regarding its desired outcome on the target site. The results of these quantitative models are expressed as efficiency scores which are used by different websites and software to predict about the efficiency of cutting of sgRNA on intended target site. A schematic representation of common steps employed by different online sgRNA designing tools is shown in Fig. 3.4.

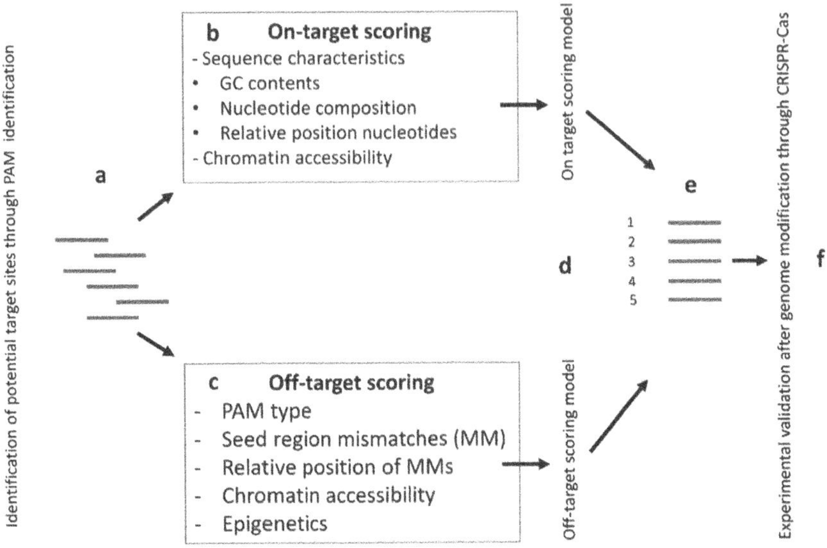

Fig. 3.4 Schematic representation of optimal sgRNA design pipeline. (**a**) Potential target sites are identified by identifying all PAM sites in target region/genome. (**b**) Factors known to affect on-target activity are taken into account which include target sequence characteristics and chromatin features. (**c**) Factors known to affect off-target activity are taken into account which include PAM type, seed region mismatches, relative position of these mismatches, chromatin accessibility, and epigenetic marks. (**d**) On-target efficiency scores and off-target activity scores are calculated by using different mathematical models. (**e**) Based on the scores of mathematical model, a ranked list of optimal targets with maximum on-target score and minimum off-target scores is generated. (**f**) Optimal sgRNA are chosen for experimental genome editing and results are validated

Online software tools and sgRNA design applications provide on-target efficacy prediction metrics based on which selection of sgRNA can be done. The analysis and scoring of on-target efficiencies is dependent on subjective approaches and previous results of experimental methodologies. The examples of such tools include Broad Institute sgRNA Designer or other tools such as CRISPR Multi-Targeter or Protospacer which employ scoring metrics from Doench et al. studies. This approach of scoring on-target efficiencies uses experimentally determined cutting efficiencies to identify sets of sgRNA with high or low cleavage activity (Doench et al. 2016). Initially, different studies used 1000s of sgRNA to determine characteristics within the sequences of protospacer of each sgRNA which may be associated with presumably high or low efficiencies. Various studies differed in their approach for calculating efficiency scores, for example, some studies only considered 20 bp target sequence, while others included PAM and flanking sequence (Chari et al. 2015). Despite differences across various studies, some key features have been repeatedly found to be important regarding the optimized gRNA design. Quantification of these key features related to sequence of gRNA and other characteristics such as thermodynamic stability of sgRNA forms the basis for calculation of on-target efficiencies scoring algorithms which are used to predict the on-target activity of every other designed sgRNA. Such prediction models usually give score in the range of 0–1, where a higher score means more effective sgRNA.

It has been shown that GC content of sgRNA is important as sgRNAs containing intermediate GC content demonstrated best activity with regard to Cas9 cleavage efficiency as compared to those containing high or low GC content (Wang et al. 2014; Doench et al. 2014). It has been reported that guanine enrichment enhances efficiency by contributing to the stability, loading, and activity of sgRNA (Moreno-Mateos et al. 2015). Presence of guanine in the position 20 next to PAM was shown to enhance cutting efficacy for SpCas9 and St1Cas9 (Chari et al. 2015). Furthermore, previous research had shown that certain nucleotides are favored in many positions of sgRNA design, including the variable nucleotide of NGG. Many predictive models, take into account the position dependent effects of specific nucleotides in the sequence of sgRNA. One previous study reported that higher frequency of T and TT is negatively associated with efficiency of sgRNA significantly (Kuan et al. 2017). Occurrence of too many PAMs (NGG) within target DNA sites could inhibit the activity of CRISPR/Cas9 (Malina et al. 2015). Majority of sequence related characteristics impacting efficiency scores are found to be distributed within 10–12 bp region proximal to PAM in sgRNA, and this important region is termed as seed region. This region is important as it is the first which binds to the DNA in the first step following recognition of PAM (Thomas et al. 2019). Melting temperature of sgRNA impacts the efficiency scores (Cui et al. 2018). While calculating the efficiency scores, different online tools for the design of sgRNA take into considerations these various factors and employ different algorithms derived from different previous experimental studies, and there might be varying results while using different algorithms. Therefore, some online tools provide users with more than one efficiency prediction models, the examples of such tools include E-CRISP, CHOPCHOP, CLD, and CRISPOR. Details of different features of various online tools for designing sgRNA along with description of their predictive models based on previous experimental studies is shown in Table 3.2. Also see Fig. 3.4 for a schematic representation of common steps used for designing gRNA.

One of the factors which can further improve the predictive models for calculating on-target efficiency of sgRNA is chromatin structure. Chromatin accessibility has been shown to be the major determinant of binding of sgRNA (Wu et al. 2014b). The factors which are involved in chromatin accessibility and remodeling include DNAase I hypersensitivity sites, transcription factor binding, DNA methylation, and histone modifications. Both DNA sequence composition of sgRNA and accessibility of chromatin structures can be incorporated in future predictive models to calculate the binding efficiency of sgRNA.

3.8.1.3 Other Useful Considerations for Designing sgRNA

The above-mentioned features of sgRNA design relate to the functioning of sgRNAs for its on-target efficacy. There are other features which can be considered and pertain to the efficient production of sgRNA through optimal transcription (Montague et al. 2014). For example, some design tools (CHOPCHOP) offer to select gRNAs with 5′ G to promote PolIII dependent transcription from U6

Table 3.2 Examples of different sgRNA design tools. A comparison of different salient features of commonly used gRNA design tools

Tool	Genomes	Input	Output	Nuclease/ PAM	Off-target scoring	On-target prediction	Websites
CCTop	Around 100 different species including animals, plants and microbes	DNA sequence, accepts annotations	Scores OTs, ranks sgRNA for OTs, gives number and position of MMs, gives efficiency scores, GC%	13 PAMs	Stemmer et al. (2015)	CRISPRater (Labuhn et al. 2017)	https://crispr.cos.uni-heidelberg.de/index.html
PROTOSPACER	Provided by user	Gene ID DNA sequence Genomic region	gives OT sites, gives efficiency scores, GC%, no position with respect to CDS of gene	SpCas9	Hsu et al. (2013)	Doench et al. (2014)	http://www.protospace.r.com/
CRISPOR	Large number of genomes of plants, animals and microbes covered (258)	DNA sequence Genomic region	Number of off-target sites listed, scores OTs, gives efficiency scores, GC%	21 PAMs	CFD (Doench et al. 2016; Hsu et al. 2013)	Rule set 2 (Doench et al. 2016, 2014)	http://crispor.tefor.net/
WU-CRISPR	Mouse and human	DNA sequence Gene ID Gene symbol	Scores OTs, gives efficiency scores, GC%, filters for selecting guides, Checks for polyT/U sequences and secondary structures	SpCas9	Wong et al. (2015)	Wong et al. (2015)	http://crispr.wustl.edu
CRISPRdirect	Many genomes from plant, animals and fungi	DNA sequence Genomic region Accession numbers	Gives table with target positions and context of sequence, Number of off-target sites listed, GC %, Checks for polyT sequences	Custom PAM	kmer + PAM (Naito et al. 2014)	No	http://crispr.dbcls.jp

E-CRISP	Many genomes from plant, animals and microbes	Gene ID Gene Symbol DNA sequence	Number of off-target sites, scores OTs, number MMs, table with target positions and Genomic context of sequence (CpG islands, efficiency scores, Introns, exons), GC%, hecks for polyT sequences, filters for selecting guides, option for double nickase	17 PAMs Alternative PAM Custom PAM	S-score (Heigwer et al. 2014)	Doench et al. (2014)	www.e-crisp.org/E-CRISP/
CHOPCHOP	Approximately 100 different species and mostly microbes including viruses	RefSeq Gene ID Genomic region	Scores OTs, ranks sgRNA for OTs, gives number and position of MMs, position with respect to CDS of gene, gives efficiency scores, filters for selecting guides	7 PAMs Custom PAM	Cong et al. (2013), Hsu et al. (2013)	Doench et al. (2014), Chari et al. (2015), Xu et al. (2015)	https://chopchop.rc.fas.harvard.edu
Cas-designer	Many animal species including fishes	CrRNA DNA sequence or FASTA	Target sequence with efficiency scores, OT list with link to Ensembl, bulge MMs, GC%, polyT	16 PAMs Alternative PAMs	Cas-OFFinder	No	www.rgenome.net/cas-designer/
CROP-IT	Mouse and human	Genomic region, DNA sequence sgRNA sequence	List of OT sites with genomic context, scores OTs, Number and Position of MMs, list of target sites with genomic context, double nickase option	SpCas9	Singh et al. (2015)	No	http://cheetah.bioch.virginia.edu/AdliLab/CROP-IT/homepage.html

(continued)

Table 3.2 (continued)

Tool	Genomes	Input	Output	Nuclease/ PAM	Off-target scoring	On-target prediction	Websites
CRISPR multi-targeter	12 genomes from animals and plants	DNA sequence Gene symbol Gene ID/transcripts ID RefSeq sequence	Connects to "GT-Scan and Cas-OFFinder" websites for predicting OTs Table of target sequences with efficiency scores, GC%, Tm, option for double nickase	SpCas9 Custom PAM	Links to websites GT-Scan Cas-OFFinder	Doench et al. (2014)	http://www.multicrispr.net/

promotors or 5′ GG for in vitro transcription using T7 polymerase. Some design tools offer the exclusion of sgRNAs with TTTT stretches which is a signal for termination of PolIII transcription.

3.8.2 Factors Influencing Specificity and Off-Target Activity

To achieve specific results in genome editing experiments using CRISPR-Cas, the most basic design criteria is to target only PAM and adjacent 20 nucleotide region. This means this 20-nucleotide region should be unique which occurs only once in genome. Even if sgRNA is designed against this unique occurrence of a target sequence in a given genome, this does not ensure that targeting will be perfectly specific as Cas nuclease activity does occur at sites where sgRNA binds to imperfectly matching sites. Despite the fact that CRISPR systems can be highly specific, still it happens that sgRNAs exhibit some off-target (OT) activity against imperfectly matched sites (Yang et al. 2014). How one can design sgRNA with minimum OT effects? Given the current knowledge, the ability to predict OT is limited. However, some factors are known which can affect specificity of the Cas9/sgRNA and thus also determine the activity at off-target sites. These factors can be broadly classified into two categories. Firstly, these factors relate to the intrinsic properties of Cas9 protein and this could be determined by the relative importance of each nucleotide position in sgRNA for target recognition and depends on sgRNA sequence. Secondly, specificity at the target site also depends upon the relative abundance of effective Cas9/sgRNA complex with respect to effective target DNA concentration (Wu et al. 2014a). The knowledge of the factors which determine the target specificity can help to estimate off-target activity of a given sgRNA. These factors which determine the target specificity at target site as well as the off-target activity of a given sgRNA the are described below.

3.8.2.1 PAM Sequence

Requirement for the PAM sequence, which is recognized by a domain in Cas9 protein (Nishimasu et al. 2014), varies greatly among different Cas9 proteins derived from different bacterial species (Garneau et al. 2010). Presumably those Cas9 proteins which recognize a longer PAM will have less targetable sites in the genome and, therefore will have fewer off-target sites in a given target DNA (Wu et al. 2014a). For an overview of different PAM requirements of different Cas9 proteins derived from different bacterial species, see Table 3.1.

3.8.2.2 Seed Region

It is known that SpCas9 cleavage efficiency is affected more if there are mismatches in the sgRNA core region or seed region, i.e. 10–12 base pairs in the PAM proximal region. In contrast mismatches in the last eight base pairs in the PAM-distal region are well tolerated and affect the cleavage efficiency to a much lesser extent (Cong et al. 2013). Though some studies have defined seed region to be only consisting of

5 base pairs in the PAM proximal region of sgRNA (Wu et al. 2014b) and generally it is considered that mismatches in near the PAM region are less tolerated.

3.8.2.3 Frequency of Seed-Matching Genomic Sites

Because of relatively short length of seed regions of sgRNA (5–12 nucleotides), there can be potentially a large number of such sites matching to the DNA in given genome for a given designed sgRNA. For example, it has been calculated that for a AAGGA+NGG sequence there are a million sites available in mouse genome as compared to CGTCG+NGG which only had an occurrence of 10,000 times (Wu et al. 2014b). Therefore, this abundance of potential seed plus PAM sites for a given sgRNA against a given genome can potentially affect the number of off-targets.

3.8.2.4 Chromatin Accessibility and Epigenetics

If the chromatin is in condensed packed form, then it is less accessible to Cas9 protein and for recognition of PAM site. Accessibility of chromatin is typically measured through DNase-I hypersensitivity (Wu et al. 2014b). There are many factors which can influence the open state of chromatin, for example, histone epigenetic modifications and CpG DNA methylation marks can control the compactness of chromatin. Cas9/sgRNA can only target DNA in open chromatin state and this can influence its activity both on-target as well as off-target sites.

3.8.2.5 Length of Target Sequence

Normally, a guide RNA of 20 nucleotide length is designed but studies have shown that decreasing the length to 17–18 nucleotides substantially increases the specificity by decreasing the number of off-targets (Fu et al. 2014). It has been hypothesized that these last three nucleotides probably play more role in stabilizing the sgRNA on off-target sites as compared to on-target site.

3.8.3 Off-Target scoring

Target or gRNA sequence is the single most important parameter for determining the specificity and off-target activity of a given gRNA. Even if there is unique occurrence of a nucleotide sequence with exact complementarity to gRNA, gRNAs do bind to other highly similar sequences in genome. Mismatches between sequence on gRNA and DNA sites, is an important criterion determining the off-target activity and scoring. Previous study have shown that the mismatch tolerance between SpCas9 complex and DNA is influenced by number, position, and distribution of mismatches (Hsu et al. 2013). For determining off-targets, protospacer gRNA sequence is compared and aligned to reference genome and each DNA sequence adjacent to every PAM sequence of a specific Cas9 endonuclease is searched for similarities between gRNA sequence and target DNA sequence. Occurrence of high similarity between gRNA sequence and target DNA sequence at places other than intended site of genome editing can result in unwanted effects. Collectively, these potential unwanted effects of a given sgRNA against a reference genome are termed as "off-target" effects (Wilson et al. 2018). Results from previous studies using unbiased sequencing approaches to detect off-targets have shown variable results but sometimes the off-targets do occur at significantly higher rate (Kim et al. 2015; Yang et al. 2014). Therefore, to be able to successfully carry out the CRISPR-Cas9 based genome editing and avoid inadvertent unwanted effects, it is important to be able to predict and select for those sgRNA which will have higher specificity and lower off-targets. Different gRNA design tools provide approaches to identify the off-target sites for designed gRNA and then assign scores to each gRNA for its potential off-target effects. The off-target scores are calculated after taking into consideration different factors which are known to affect binding of sgRNA to DNA and cutting efficiencies of Cas9 nuclease. Some of these factors have already been described in paragraphs above in Sect. 3.8.2. A schematic representation of common steps employed by different online sgRNA designing tools is shown in Fig. 3.4.

The techniques for detecting off-targets rely on sequence comparison methods mostly based on BWA (Burrows Wheeler Aligner) algorithms, Bowtie or pair wise sequence comparisons. After sequence comparison, off-target values are represented in two forms, i.e. either in the form of scoring matrix or in the form of specificity scores. Scoring matrices display the number of genomic sites for a given number of nucleotide mismatches between protospacer region of potential sgRNA and DNA sequences adjacent to all PAM sequences (Thomas et al. 2019). While specificity scoring takes into account the frequency of mismatch for each nucleotide and its positional weighing. This weighing is calculated based on position of individual mismatches within protospacer region of sgRNA sequence and distance from other mismatches. Mismatches that occur in close proximity to PAM as well as each other have greater potential to disrupt the DNA:RNA heteroduplex and will have greater impact to affect Cas endonuclease activity. Mismatches which occur in the seed region have greater potential to disrupt cleavage and impact specificity more significantly (Hsu et al. 2013; Thomas et al. 2019). After taking into account all these factors, potential off-target scores are calculated by using mathematical models and often expressed in the form of penalty matrix scores. A score can be assigned between 0 and 1, where a 0 value means no mismatches while a higher value means more mismatches and bigger impact on cleavage activity of Cas9 protein. These off-target scores are calculated for each potential off-target position and each gRNA can be assigned a score according to its potential off-target sites. This score is used to select the best gRNA with minimum off-target score from among a number of available gRNAs (Cui et al. 2018).

Many different tools use a combination of different factors known to affect specificity to calculate off-target scores in different mathematical models. Different mathematical models are being applied by different tools. Cutting frequency determination score (CFD) (Doench et al. 2016), CASPER (Xu et al. 2015), Hsu-Zhang matrix (Hsu et al. 2013; Doench et al. 2016) are the names of few examples of algorithms used by various online tools for designing sgRNA. Many other tools have devised their own algorithms for calculating off-targets. Given the fact that no one algorithm is always effective for efficient designing of sgRNA, researchers are choosing to use more than one algorithm for designing sgRNA. Different sgRNA online tools are providing with option of choosing for more than one off-target scoring methods. For example, CHOPCHOP (Montague et al. 2014), an online sgRNA design tool, provides the option of off-target score calculation separately by two methods, namely Hsu et al. (2013) and Cong et al. (2013). Another sgRNA design tool, CRISPOR, provides option of off-target scoring from Hsu et al. (2013) and Doench et al. (2016) algorithms. For a detailed overview of different salient features of various online sgRNA design tools, see Table 3.2.

Despite implementation of different mathematical algorithms and scoring methods, designed sgRNA can still result in false positive and false negative off-target sites. Comparison of experimentally validated off-targets with off-targets sites predicted by sgRNA design tool, showed that prediction tools can overestimate the number of potential off-target sites by a factor of tenfold (Cameron et al. 2017). Such false positive predictions can potentially lead to erroneous disqualification of optimal sgRNA with good on-target efficiency scores. To minimize overcalculation of false positive off-target sites, design tools often limit the potential off-targets to a maximum number of certain mismatches and calculate off-targets with only very specific PAMs (Bae et al. 2014). Previous studies have also shown that large number of false negative off-target sites also occur (Tsai et al. 2017, 2015). Therefore, a good off-target predictor balances between erroneous off-target false negative and false positives results.

Following mechanisms have been proposed through which a change in sequence of sgRNA can play a role in determining the activity at target site as well as off-target sites. The mechanisms include (1) changes that alter number of seed-matching sites in the genome, (2) changes that alter effective concentration of sgRNA at target site through modulation of transcription of sgRNA (changes in sgRNA sequence have been shown to affect concentration of U6 promoter transcribed sgRNA) (Wu et al. 2014b), stability of sgRNA and loading of sgRNA onto Cas9, (3) changes that depend upon the chromatin accessibility, and (4) changes that depend upon thermodynamic stability of guide RNA-DNA duplex (Wu et al. 2014a).

3.8.4 Guide RNA Design for Different Applications of CRISPR

There are many experimental applications of CRISPR-Cas system. These applications include gene knockout, gene knockin applications, base editing, and gene expression activation and inhibition. Not all of these applications depend upon the cleavage activity of Cas endonuclease but all of these depend on the ability of gRNA to recognize target sequence. Though there are differences between these applications in the mechanisms at molecular level but overall principles for the design of guide RNA are same for different applications but the positioning of gRNA may differ between these applications. The basic principle relies on the selecting suitable target region, selecting suitable gRNAs after considering the positioning of gRNAs in relation to target region, considering off-targets and

Table 3.3 Different applications of CRISPR-Cas for genome modification and gRNA design

Application of sgRNA	Purpose	Target region	gRNAs required
Knockout	Disrupting a gene or other genomic element	Coding exons, or genomic regions	One or more
Knockin	Insertion of DNA on a specific region	Any genomic region	One or more
Point mutation	Specifically altering a single nucleotide	Within 20 nucleotides of intended mutation	Single gRNA
Base editing	Specific conversion of nucleotides; either C to T or A to G through cytosine base editor or adenine base editor respectively	All relevant nucleotides, occurring within base editing active window	Single gRNA
Regulation—CRISPRi	Transcriptional repression of a target gene	Promoter region, −50 nt to +400 nt relative to the transcription start site (TSS)	Multiple
Regulation—CRISPRa	Transcriptional activation of a target gene	Promoter region, −400 to −50 nt relative to the transcription start site (TSS)	Multiple

on-target scoring. While employing CRISPR-Cas for different applications, important points which should be considered for the design of gRNA are described below and summarized in Table 3.3.

3.8.4.1 Knockouts and sgRNA Design

The aim is to ablate or disrupt the function of a gene. This is done by introducing one or more double stranded breaks (DSBs) in the selected genomic region or gene. Frameshift mutations can be introduced when one double stranded break is generated in the coding sequence of gene within exon. In this case, DSB will stimulate the DNA repair mechanisms and in the absence of repair template, non-homologous end joining (NHEJ) pathway is initiated which leads to indels within coding sequence, thus introducing frameshift mutations which disrupt protein translation and results in gene knockout. When using this technique for gene knockout, one should consider the size and relative positioning of exon within target gene. Larger exons are preferred as these give more choice for designing gRNA and therefore, one can select suitable gRNA from wider range of available choice. Relative positioning of exon within transcript is also important as targeting early codon can result in early frameshift with higher chance of introduction of stop codon and thus leading to gene knockout. Apart from ablation of gene function using induced frameshift mutations via single DSBs, gene function can also be disrupted by deleting an entire gene region and this is done by inducing two or more DSBs. For gene knockout applications, one may consider other parameters, such as location of functional domains of proteins.

3.8.4.2 Knockins and sgRNA Design

Gene knockin applications involve the insertion of exogenous DNA or introducing a desired change in the nucleotide sequence through point mutations. A DSB is introduced using CRISPR-Cas and a suitable DNA repair template is also introduced into the cell. In this case, cell repairs its DNA using homology directed repair using the DNA repair template, thus leading to insertion or knock-in of any exogenous DNA included within the introduced DNA template. For the purpose of introducing point mutations, the distance between induced DSB and desired point mutation insertion point should be less than 20 bp, as the polymerase might not extend far enough while repairing DNA template to include desired point mutation. Shorter DNA repair template usually contain 50 bases of homology on either side of induced DSB and these are incorporated into DNA using single DSB and one gRNA. Shorter DNA repair templates are suitable for applications involving point mutation. For inducing larger DNA insertions, such as plasmids or BACs, multiple DSBs are introduced using more than one gRNAs.

3.8.4.3 Transcriptional Gene Activation or Repression and sgRNA Design

Gene activation or gene repression can be achieved using catalytically inactivated Cas endonuclease to target transcriptional activator or repressor complexes to specific site within gene promoter region. CRISPRa technique is used to activate gene expression by targeting promotor region upstream of transcription start site (TSS), while CRISPRi technique represses gene expression by targeting downstream of TSS. For the application CRISPRa/i, accurate identification of TSS start site is required. This identification of TSS have been done by using Rapid Amplification of cDNA End PCR (RACE-PCR) and Cap Analysis of Gene Expression Sequencing (CAGE-seq) methods, and CAGE-seq data is available for both human and mouse genome which can be displayed in genome browser (Davis et al. 2018; Noguchi et al. 2017). Arbitrary values of 400 bp upstream and 300 bp downstream of TSS are chosen for the purpose of selecting target regions of CRISPRa and CRISPRi experiments, respectively, but in actual practice promotor regions may be significantly larger. After defining target regions for CRISPRa and CRISPRi, it may be necessary to select multiple gRNA to achieve regulatory effects (Thomas et al. 2019).

3.8.5 Computer Tools for sgRNA Design

There are various computer tools available which assist the researcher with sgRNA selection, having good on-target score and less off-target activity. These computer tools are either online websites or installable software packages such as Cas-OT (Xiao et al. 2014) and sgRNAcas9 (Xie et al. 2014). For a comprehensive review of these tools, regarding their various functions and available options, see Table 3.2. The software packages allow users to input any genome of interest but this might involve the preliminary work of preparing and formatting input sequence files before being able to use these on such software packages. Alternatively, gene identification, gene symbol can be given or a genomic region can be defined. These tools help the researchers in designing the sgRNA by taking into account many factors. All of these tools firstly apply the most basic criterion for the selection of sgRNA by identifying all PAM sites in the target DNA for a specified Cas9. Many different online tools offer many possibilities regarding PAM site selection. For example, CCTOP, PROTOSPACER, CRISPOR, Cas-Designer, CHOPCHOP, and E-CRISP give the output of list of available PAM sequence options from which the users can make a selection while designing sgRNA (Upadhyay and Sharma 2014). Some design tools also provide the option of giving a user defined custom PAM sequence, the examples of such tools are CRISPRdirect, E-CRISP, CHOPCHOP, and CRISPR Multi-Targeter. When the traditional SpCas9 PAM (NGG) sites are absent in a given genomic region or when one is looking for greater target specificity, then there are possibilities of selection of other PAM sites coming from new altered forms of Cas9 or Cas9 from other bacterial species (see Table 3.1 for the occurrence of naturally available Cas proteins from different bacterial species and their corresponding PAM sites) (Adli 2018).

After PAM site consideration, the next most important thing in the design of sgRNA is the location of cut site in a given gene or piece of DNA. Some design tools output a graphical representation of target gene in conjunction with overlayed sgRNA sites to help visualize the optimal sites for genetic manipulations (Montague et al. 2014). The sgRNA sequences targeting intended site of genome modification are given with predicted on-target and off-target effects. Different parameters are provided by these tools to calculate on-target and off-target effects. User input parameters are used to score off-targets based on location and number of mismatches to the guide RNA. Some tools provide option of determining off-targets with less than certain number of mismatches over entire guide RNA, or within certain PAM proximal or distal region of guide RNA (Ma et al. 2013; Bae et al. 2014; Xiao et al. 2014; Montague et al. 2014; Xie et al. 2014). Guides are ranked either by number of off-targets or by weighted sum of off-targets score. Regarding the evaluation of the on-target efficiency of guide RNA, some tools provide the information on the presence of SNPs (Ma et al. 2013) which can impact targeting as well as genomic features of the on-target site which can influence the intended purpose of using guide RNA such as position of exons and introns of a gene, slice variants of a gene and presence of epigenetic marks such as CpG DNA (Heigwer et al. 2014; Montague et al. 2014). Some tools provide the GC content of guide RNA which could affect the effectiveness of guide RNA (Heigwer et al. 2014; Montague et al. 2014; Xie et al. 2014; Wang et al. 2014). In addition to providing support for the design of guide RNA, some tools also provide further support in providing oligonucleotide sequences for construction of guide RNA or for designing primers for confirming cleavage at the intended target site.

One important aspect to consider while choosing for an appropriate tool for the design of guide RNA is the purpose of designing the guide. For example, when the intention is to design a guide RNA with high on-target activity, some tools may be more appropriate for this than others (Montague et al. 2014). When minimum off-targets are desired then those tools may be more helpful which score predicted off-targets quantitatively (Hsu et al. 2013; Heigwer et al. 2014). These tools vary in their methods of scoring off-targets, for example, most of the tools search for off-targets with a total of three mismatches or less in either the PAM proximal region or in the total length of guide RNA. In designing guide RNA with such tools, one may miss many off-targets as compared to tools which detect more mismatches in off-target sites of a guide RNA. Therefore, it is more convenient to use multiple tools while designing sgRNA since every tool has its own unique features and advantages. A comparison of different sgRNA design tools regarding their different features is given in Table 3.2.

3.9 Tracking Tools for CRISPR/Cas Mutations

CRISPR/Cas platform is adopted in a very short time and it revolutionized the practices of molecular biology and biotechnology. CRISPR/Cas is a simple and precise suite of DNA sequences within the genome of archaea and bacteria which allowed genome manipulation at cellular level never possible before (Brinkman 2019). Moreover, applications of CRISPR/Cas are not limited to dsDNA but also to ssDNA and RNA as well. CRISPR/Cas creates double stranded breaks at a precise location in the genome, which can be repaired through NHEJ, HDR, or MMEJ (Micro-homology mediated end joining) repair system. Although CRISPR is an editing tool used for creating mutations, however, repurposing of CRISPR/Cas has made possible to delete, replace, modify, and regulate genes in a precise manner. Generally, CRISPR/Cas is recognized as a precise technique, however, non-specific indels may also occur due to the binding of sgRNA at non-target sites in the genome. Therefore, quantification of indels plays vital role in evaluating efficiency and specificity of CRISPR/Cas system (Brinkman 2019). Furthermore, quantification of indels and specificity of CRISPR/Cas outcomes are also critical for regulatory authorities (Duensing et al. 2018).

Outcomes of CRISPR/Cas mediated genome editing are not random and largely depend upon the type of intended modifications in the genome (Bi and Yang 2017). Conventional genetic engineering mostly results in stable integration of foreign DNA such as 35S promoters, NOS terminators, and antibiotic resistance genes, in the host genome. Consequently, the outcomes of these elements can be easily confirmed with event-specific primers using PCR (Cho et al. 2014). In contrast, sometimes CRISPR/Cas system integrates single base modification at the target site even without integration of foreign DNA. It is identified in the genome-edited organisms that they do not possess foreign DNA (CamV 35S promoter, NOS terminator or stable integration of Cas gene) if produced from in vitro transcribed CRISPR/Cas system or purified RNPs. Therefore, such changes are difficult to recognize as foreign because they highly resemble natural random mutations, chemical mutagenesis or radiation mutations (Kim and Kim 2014). Hence, it becomes difficult and challenging for scientists and regulatory authorities to recognize and quantify these mutations. Such as, European countries demand event-specific identification of GMOs derived food or feed before their marketing and commercialization. Organisms produced through random mutations such as chemicals or radiation do not need any identification or post mutation identification as they are not regulated (Fellmann et al. 2017). In contrast, genome-edited crops are not exempted from regulation and need to quantify and detect their events before commercialization. The marketing and commercialization of these plants, therefore, require clear detection with rigorous regulations and environmental risk assessment (Liang et al. 2018). German government recommendations for identification of genome-edited plants are described as: (1) modifications must be detectable; (2) it should be analytically possible to confirm that the given modifications were introduced and specifically through genome editing; (3) identification of genome editing should be unambiguous (Schaeffer and Nakata 2015). However, in the USA, Argentina, Australia, and Canada, genome-edited organisms are not regulated under conventional genetic engineering regulations but still quantification of specificity and off-target evaluation is compulsory.

Different detection methods are available for CRISPR mediated genome editing organisms, ranging from simple PCR to Next Generation Sequencing (NGS) (Stasik et al. 2018). However, the choice of the quantification method depends upon the type of predicted modifications such as mutation, insertions, regulation, or epigenetic modifications. PCR methods are inexpensive, efficient, and sensitive while sequencing-based methods especially WGS (whole genome sequencing) is expensive require expert skills to analyze data. In the following section, different analytical methods commonly used for the detection of genome-edited organisms are described (Tang et al. 2018).

3.9.1 Tracking CRISPR Mutations with Surveyor or T7E1A Assay

The Surveyor or T7E1A assay is an extensively used method and represents a gold standard for the detection of CRISPR induced mutations. It is a simple, PCR based method that can be performed a few days after the experiment. The entire procedure is based on-target-specific PCR amplification using a proofreading polymerase, denaturation, and reannealing of amplicons to generate heteroduplex and finally cleavage of mismatches with T7 endonuclease (Gwiazda 2014). The cleavage products will be resolved on agarose gel electrophoresis to analyze the results. Although the assay is simple and straightforward, however, it requires optimization of amplification and generation of heteroduplex. Moreover, T7E1A would not always produce heteroduplex as increasing evidence suggest that mutation-induced by CRISPR is not random. Subsequently, T7E1 assay could underestimate the CRISPR efficiency and T7E1A could only be useful for the activity of CRISPR/Cas (Miraglia et al. 2004).

3.9.2 PCR-Restriction Enzyme

Analogous to the tracking CRISPR mutations with a surveyor or T7E1A assay CRISPR indels, PCR-Restriction Enzyme (PCR-RE) is simple and fast method to analyze CRISPR indels, however, it is based on compulsory presence of restriction site in the gRNA (Tsai et al. 2017). Positive CRISPR editing will result in mutation and consequently impair the restriction site. The target DNA will be amplified with PCR and amplicon will be analyzed for loss in restriction site with a restriction enzyme having site in gRNA (Cho et al. 2014). The wild-type fragment will be digested with enzyme while mutated fragment will remain undigested due to loss in restriction site. The PCR-RE can differentiate between monoallelic and biallelic mutations. However, it is not possible to track homo and heterozygous mutants (Lin and Luo 2019).

3.9.3 Restriction Guided Endonuclease-Restriction Fragment Length Polymorphism Method

The process of detecting mutation through RGEN-RFLP (Restriction guided endonuclease-restriction fragment length polymorphism) is based on cleavage of PCR products in the presences of in-vitro-transcribed gRNA and purified Cas protein (Kleinstiver et al. 2016). Wild-type amplicons will be digested whereas mutated fragment will remain intact due to impaired gRNA binding site. Similar to PCR-RE this method cannot distinguish between biallelic homozygous and biallelic heterozygous mutants (Thomsen et al. 2012).

3.9.4 PCR-PAGE

PCR-PAGE (PCR-polyacrylamide gel electrphoresis) is an efficient and less expensive method for the detection of CRISPR indels by resolving heteroduplex and homoduplex amplicons on polyacrylamide gel electrophoresis. It is based on amplification of cleavage site by specific primers followed by denaturation and annealing of the amplicons (Zhu et al. 2014). Denaturation and annealing of the amplicons will produce a mixture of heteroduplex and homoduplex amplicons which can be resolved on native PAGE (Thomsen et al. 2012). This method can easily distinguish between mono and biallelic mutations, however, it is unable to differentiate between wild-type and mutated homoduplex.

3.9.5 High-Resolution Melt Curve Analysis (HRMA)

HRMA is based on base composition and size dependent analysis of the amplified products. HRMA involves sequence-specific amplification of about 90–200 bp around the target site and subsequent melt curve analysis of the amplicon. HRMA is a simple, efficient, and high throughput method of CRISPR-based mutation analysis (Kc et al. 2016). The entire procedure can be performed in a single tube and usually takes less than 2 h. As compared to the T7E1A and TIDE (Tracking of indels by decomposition) HRMA does not require purification, sequencing, and electrophoresis of PCR products. However, electrophoresis and sequencing can be performed if required. Although HRMA has been used for CRISPR indels, epigenetics and SNP genotyping but it also possesses some limitations. HRMA cannot quantify large scale mutations, in addition to this it is not possible to quantify the frequency and base composition of indels using HRMA (Thomsen et al. 2012).

3.9.6 qPCR

qPCR has been one of the easiest and rapid methods for quantification analysis of CRISPR mutations (Tycko et al. 2016). It is based on PCR amplification of the target region followed by Sanger sequencing. It can distinguish between homozygous and heterozygous mutations and have been used to identify CRISPR mutation in several plant species such as *Arabidopsis*, sorghum, maize, and rice (Dabrowska et al. 2018).

3.9.7 TIDE and TIDER

TIDE and TIDER (Tracking of insertions, deletions and recombination events) have been used as a fast and effective method for quantification of CRISPR indels in the pool of cells (Zischewski et al. 2017). These methods are based on two PCRs followed by sequencing of the amplicons. Furthermore, these are analysed through an easy to use webtool. Although TIDE is simple and detect mutations from NHEJ repair pathway but it is unable to detect base substitutions and template-directed mutations (Brinkman 2019). In contrast, TIDER can detect point mutations arising from NHEJ and template-directed mutations (Brinkman et al. 2018).

3.9.8 IDAA

Next Generation Sequencing (NGS) is the most widely adopted method to evaluate indels of CRISPR (Jacobi et al. 2017). Whereas, the software and the algorithms used for NGS have some limitations as well along with its advantages. The overall detection accuracy of NGS depends upon the platform used for data analysis (Brocal et al. 2016). TIDE, TIDER, and ICE-D offer an alternate solution to NGS, using Sanger sequencing but Sanger sequencing limits detection efficiency to 5–10%. In addition to this, the sensitivity of TIDE and ICE-D (Inference of CRISPR edits) for profiling complex cell pools is not yet evaluated (Stasik et al. 2018). IDAA offers a fast and accurate quantitative method of profiling CRISPR indels based on tri-primer amplification and capillary electrophoresis. IDAA (Indel detection by amplicon analysis) uses a unique FAM6 labeled primer that allows uniform labeling of amplicon with fluorophore. TIDE and IDAA have similar cost-effectiveness, efficiency, and reproducibility. However, IDAA can only quantify the type of indels, whereas TIDE can also provide information about indels such as the constitution of bases (Zischewski et al. 2017).

3.9.9 Droplet Digital PCR (ddPCR)

ddPCR offers an ultrasensitive quantification of both HDR and NHEJ based mutations in CRISPR outcomes (Hindson et al. 2011). For the detection and quantification of indels, labeled duplex primer-probe is used. In HDR based genome editing, ddPCR uses a FAM-labeled probe that binds HDR editing site while reference assay will use a HEX probe (Miyaoka et al. 2018). FAM positive will indicate the HDR edits and HEX positive will represent the total number of alleles. ddPCR can detect single bp insertion, deletion, and substitution up to 25 bp. In NHEJ mutations, the FAM-labeled probe binds a reference sequence distant from cleavage site, while HEX probe binds wild-type sequence at the cleavage site. Droplets with signals from HEX and FAM will contain wildtype while FAM positive and HEX negative contain mutated alleles. ddPCR can detect indels up to 40 bp on either side of the cleavage site (Liang et al. 2018). Dropoff assay can distinguish monoallelic and biallelic mutations with high sensitivity.

3.9.10 TA-NGS

TA-NGS is a highly sensitive method based on the amplification of the target site followed by NGS (Molinero and Vandenabeele 2016). Compared with other tools such as T7E1A, TIDE, and IDAA, TA-NGS is highly sensitive and gives detailed information on indels such as the composition of indels and their percentage. However, TA-NGS is expensive, time-consuming and require expert skills to handle the data (Kim and Kim 2014). Nevertheless, several web tools such as Batch-GE, CRISPR-DAV, and CRISPResso have been developed to facilitate the analysis of TA-NGS (Brinkman 2019).

3.9.11 Genome-wide Detection of CRISPR Footprints

Entire genome sequencing of genome-edited organisms is important for identification of genome-wide indels and evaluation of specificity of CRISPR/Cas system (Shui et al. 2016). Sanger sequencing is suitable for detection of mutations especially in homogenous samples whereas it is not suitable for heterogeneous samples. For heterogeneous samples targeted deep sequencing through NGS is an appropriate approach. WGS is becoming an increasingly important method for global detection of genetic alterations and specificity of CRISPR/Cas system (Bassett 2017). It does not require any prior information of a specific locus and it can be used to detect unknown mutations in the genome. WGS through NGS generate bulk amount of data of small sequences which require bioinformatics toolkit and the experts to evaluate it. Therefore, a reference genome sequence is necessary to compare the results. The applications of NGS become challenging in polyploidy crops and genomes containing more repetitive regions. For specificity of CRISPR/Cas it is necessary to observe non-target binding and cleavage sites, collectively called CRISPR footprints.

Specificity is a major concern in genome editing experiments as CRISPR suite has been re-engineered for a wide range of applications such as mutations, niching, insertion, base editing, and epigenetic modifications (Liang et al. 2018). All other platforms except WGS can detect target site-specific indels. However, WGS-NGS can detect unbiased global footprints of CRISPR/Cas using computerized algorithms. Using computerized platforms, off-target sites can be predicted and researchers can use that information to detect off-targets through targeted sequencing. In-silico prediction is relatively quick and cheap but each platform has its parameters and such predictions are not guaranteed. Moreover, experimental confirmation of the predicted sites is not an easy task as they are in large numbers (Lin and Luo 2019). To overcome these limitations researchers have been using in vitro and in vivo based whole genome sequencing of genome-edited experiments.

Whole genome sequencing is crucial for the development of patents, clinical trials, and commercialization of genome-edited plants. WGS can detect any unforeseen mutations which have the potential to have harmful effects (Tycko et al. 2016). Furthermore, it would also be important for regulatory process of plants and animals. The commonly used in vitro methods of WGS are Digenome-Seq and Circle-seq. Digenome-seq is based on in vitro nuclease digestion of genome while Circle-seq is based on circularization of fragments after nuclease digestion (Dabrowska et al. 2018). One disadvantage of in vitro methods is that they are unable to capture the effect of chromatin on CRISPR cleavage (Bi and Yang 2017). In vivo methods include high throughput genome-wide translocation sequencing (HTGTS), capturing of double stranded oligonucleotides into DSBs (Guide-seq), capturing of insertion defective lentiviral into DSBs (IDVL-Seq), and BLESS (directly labeling of DSBs in situ with biotinylated and streptavidin enrichment). All these methods capture DSBs chemically or by inserting tags. WGS methods are also used to evaluate off-target effects introduced by CRISPR (Bassett 2017).

3.10 Summary and Conclusion

In summary, CRRISPR/Cas system can be used for diverse applications of genome editing such as gene knockin and knockout, gene upregulation and downregulation, point mutations, and base editing. Researcher starts with a hypothesis with intention of genome perturbation for any of these described in previous line. A CRSIPR/Cas system is selected for this purpose. The most widely used system which have been used for experimental genome editing consists of two components, i.e. Cas9

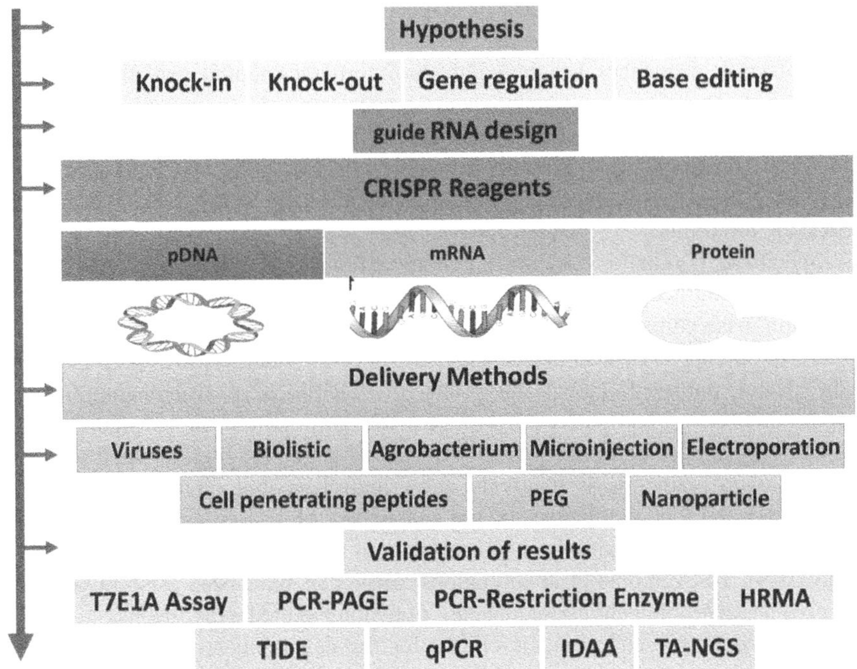

Fig. 3.5 Schematic representation of different involved in CRISPR-Cas genome editing experiment. Steps are shown starting from hypothesis and purpose of experiment, for example, gene knockin, knockout or gene regulation. Accordingly, gRNA is designed using online software with maximum on-target score and minimum off-target score. This guide RNA along with Cas protein is delivered to cells in the form of plasmid DNA or mRNA for guide RNA and Cas or guide RNA and Cas itself. Most commonly used delivery methods include viruses, biolistic, Agrobacterium, Microinjection, Electroporation, and Nanoparticles. CRISPR-Cas induced genetic changes are validated and confirmed

nuclease protein and sgRNA (a 20 nucleotide (nt) long RNA molecule). Accordingly, a guide RNA is designed using online software described above. The best guide RNA is selected based on their maximum on-target scores and minimum off-target scores. These two components can either be delivered in the form of ribonucleoprotein complex (Cas9 and sgRNA) or in the form of DNA which is transcribed in vivo to produce Cas9 and sgRNA. Most commonly used delivery methods include viruses, biolistic, Agrobacterium, microinjection, electroporation, and nanoparticles. CRISPR-Cas induced genetic changes are validated and confirmed. Steps of a CRISPR/Cas based genome editing experiments are shown in Fig. 3.5.

In conclusion this a tool which enabled breakthrough discoveries and methodological advancements having impact across diverse fields of biology such as molecular biology, medicine and biotechnology, to name only a few. Stunning progress have been made in recent years regarding the development of various CRISPR-based technologies. Biotechnological and therapeutic applications of this system been hailed widely in mass media and social media. Growing scope and power of CRSIPR to impact scientific developments is leading to social and ethical concerns. CRISPR-Cas technology is being used for genetic modification of genes for improving major agronomic traits in crops of economic importance. The crop improvements have been previously done using traditional breeding approaches and this will take usually take 7–10 years to get desired improved changes, whereas the CRISPR technology gives results in much shorter time. Another application of CRISPR is "gene drive." Using this technique, a particular gene can be propagated throughout an entire population or species. Due to global impact of such technique, regulatory issues and safety aspects should be carefully considered. Due to simplicity, flexibility, and accuracy of CRISPR/Cas9 in gene targeting, it is being widely used in gene editing experiments and it is expected that numerous breakthroughs will be made using the technology in the coming times.

3.11 Future Perspective

With the discovery CRISPR-Cas, a programmable tool for genetic modifications, researchers have been equipped with an unprecedented toolbox for genetic modifications. This technique holds tremendous advantages as compared to other ways of genetic manipulations due to ease of use, specificity, efficacy, and versatility. Many efforts are being employed to improve and expand CRISPR technology on many aspects. The common target for these efforts is to achieve more efficient and predictable gene editing. This will make it possible to make gene editing more efficient with multiple edits. To achieve these goals, re-engineering Cas9 for desirable characteristics can be considered, better delivery tools with packing Cas9 are needed, better binding and cutting efficacy of with higher specificity of Cas9 protein is needed. Better and new Cas9 proteins or other type II CRISPR proteins with better performance advantages might also be discovered. Discovery of delivery tools to be able to carry CRISPR-Cas system to target cells with higher efficiencies will make it possible to convert every cell in a population to desired genotype without extra work of selecting and characterizing every individual clone. There are a large number of sgRNA design tools are available which employ different mathematical models and algorithms to identify all available sgRNA and select most optimal sgRNA based on target specificity, cutting efficiency scores, and position of sgRNA in genomic context. Rules governing the specificity of Cas9/sgRNA targeting are not completely understood yet, as it is challenging to predict genome-wide off-targets and this results in false positive and false negative results. Therefore, better data driven models to predict on-target efficiencies and off-target effects are needed which should also take into account other possible factors influencing CRIPSR-Cas activity such as the higher level of chromatin structures limiting binding of Cas9 protein.

References

Adli M (2018) The CRISPR tool kit for genome editing and beyond. Nat Commun 9(1):1–13

Ali Z, Abul-Faraj A, Li L, Ghosh N, Piatek M, Mahjoub A, Aouida M, Piatek A, Baltes NJ, Voytas DF (2015) Efficient virus-mediated genome editing in plants using the CRISPR/Cas9 system. Mol Plant 8(8):1288–1291

Altenbuchner J (2016) Editing of the Bacillus subtilis genome by the CRISPR-Cas9 system. Appl Environ Microbiol 82(17):5421–5427

Anders C, Niewoehner O, Duerst A, Jinek M (2014) Structural basis of PAM-dependent target DNA recognition by the Cas9 endonuclease. Nature 513(7519):569

Andersson M, Turesson H, Nicolia A, Fält A-S, Samuelsson M, Hofvander P (2017) Efficient targeted multiallelic mutagenesis in tetraploid potato (Solanum tuberosum) by transient CRISPR-Cas9 expression in protoplasts. Plant Cell Rep 36(1):117–128

Aubrey BJ, Kelly GL, Kueh AJ, Brennan MS, O'Connor L, Milla L, Wilcox S, Tai L, Strasser A, Herold MJ (2015) An inducible lentiviral guide RNA platform enables the identification of tumor-essential genes and tumor-promoting mutations in vivo. Cell Rep 10(8):1422–1432

Bae S, Park J, Kim JS (2014) Cas-OFFinder: a fast and versatile algorithm that searches for potential off-target sites of Cas9 RNA-guided endonucleases. Bioinformatics 30 (10):1473–1475. https://doi.org/10.1093/bioinformatics/btu048

Baltes NJ, Gil-Humanes J, Cermak T, Atkins PA, Voytas DF (2014) DNA replicons for plant genome engineering. Plant Cell 26(1):151–163

Barrangou R, Fremaux C, Deveau H, Richards M, Boyaval P, Moineau S, Romero DA, Horvath P (2007) CRISPR provides acquired resistance against viruses in prokaryotes. Science 315 (5819):1709–1712

Bassett AR (2017) Editing the genome of hiPSC with CRISPR/Cas9: disease models. Mamm Genome 28(7-8):348–364

Bates K, Kostarelos K (2013) Carbon nanotubes as vectors for gene therapy: past achievements, present challenges and future goals. Adv Drug Deliv Rev 65(15):2023–2033

Bhakdi S, Tranum-Jensen J, Sziegoleit A (1985) Mechanism of membrane damage by streptolysin-O. Infect Immun 47(1):52–60

Bi H, Yang B (2017) Gene editing with TALEN and CRISPR/Cas in rice. In: Prog mol biol transl sci, vol 149. Elsevier, London, pp 81–98

Bolotin A, Quinquis B, Sorokin A, Ehrlich SD (2005) Clustered regularly interspaced short palindrome repeats (CRISPRs) have spacers of extrachromosomal origin. Microbiology 151 (8):2551–2561

Bortesi L, Fischer R (2015) The CRISPR/Cas9 system for plant genome editing and beyond. Biotechnol Adv 33(1):41–52

Breakefield XO, Frederickson RM, Simpson RJ (2011) Gesicles: Microvesicle "cookies" for transient information transfer between cells. Mol Ther 19(9):1574–1576

Brinkman E (2019) Ride the tide: observing CRISPR/Cas9 genome editing by the numbers

Brinkman EK, Kousholt AN, Harmsen T, Leemans C, Chen T, Jonkers J, van Steensel B (2018) Easy quantification of template-directed CRISPR/Cas9 editing. Nucleic Acids Res 46(10):e58–e58

Brito JL, Davies FE, Gonzalez D, Morgan GJ (2008) Streptolysin-O reversible permeabilisation is an effective method to transfect siRNAs into myeloma cells. J Immunol Methods 333 (1-2):147–155

Brocal I, White RJ, Dooley CM, Carruthers SN, Clark R, Hall A, Busch-Nentwich EM, Stemple DL, Kettleborough RN (2016) Efficient identification of CRISPR/Cas9-induced insertions/deletions by direct germline screening in zebrafish. BMC Genomics 17(1):259

Cameron P, Fuller CK, Donohoue PD, Jones BN, Thompson MS, Carter MM, Gradia S, Vidal B, Garner E, Slorach EM, Lau E, Banh LM, Lied AM, Edwards LS, Settle AH, Capurso D, Llaca V, Deschamps S, Cigan M, Young JK, May AP (2017) Mapping the genomic landscape of CRISPR-Cas9 cleavage. Nat Methods 14(6):600–606. https://doi.org/10.1038/nmeth.4284

Carroll KJ, Makarewich CA, McAnally J, Anderson DM, Zentilin L, Liu N, Giacca M, Bassel-Duby R, Olson EN (2016) A mouse model for adult cardiac-specific gene deletion with CRISPR/Cas9. Proc Natl Acad Sci 113(2):338–343

Čermák T, Baltes NJ, Čegan R, Zhang Y, Voytas DF (2015) High-frequency, precise modification of the tomato genome. Genome Biol 16(1):232

Chandrasekaran J, Brumin M, Wolf D, Leibman D, Klap C, Pearlsman M, Sherman A, Arazi T, Gal-On A (2016) Development of broad virus resistance in non-transgenic cucumber using CRISPR/Cas9 technology. Mol Plant Pathol 17(7):1140–1153

Chandrasekaran AP, Song M, Kim K-S, Ramakrishna S (2018) Different methods of delivering CRISPR/Cas9 into cells. In: Prog mol biol transl sci, vol 159. Elsevier, London, pp 157–176

Chang F-P, Kuang L-Y, Huang C-A, Jane W-N, Hung Y, Yue-ie CH, Mou C-Y (2013) A simple plant gene delivery system using mesoporous silica nanoparticles as carriers. J Mater Chem B 1 (39):5279–5287

Char SN, Neelakandan AK, Nahampun H, Frame B, Main M, Spalding MH, Becraft PW, Meyers BC, Walbot V, Wang K (2017) An agrobacterium-delivered CRISPR/Cas9 system for high-frequency targeted mutagenesis in maize. Plant Biotechnol J 15(2):257–268

Chari R, Mali P, Moosburner M, Church GM (2015) Unraveling CRISPR-Cas9 genome engineering parameters via a library-on-library approach. Nat Methods 12(9):823

Chew WL, Tabebordbar M, Cheng JK, Mali P, Wu EY, Ng AH, Zhu K, Wagers AJ, Church GM (2016) A multifunctional AAV–CRISPR–Cas9 and its host response. Nat Methods 13(10):868

Cho SW, Kim S, Kim Y, Kweon J, Kim HS, Bae S, Kim J-S (2014) Analysis of off-target effects of CRISPR/Cas-derived RNA-guided endonucleases and nickases. Genome Res 24(1):132–141

Choi J, Dang Y, Abraham S, Ma H, Zhang J, Guo H, Cai Y, Mikkelsen J, Wu H, Shankar P (2016) Lentivirus pre-packed with Cas9 protein for safer gene editing. Gene Ther 23(7):627

Chuang C-K, Chen C-H, Huang C-L, Su Y-H, Peng S-H, Lin T-Y, Tai H-C, Yang T-S, Tu C-F (2017) Generation of GGTA1 mutant pigs by direct pronuclear microinjection of CRISPR/Cas9 plasmid vectors. Anim Biotechnol 28(3):174–181

Cong L, Ran FA, Cox D, Lin S, Barretto R, Habib N, Hsu PD, Wu X, Jiang W, Marraffini LA (2013) Multiplex genome engineering using CRISPR/Cas systems. Science 339(6121):819–823

Crawley AB, Henriksen JR, Barrangou R (2018) CRISPRdisco: an automated pipeline for the discovery and analysis of CRISPR-Cas systems. CRISPR J 1(2):171–181

Cui Y, Xu J, Cheng M, Liao X, Peng S (2018) Review of CRISPR/Cas9 sgRNA design tools. Interdiscip Sci 10(2):455–465. https://doi.org/10.1007/s12539-018-0298-z

Cunningham FJ, Goh NS, Demirer GS, Matos JL, Landry MP (2018) Nanoparticle-mediated delivery towards advancing plant genetic engineering. Trends Biotechnol 36(9):882–897

Dabrowska M, Czubak K, Juzwa W, Krzyzosiak WJ, Olejniczak M, Kozlowski P (2018) qEva CRISPR: a method for quantitative evaluation of CRISPR/Cas-mediated genome editing in target and off-target sites. Nucleic Acids Res 46(17):e101

Dalby B, Cates S, Harris A, Ohki EC, Tilkins ML, Price PJ, Ciccarone VC (2004) Advanced transfection with Lipofectamine 2000 reagent: primary neurons, siRNA, and high-throughput applications. Methods 33(2):95–103. https://doi.org/10.1016/j.ymeth.2003.11.023

Davis CA, Hitz BC, Sloan CA, Chan ET, Davidson JM, Gabdank I, Hilton JA, Jain K, Baymuradov UK, Narayanan AK, Onate KC, Graham K, Miyasato SR, Dreszer TR, Strattan JS, Jolanki O, Tanaka FY, Cherry JM (2018) The encyclopedia of DNA elements (ENCODE): data portal update. Nucleic Acids Res 46(D1):D794–D801. https://doi.org/10.1093/nar/gkx1081

de Figueiredo IR, Freire JM, Flores L, Veiga AS, Castanho MA (2014) Cell-penetrating peptides: a tool for effective delivery in gene-targeted therapies. IUBMB Life 66(3):182–194

Demirer G, Zhang H, Matos J, Chang R, Chio L, Staskawicz B, Landry M (2018) High aspect ratio nanomaterials enable biomolecule delivery and transgene expression or silencing in mature plants. bioRxiv 2018

Deng W, Shi X, Tjian R, Lionnet T, Singer RH (2015) CASFISH: CRISPR/Cas9-mediated in situ labeling of genomic loci in fixed cells. Proc Natl Acad Sci 112(38):11870–11875

Deyle DR, Russell DW (2009) Adeno-associated virus vector integration. Curr Opin Mol Ther 11 (4):442–447
Doench JG, Hartenian E, Graham DB, Tothova Z, Hegde M, Smith I, Sullender M, Ebert BL, Xavier RJ, Root DE (2014) Rational design of highly active sgRNAs for CRISPR-Cas9–mediated gene inactivation. Nat Biotechnol 32(12):1262
Doench JG, Fusi N, Sullender M, Hegde M, Vaimberg EW, Donovan KF, Smith I, Tothova Z, Wilen C, Orchard R (2016) Optimized sgRNA design to maximize activity and minimize off-target effects of CRISPR-Cas9. Nat Biotechnol 34(2):184
Du X, Shi B, Tang Y, Dai S, Qiao SZ (2014) Label-free dendrimer-like silica nanohybrids for traceable and controlled gene delivery. Biomaterials 35(21):5580–5590
Duensing N, Sprink T, Parrott WA, Fedorova M, Lema MA, Wolt JD, Bartsch D (2018) Novel features and considerations for ERA and regulation of crops produced by genome editing. Front Bioeng Biotechnol 6:e79
Durfee PN, Lin Y-S, Dunphy DR, Muñiz AJ, Butler KS, Humphrey KR, Lokke AJ, Agola JO, Chou SS, Chen I-M (2016) Mesoporous silica nanoparticle-supported lipid bilayers (protocells) for active targeting and delivery to individual leukemia cells. ACS Nano 10(9):8325–8345
Fellmann C, Gowen BG, Lin P-C, Doudna JA, Corn JE (2017) Cornerstones of CRISPR–Cas in drug discovery and therapy. Nat Rev Drug Discov 16(2):89
Feng C, Yuan J, Wang R, Liu Y, Birchler JA, Han F (2016) Efficient targeted genome modification in maize using CRISPR/Cas9 system. J Genet Genomics 43(1):37–43
Fu Y, Sander JD, Reyon D, Cascio VM, Joung JK (2014) Improving CRISPR-Cas nuclease specificity using truncated guide RNAs. Nat Biotechnol 32(3):279–284. https://doi.org/10.1038/nbt.2808
Gaj T, Gersbach CA, Barbas CF (2013) ZFN, TALEN, and CRISPR/Cas-based methods for genome engineering. Trends Biotechnol 31(7):397–405
Garneau JE, Dupuis ME, Villion M, Romero DA, Barrangou R, Boyaval P, Fremaux C, Horvath P, Magadan AH, Moineau S (2010) The CRISPR/Cas bacterial immune system cleaves bacteriophage and plasmid DNA. Nature 468(7320):67–71. https://doi.org/10.1038/nature09523
Gasiunas G, Barrangou R, Horvath P, Siksnys V (2012) Cas9–crRNA ribonucleoprotein complex mediates specific DNA cleavage for adaptive immunity in bacteria. Proc Natl Acad Sci 109(39): E2579–E2586
Gasiunas G, Sinkunas T, Siksnys V (2014) Molecular mechanisms of CRISPR-mediated microbial immunity. Cell Mol Life Sci 71(3):449–465. https://doi.org/10.1007/s00018-013-1438-6
Gaudelli NM, Komor AC, Rees HA, Packer MS, Badran AH, Bryson DI, Liu DR (2017) Programmable base editing of A• T to G• C in genomic DNA without DNA cleavage. Nature 551(7681):464
Gil-Humanes J, Wang Y, Liang Z, Shan Q, Ozuna CV, Sánchez-León S, Baltes NJ, Starker C, Barro F, Gao C (2017) High-efficiency gene targeting in hexaploid wheat using DNA replicons and CRISPR/Cas9. Plant J 89(6):1251–1262
Guidotti G, Brambilla L, Rossi D (2017) Cell-penetrating peptides: from basic research to clinics. Trends Pharmacol Sci 38(4):406–424
Gwiazda K (2014) Increasing the efficiency of targeted genome engineering by directly manipulating DNA double strand break repair pathways
Heigwer F, Kerr G, Boutros M (2014) E-CRISP: fast CRISPR target site identification. Nat Methods 11(2):122–123. https://doi.org/10.1038/nmeth.2812
Hiei Y, Komari T (2008) Agrobacterium-mediated transformation of rice using immature embryos or calli induced from mature seed. Nat Protoc 3(5):824
Hilton IB, D'ippolito AM, Vockley CM, Thakore PI, Crawford GE, Reddy TE, Gersbach CA (2015) Epigenome editing by a CRISPR-Cas9-based acetyltransferase activates genes from promoters and enhancers. Nat Biotechnol 33(5):510
Hindson BJ, Ness KD, Masquelier DA, Belgrader P, Heredia NJ, Makarewicz AJ, Bright IJ, Lucero MY, Hiddessen AL, Legler TC (2011) High-throughput droplet digital PCR system for absolute quantitation of DNA copy number. Anal Chem 83(22):8604–8610

Hirano H, Gootenberg JS, Horii T, Abudayyeh OO, Kimura M, Hsu PD, Nakane T, Ishitani R, Hatada I, Zhang F (2016) Structure and engineering of Francisella novicida Cas9. Cell 164 (5):950–961

Horii T, Arai Y, Yamazaki M, Morita S, Kimura M, Itoh M, Abe Y, Hatada I (2014) Validation of microinjection methods for generating knockout mice by CRISPR/Cas-mediated genome engineering. Sci Rep 4:4513

Horvath P, Romero DA, Coûté-Monvoisin A-C, Richards M, Deveau H, Moineau S, Boyaval P, Fremaux C, Barrangou R (2008) Diversity, activity, and evolution of CRISPR loci in Streptococcus thermophilus. J Bacteriol 190(4):1401–1412

Hou Z, Zhang Y, Propson NE, Howden SE, Chu L-F, Sontheimer EJ, Thomson JA (2013) Efficient genome engineering in human pluripotent stem cells using Cas9 from Neisseria meningitidis. Proc Natl Acad Sci 110(39):15644–15649

Hsu PD, Scott DA, Weinstein JA, Ran FA, Konermann S, Agarwala V, Li Y, Fine EJ, Wu X, Shalem O, Cradick TJ, Marraffini LA, Bao G, Zhang F (2013) DNA targeting specificity of RNA-guided Cas9 nucleases. Nat Biotechnol 31(9):827–832. https://doi.org/10.1038/nbt.2647

Hua K, Tao X, Yuan F, Wang D, Zhu J-K (2018) Precise A· T to G· C base editing in the rice genome. Mol Plant 11(4):627–630

Hung SS, Chrysostomou V, Li F, Lim JK, Wang J-H, Powell JE, Tu L, Daniszewski M, Lo C, Wong RC (2016) AAV-mediated CRISPR/Cas gene editing of retinal cells in vivo. Invest Ophthalmol Vis Sci 57(7):3470–3476

Ishino Y, Shinagawa H, Makino K, Amemura M, Nakata A (1987) Nucleotide sequence of the iap gene, responsible for alkaline phosphatase isozyme conversion in Escherichia coli, and identification of the gene product. J Bacteriol 169(12):5429–5433

Jacobi AM, Rettig GR, Turk R, Collingwood MA, Zeiner SA, Quadros RM, Harms DW, Bonthuis PJ, Gregg C, Ohtsuka M (2017) Simplified CRISPR tools for efficient genome editing and streamlined protocols for their delivery into mammalian cells and mouse zygotes. Methods 121:16–28

Jansen R, Embden JD, Gaastra W, Schouls LM (2002) Identification of genes that are associated with DNA repeats in prokaryotes. Mol Microbiol 43(6):1565–1575

Jiang L, Ding L, He B, Shen J, Xu Z, Yin M, Zhang X (2014) Systemic gene silencing in plants triggered by fluorescent nanoparticle-delivered double-stranded RNA. Nanoscale 6 (17):9965–9969

Jinek M, Chylinski K, Fonfara I, Hauer M, Doudna JA, Charpentier E (2012) A programmable dual-RNA–guided DNA endonuclease in adaptive bacterial immunity. Science 337 (6096):816–821

Kaeppler HF, Gu W, Somers DA, Rines HW, Cockburn AF (1990) Silicon carbide fiber-mediated DNA delivery into plant cells. Plant Cell Rep 9(8):415–418

Kaeppler H, Somers D, Rines H, Cockburn A (1992) Silicon carbide fiber-mediated stable transformation of plant cells. Theor Appl Genet 84(5-6):560–566

Kc R, Srivastava A, Wilkowski J, Richter C, Shavit J, Burke D, Bielas S (2016) Detection of nucleotide-specific CRISPR/Cas9 modified alleles using multiplex ligation detection. Sci Rep 6:32048

Kim H, Kim J-S (2014) A guide to genome engineering with programmable nucleases. Nat Rev Genet 15(5):321–334

Kim D, Bae S, Park J, Kim E, Kim S, Yu HR, Hwang J, Kim JI, Kim JS (2015) Digenome-seq: genome-wide profiling of CRISPR-Cas9 off-target effects in human cells. Nat Methods 12 (3):237–243. https://doi.org/10.1038/nmeth.3284

Kim E, Koo T, Park SW, Kim D, Kim K, Cho H-Y, Song DW, Lee KJ, Jung MH, Kim S (2017) In vivo genome editing with a small Cas9 orthologue derived from Campylobacter jejuni. Nat Commun 8:14500

Kleinstiver BP, Prew MS, Tsai SQ, Topkar VV, Nguyen NT, Zheng Z, Gonzales AP, Li Z, Peterson RT, Yeh J-RJ (2015) Engineered CRISPR-Cas9 nucleases with altered PAM specificities. Nature 523(7561):481

Kleinstiver BP, Pattanayak V, Prew MS, Tsai SQ, Nguyen NT, Zheng Z, Joung JK (2016) High-fidelity CRISPR–Cas9 nucleases with no detectable genome-wide off-target effects. Nature 529 (7587):490

Kogure K, Moriguchi R, Sasaki K, Ueno M, Futaki S, Harashima H (2004) Development of a non-viral multifunctional envelope-type nano device by a novel lipid film hydration method. J Control Release 98(2):317–323

Koike-Yusa H, Li Y, Tan E-P, Velasco-Herrera MDC, Yusa K (2014) Genome-wide recessive genetic screening in mammalian cells with a lentiviral CRISPR-guide RNA library. Nat Biotechnol 32(3):267

Komor AC, Zhao KT, Packer MS, Gaudelli NM, Waterbury AL, Koblan LW, Kim YB, Badran AH, Liu DR (2017) Improved base excision repair inhibition and bacteriophage Mu Gam protein yields C: G-to-T: A base editors with higher efficiency and product purity. Sci Adv 3 (8):eaao4774

Kuan PF, Powers S, He S, Li K, Zhao X, Huang B (2017) A systematic evaluation of nucleotide properties for CRISPR sgRNA design. BMC Bioinf 18(1):297

Labuhn M, Adams FF, Ng M, Knoess S, Schambach A, Charpentier EM, Schwarzer A, Mateo JL, Klusmann J-H, Heckl D (2017) Refined sgRNA efficacy prediction improves large-and small-scale CRISPR–Cas9 applications. Nucleic Acids Res 46(3):1375–1385

Larue C, Castillo-Michel H, Sobanska S, Cécillon L, Bureau S, Barthès V, Ouerdane L, Carrière M, Sarret G (2014) Foliar exposure of the crop Lactuca sativa to silver nanoparticles: evidence for internalization and changes in Ag speciation. J Hazard Mater 264:98–106

Lawrenson T, Shorinola O, Stacey N, Li C, Østergaard L, Patron N, Uauy C, Harwood W (2015) Induction of targeted, heritable mutations in barley and Brassica oleracea using RNA-guided Cas9 nuclease. Genome Biol 16(1):258

Ledford H (2016) CRISPR: gene editing is just the beginning. Nat News 531(7593):156

Li J-F, Norville JE, Aach J, McCormack M, Zhang D, Bush J, Church GM, Sheen J (2013) Multiplex and homologous recombination–mediated genome editing in Arabidopsis and Nicotiana benthamiana using guide RNA and Cas9. Nat Biotechnol 31:688. https://doi.org/10.1038/nbt.2654

Li Z, Liu Z-B, Xing A, Moon BP, Koellhoffer JP, Huang L, Ward RT, Clifton E, Falco SC, Cigan AM (2015) Cas9-guide RNA directed genome editing in soybean. Plant Physiol 169 (2):960–970

Li J, Meng X, Zong Y, Chen K, Zhang H, Liu J, Li J, Gao C (2016) Gene replacements and insertions in rice by intron targeting using CRISPR–Cas9. Nat Plants 2(10):16139

Li X, Wang Y, Chen S, Tian H, Fu D, Zhu B, Luo Y, Zhu H (2018) Lycopene is enriched in tomato fruit by CRISPR/Cas9-mediated multiplex genome editing. Front Plant Sci 9:559. https://doi.org/10.3389/fpls.2018.00559

Liang X, Potter J, Kumar S, Zou Y, Quintanilla R, Sridharan M, Carte J, Chen W, Roark N, Ranganathan S (2015) Rapid and highly efficient mammalian cell engineering via Cas9 protein transfection. J Biotechnol 208:44–53

Liang Z, Chen K, Li T, Zhang Y, Wang Y, Zhao Q, Liu J, Zhang H, Liu C, Ran Y (2017) Efficient DNA-free genome editing of bread wheat using CRISPR/Cas9 ribonucleoprotein complexes. Nat Commun 8:14261

Liang P, Sun H, Zhang X, Xie X, Zhang J, Bai Y, Ouyang X, Zhi S, Xiong Y, Ma W (2018) Effective and precise adenine base editing in mouse zygotes. Protein Cell 9(9):808–813

Lin L, Luo Y (2019) Tracking CRISPR's footprints. In: CRISPR gene editing. Springer, Cham, pp 13–28

Lino CA, Harper JC, Carney JP, Timlin JA (2018) Delivering CRISPR: a review of the challenges and approaches. Drug Deliv 25(1):1234–1257

Liu J, Stace-Naughton A, Jiang X, Brinker CJ (2009) Porous nanoparticle supported lipid bilayers (protocells) as delivery vehicles. J Am Chem Soc 131(4):1354–1355

Lo T-W, Pickle CS, Lin S, Ralston EJ, Gurling M, Schartner CM, Bian Q, Doudna JA, Meyer BJ (2013) Precise and heritable genome editing in evolutionarily diverse nematodes using TALENs and CRISPR/Cas9 to engineer insertions and deletions. Genetics 195(2):331–348

Long C, Amoasii L, Mireault AA, McAnally JR, Li H, Sanchez-Ortiz E, Bhattacharyya S, Shelton JM, Bassel-Duby R, Olson EN (2016) Postnatal genome editing partially restores dystrophin expression in a mouse model of muscular dystrophy. Science 351(6271):400–403

Luo D, Saltzman WM (2000) Enhancement of transfection by physical concentration of DNA at the cell surface. Nat Biotechnol 18(8):893

Luo G-F, Chen W-H, Liu Y, Lei Q, Zhuo R-X, Zhang X-Z (2014) Multifunctional enveloped mesoporous silica nanoparticles for subcellular co-delivery of drug and therapeutic peptide. Sci Rep 4:6064

Ma M, Ye AY, Zheng W, Kong L (2013) A guide RNA sequence design platform for the CRISPR/Cas9 system for model organism genomes. Biomed Res Int 2013:270805. https://doi.org/10.1155/2013/270805

Ma H, Dang Y, Wu Y, Jia G, Anaya E, Zhang J, Abraham S, Choi J-G, Shi G, Qi L (2015a) A CRISPR-based screen identifies genes essential for West-Nile-virus-induced cell death. Cell Rep 12(4):673–683

Ma X, Zhang Q, Zhu Q, Liu W, Chen Y, Qiu R, Wang B, Yang Z, Li H, Lin Y (2015b) A robust CRISPR/Cas9 system for convenient, high-efficiency multiplex genome editing in monocot and dicot plants. Mol Plant 8(8):1274–1284

Ma X, Zhu Q, Chen Y, Liu Y-G (2016) CRISPR/Cas9 platforms for genome editing in plants: developments and applications. Mol Plant 9(7):961–974

Mackowiak SA, Schmidt A, Weiss V, Argyo C, von Schirnding C, Bein T, Bräuchle C (2013) Targeted drug delivery in cancer cells with red-light photoactivated mesoporous silica nanoparticles. Nano Lett 13(6):2576–2583

Macovei A, Sevilla NR, Cantos C, Jonson GB, Slamet-Loedin I, Čermák T, Voytas DF, Choi IR, Chadha-Mohanty P (2018) Novel alleles of rice eIF4G generated by CRISPR/Cas9-targeted mutagenesis confer resistance to Rice tungro spherical virus. Plant Biotechnol J 16 (11):1918–1927

Maggio I, Holkers M, Liu J, Janssen JM, Chen X, Gonçalves MA (2014) Adenoviral vector delivery of RNA-guided CRISPR/Cas9 nuclease complexes induces targeted mutagenesis in a diverse array of human cells. Sci Rep 4:5105

Mahas A, Ali Z, Tashkandi M, Mahfouz MM (2019) Virus-mediated genome editing in plants using the CRISPR/Cas9 system. In: Plant genome editing with CRISPR systems. Springer, Cham, pp 311–326

Mali P, Aach J, Stranges PB, Esvelt KM, Moosburner M, Kosuri S, Yang L, Church GM (2013a) CAS9 transcriptional activators for target specificity screening and paired nickases for cooperative genome engineering. Nat Biotechnol 31:833. https://doi.org/10.1038/nbt.2675

Mali P, Yang L, Esvelt KM, Aach J, Guell M, DiCarlo JE, Norville JE, Church GM (2013b) RNA-guided human genome engineering via Cas9. Science 339(6121):823–826

Malina A, Cameron CJ, Robert F, Blanchette M, Dostie J, Pelletier J (2015) PAM multiplicity marks genomic target sites as inhibitory to CRISPR-Cas9 editing. Nat Commun 6:10124

Malnoy M, Viola R, Jung M-H, Koo O-J, Kim S, Kim J-S, Velasco R, Nagamangala Kanchiswamy C (2016) DNA-free genetically edited grapevine and apple protoplast using CRISPR/Cas9 ribonucleoproteins. Front Plant Sci 7:1904. https://doi.org/10.3389/fpls.2016.01904

Mangeot P-E, Dollet S, Girard M, Ciancia C, Joly S, Peschanski M, Lotteau V (2011) Protein transfer into human cells by VSV-G-induced nanovesicles. Mol Ther 19(9):1656–1666. https://doi.org/10.1038/mt.2011.138

Martin-Ortigosa S, Wang K (2014) Proteolistics: a biolistic method for intracellular delivery of proteins. Transgenic Res 23(5):743–756

Martin-Ortigosa S, Peterson DJ, Valenstein JS, Lin VSY, Trewyn BG, Lyznik LA, Wang K (2014) Mesoporous silica nanoparticle-mediated intracellular Cre protein delivery for maize genome editing via *loxP* site excision. Plant Physiol 164(2):537. https://doi.org/10.1104/pp.113.233650

Matano M, Date S, Shimokawa M, Takano A, Fujii M, Ohta Y, Watanabe T, Kanai T, Sato T (2015) Modeling colorectal cancer using CRISPR-Cas9–mediated engineering of human intestinal organoids. Nat Med 21(3):256

McGinn J, Marraffini LA (2018) Molecular mechanisms of CRISPR–Cas spacer acquisition. Nat Rev Microbiol 2018:1

Miraglia M, Berdal K, Brera C, Corbisier P, Holst-Jensen A, Kok E, Marvin H, Schimmel H, Rentsch J, Van Rie J (2004) Detection and traceability of genetically modified organisms in the food production chain. Food Chem Toxicol 42(7):1157–1180

Miyaoka Y, Mayerl SJ, Chan AH, Conklin BR (2018) Detection and quantification of HDR and NHEJ induced by genome editing at endogenous gene loci using droplet digital PCR. In: Digital PCR. Springer, Cham, pp 349–362

Mojica FJ, Juez G, Rodriguez-Valera F (1993) Transcription at different salinities of Haloferax mediterranei sequences adjacent to partially modified PstI sites. Mol Microbiol 9(3):613–621

Mojica FJ, García-Martínez J, Soria E (2005) Intervening sequences of regularly spaced prokaryotic repeats derive from foreign genetic elements. J Mol Evol 60(2):174–182

Mojica FJ, Díez-Villaseñor C, García-Martínez J, Almendros C (2009) Short motif sequences determine the targets of the prokaryotic CRISPR defence system. Microbiology 155 (3):733–740

Molinero AIS, Vandenabeele S (2016) Plants having enhanced yield-related traits and a method for making the same. Google Patents

Montague TG, Cruz JM, Gagnon JA, Church GM, Valen E (2014) CHOPCHOP: a CRISPR/Cas9 and TALEN web tool for genome editing. Nucleic Acids Res 42:401–407. https://doi.org/10.1093/nar/gku410

Moreno-Mateos MA, Vejnar CE, Beaudoin J-D, Fernandez JP, Mis EK, Khokha MK, Giraldez AJ (2015) CRISPRscan: designing highly efficient sgRNAs for CRISPR-Cas9 targeting in vivo. Nat Methods 12(10):982

Morris MC, Deshayes S, Heitz F, Divita G (2008) Cell-penetrating peptides: from molecular mechanisms to therapeutics. Biol Cell 100(4):201–217

Mout R, Ray M, Yesilbag Tonga G, Lee Y-W, Tay T, Sasaki K, Rotello VM (2017) Direct cytosolic delivery of CRISPR/Cas9-ribonucleoprotein for efficient gene editing. ACS Nano 11 (3):2452–2458

Naito Y, Hino K, Bono H, Ui-Tei K (2014) CRISPRdirect: software for designing CRISPR/Cas guide RNA with reduced off-target sites. Bioinformatics 31(7):1120–1123

Nakagawa Y, Sakuma T, Sakamoto T, Ohmuraya M, Nakagata N, Yamamoto T (2015) Production of knockout mice by DNA microinjection of various CRISPR/Cas9 vectors into freeze-thawed fertilized oocytes. BMC Biotechnol 15(1):33

Nakamura T, Akita H, Yamada Y, Hatakeyama H, Harashima H (2012) A multifunctional envelope-type nanodevice for use in nanomedicine: concept and applications. Acc Chem Res 45(7):1113–1121

Nakayasu M, Akiyama R, Lee HJ, Osakabe K, Osakabe Y, Watanabe B, Sugimoto Y, Umemoto N, Saito K, Muranaka T, Mizutani M (2018) Generation of α-solanine-free hairy roots of potato by CRISPR/Cas9 mediated genome editing of the St16DOX gene. Plant Physiol Biochem 131:70–77. https://doi.org/10.1016/j.plaphy.2018.04.026

Naldini L (2015) Gene therapy returns to centre stage. Nature 526(7573):351

Naldini L, Blömer U, Gage FH, Trono D, Verma IM (1996a) Efficient transfer, integration, and sustained long-term expression of the transgene in adult rat brains injected with a lentiviral vector. Proc Natl Acad Sci 93(21):11382–11388. https://doi.org/10.1073/pnas.93.21.11382

Naldini L, Blömer U, Gallay P, Ory D, Mulligan R, Gage FH, Verma IM, Trono D (1996b) In vivo gene delivery and stable transduction of nondividing cells by a lentiviral vector. Science 272 (5259):263–267

Nekrasov V, Staskawicz B, Weigel D, Jones JD, Kamoun S (2013) Targeted mutagenesis in the model plant Nicotiana benthamiana using Cas9 RNA-guided endonuclease. Nat Biotechnol 31 (8):691

Nester EW (2015) Agrobacterium: nature's genetic engineer. Front Plant Sci 5:730

Nihongaki Y, Kawano F, Nakajima T, Sato M (2015) Photoactivatable CRISPR-Cas9 for optogenetic genome editing. Nat Biotechnol 33(7):755

Nishimasu H, Ran FA, Hsu PD, Konermann S, Shehata SI, Dohmae N, Ishitani R, Zhang F, Nureki O (2014) Crystal structure of Cas9 in complex with guide RNA and target DNA. Cell 156 (5):935–949. https://doi.org/10.1016/j.cell.2014.02.001

Noguchi S, Arakawa T, Fukuda S, Furuno M, Hasegawa A, Hori F, Ishikawa-Kato S, Kaida K, Kaiho A, Kanamori-Katayama M, Kawashima T, Kojima M, Kubosaki A, Manabe RI, Murata M, Nagao-Sato S, Nakazato K, Ninomiya N, Nishiyori-Sueki H, Noma S, Saijyo E, Saka A, Sakai M, Simon C, Suzuki N, Tagami M, Watanabe S, Yoshida S, Arner P, Axton RA, Babina M, Baillie JK, Barnett TC, Beckhouse AG, Blumenthal A, Bodega B, Bonetti A, Briggs J, Brombacher F, Carlisle AJ, Clevers HC, Davis CA, Detmar M, Dohi T, Edge ASB, Edinger M, Ehrlund A, Ekwall K, Endoh M, Enomoto H, Eslami A, Fagiolini M, Fairbairn L, Farach-Carson MC, Faulkner GJ, Ferrai C, Fisher ME, Forrester LM, Fujita R, Furusawa JI, Geijtenbeek TB, Gingeras T, Goldowitz D, Guhl S, Guler R, Gustincich S, Ha TJ, Hamaguchi M, Hara M, Hasegawa Y, Herlyn M, Heutink P, Hitchens KJ, Hume DA, Ikawa T, Ishizu Y, Kai C, Kawamoto H, Kawamura YI, Kempfle JS, Kenna TJ, Kere J, Khachigian LM, Kitamura T, Klein S, Klinken SP, Knox AJ, Kojima S, Koseki H, Koyasu S, Lee W, Lennartsson A, Mackay-Sim A, Mejhert N, Mizuno Y, Morikawa H, Morimoto M, Moro K, Morris KJ, Motohashi H, Mummery CL, Nakachi Y, Nakahara F, Nakamura T, Nakamura Y, Nozaki T, Ogishima S, Ohkura N, Ohno H, Ohshima M, Okada-Hatakeyama M, Okazaki Y, Orlando V, Ovchinnikov DA, Passier R, Patrikakis M, Pombo A, Pradhan-Bhatt S, Qin XY, Rehli M, Rizzu P, Roy S, Sajantila A, Sakaguchi S, Sato H, Satoh H, Savvi S, Saxena A, Schmidl C, Schneider C, Schulze-Tanzil GG, Schwegmann A, Sheng G, Shin JW, Sugiyama D, Sugiyama T, Summers KM, Takahashi N, Takai J, Tanaka H, Tatsukawa H, Tomoiu A, Toyoda H, van de Wetering M, van den Berg LM, Verardo R, Vijayan D, Wells CA, Winteringham LN, Wolvetang E, Yamaguchi Y, Yamamoto M, Yanagi-Mizuochi C, Yoneda M, Yonekura Y, Zhang PG, Zucchelli S, Abugessaisa I, Arner E, Harshbarger J, Kondo A, Lassmann T, Lizio M, Sahin S, Sengstag T, Severin J, Shimoji H, Suzuki M, Suzuki H, Kawai J, Kondo N, Itoh M, Daub CO, Kasukawa T, Kawaji H, Carninci P, Forrest ARR, Hayashizaki Y (2017) FANTOM5 CAGE profiles of human and mouse samples. Sci Data 4:170112. https://doi.org/10.1038/sdata.2017.112

Philippe S, Sarkis C, Barkats M, Mammeri H, Ladroue C, Petit C, Mallet J, Serguera C (2006) Lentiviral vectors with a defective integrase allow efficient and sustained transgene expression in vitro and in vivo. Proc Natl Acad Sci 103(47):17684–17689

Polstein LR, Gersbach CA (2015) A light inducible CRISPR-Cas9 system for control of endogenous gene activation. Nat Chem Biol 11(3):198

Pourcel C, Salvignol G, Vergnaud G (2005) CRISPR elements in Yersinia pestis acquire new repeats by preferential uptake of bacteriophage DNA, and provide additional tools for evolutionary studies. Microbiology 151(3):653–663

Puchta H (2005) The repair of double-strand breaks in plants: mechanisms and consequences for genome evolution. J Exp Bot 56(409):1–14

Qi LS, Larson MH, Gilbert LA, Doudna JA, Weissman JS, Arkin AP, Lim WA (2013) Repurposing CRISPR as an RNA-guided platform for sequence-specific control of gene expression. Cell 152 (5):1173–1183

Quinn T, Morell M, Garachtchenko T, Lee L, Fong M, Haugwitz M, Farmer A (2016) Gesicle mediated delivery of a Cas9-sgRNA protein complex. Mol Ther 24:S126

Rakoczy-Trojanowska M (2002) Alternative methods of plant transformation-a short review. Cell Mol Biol Lett 7(3):849–858

Ramakrishna S, Dad A-BK, Beloor J, Gopalappa R, Lee S-K, Kim H (2014) Gene disruption by cell-penetrating peptide-mediated delivery of Cas9 protein and guide RNA. Genome Res 24 (6):1020–1027

Ran FA, Hsu Patrick D, Lin C-Y, Gootenberg Jonathan S, Konermann S, Trevino AE, Scott David A, Inoue A, Matoba S, Zhang Y, Zhang F (2013) Double nicking by RNA-guided CRISPR Cas9 for enhanced genome editing specificity. Cell 154(6):1380–1389. https://doi.org/10.1016/j.cell.2013.08.021

Ran FA, Cong L, Yan WX, Scott DA, Gootenberg JS, Kriz AJ, Zetsche B, Shalem O, Wu X, Makarova KS (2015) In vivo genome editing using Staphylococcus aureus Cas9. Nature 520 (7546):186

Ran Y, Liang Z, Gao C (2017) Current and future editing reagent delivery systems for plant genome editing. Science 60(5):490–505. https://doi.org/10.1007/s11427-017-9022-1

Richter KS, Serra H, White CI, Jeske H (2016) The recombination mediator RAD51D promotes geminiviral infection. Virology 493:113–127

Roointan A, Morowvat MH (2016) Road to the future of systems biotechnology: CRISPR-Cas-mediated metabolic engineering for recombinant protein production. Biotechnol Genet Eng Rev 32(1-2):74–91

Roy P, Noad R (2008) Virus-like particles as a vaccine delivery system: myths and facts. Hum Vaccin 4(1):5–12

Sahab S, Hayden MJ, Mason J, Spangenberg G (2019) Mesophyll protoplasts and PEG-mediated transfections: transient assays and generation of stable transgenic canola plants. In: Transgenic plants. Springer, Cham, pp 131–152

Sakuma T, Nishikawa A, Kume S, Chayama K, Yamamoto T (2014) Multiplex genome engineering in human cells using all-in-one CRISPR/Cas9 vector system. Sci Rep 4:5400

Samulski RJ, Muzyczka N (2014) AAV-mediated gene therapy for research and therapeutic purposes. Annu Rev Virol 1:427–451

Sanjana NE, Shalem O, Zhang F (2014) Improved vectors and genome-wide libraries for CRISPR screening. Nat Methods 11(8):783

Schaeffer SM, Nakata PA (2015) CRISPR/Cas9-mediated genome editing and gene replacement in plants: transitioning from lab to field. Plant Sci 240:130–142

Schumann K, Lin S, Boyer E, Simeonov DR, Subramaniam M, Gate RE, Haliburton GE, Chun JY, Bluestone JA, Doudna JA (2015) Generation of knock-in primary human T cells using Cas9 ribonucleoproteins. Proc Natl Acad Sci 112(33):10437–10442

Schwank G, Koo B-K, Sasselli V, Dekkers JF, Heo I, Demircan T, Sasaki N, Boymans S, Cuppen E, van der Ent CK (2013) Functional repair of CFTR by CRISPR/Cas9 in intestinal stem cell organoids of cystic fibrosis patients. Cell Stem Cell 13(6):653–658

Senthil-Kumar M, Mysore KS (2014) Tobacco rattle virus–based virus-induced gene silencing in Nicotiana benthamiana. Nat Protoc 9(7):1549

Shalem O, Sanjana NE, Hartenian E, Shi X, Scott DA, Mikkelsen TS, Heckl D, Ebert BL, Root DE, Doench JG (2014) Genome-scale CRISPR-Cas9 knockout screening in human cells. Science 343(6166):84–87

Shan Q, Wang Y, Chen K, Liang Z, Li J, Zhang Y, Zhang K, Liu J, Voytas DF, Zheng X (2013a) Rapid and efficient gene modification in rice and Brachypodium using TALENs. Mol Plant 6 (4):1365–1368

Shan Q, Wang Y, Li J, Zhang Y, Chen K, Liang Z, Zhang K, Liu J, Xi JJ, Qiu J-L (2013b) Targeted genome modification of crop plants using a CRISPR-Cas system. Nat Biotechnol 31(8):686

Shan Q, Wang Y, Li J, Gao C (2014) Genome editing in rice and wheat using the CRISPR/Cas system. Nat Protoc 9(10):2395

Shen B, Zhang J, Wu H, Wang J, Ma K, Li Z, Zhang X, Zhang P, Huang X (2013) Generation of gene-modified mice via Cas9/RNA-mediated gene targeting. Cell Res 23(5):720

Shui B, Hernandez Matias L, Guo Y, Peng Y (2016) The rise of CRISPR/Cas for genome editing in stem cells. Stem Cells Int 2016:17

Singh R, Kuscu C, Quinlan A, Qi Y, Adli M (2015) Cas9-chromatin binding information enables more accurate CRISPR off-target prediction. Nucleic Acids Res 43(18):e118

Stasik S, Schuster C, Ortlepp C, Platzbecker U, Bornhäuser M, Schetelig J, Ehninger G, Folprecht G, Thiede C (2018) An optimized targeted next-generation sequencing approach for sensitive detection of single nucleotide variants. Biomol Detect Quantif 15:6–12

Stemmer M, Thumberger T, del Sol KM, Wittbrodt J, Mateo JL (2015) CCTop: an intuitive, flexible and reliable CRISPR/Cas9 target prediction tool. PLoS One 10(4):e0124633

Strutt SC, Torrez RM, Kaya E, Negrete OA, Doudna JA (2018) RNA-dependent RNA targeting by CRISPR-Cas9. elife 7:e32724

Su J, Sun H, Meng Q, Zhang P, Yin Q, Li Y (2017) Enhanced blood suspensibility and laser-activated tumor-specific drug release of theranostic mesoporous silica nanoparticles by functionalizing with erythrocyte membranes. Theranostics 7(3):523

Sun X, Hu Z, Chen R, Jiang Q, Song G, Zhang H, Xi Y (2015) Targeted mutagenesis in soybean using the CRISPR-Cas9 system. Sci Rep 5(1):10342. https://doi.org/10.1038/srep10342

Sun Y, Zhang X, Wu C, He Y, Ma Y, Hou H, Guo X, Du W, Zhao Y, Xia L (2016) Engineering herbicide-resistant rice plants through CRISPR/Cas9-mediated homologous recombination of acetolactate synthase. Mol Plant 9(4):628–631

Svitashev S, Young JK, Schwartz C, Gao H, Falco SC, Cigan AM (2015) Targeted mutagenesis, precise gene editing, and site-specific gene insertion in maize using Cas9 and guide RNA. Plant Physiol 169(2):931–945

Svitashev S, Schwartz C, Lenderts B, Young JK, Cigan AM (2016) Genome editing in maize directed by CRISPR–Cas9 ribonucleoprotein complexes. Nat Commun 7:13274

Swiech L, Heidenreich M, Banerjee A, Habib N, Li Y, Trombetta J, Sur M, Zhang F (2015) In vivo interrogation of gene function in the mammalian brain using CRISPR-Cas9. Nat Biotechnol 33 (1):102

Tang X, Liu G, Zhou J, Ren Q, You Q, Tian L, Xin X, Zhong Z, Liu B, Zheng X (2018) A large-scale whole-genome sequencing analysis reveals highly specific genome editing by both Cas9 and Cpf1 (Cas12a) nucleases in rice. Genome Biol 19(1):84

Tashkandi M, Ali Z, Aljedaani F, Shami A, Mahfouz MM (2018) Engineering resistance against Tomato yellow leaf curl virus via the CRISPR/Cas9 system in tomato. Plant Signal Behav 13 (10):e1525996

Teng KW, Ishitsuka Y, Ren P, Youn Y, Deng X, Ge P, Lee SH, Belmont AS, Selvin PR (2017) Labeling proteins inside living cells using external fluorophores for fluorescence microscopy. elife 6:e25460

Thi EP, Mire CE, Lee AC, Geisbert JB, Ursic-Bedoya R, Agans KN, Robbins M, Deer DJ, Cross RW, Kondratowicz AS (2017) siRNA rescues nonhuman primates from advanced Marburg and Ravn virus disease. J Clin Invest 127(12):4437–4448

Thomas M, Parry-Smith D, Iyer V (2019) Best practice for CRISPR design using current tools and resources. Methods 164:3–17

Thomsen N, Ali RG, Ahmed JN, Arkell RM (2012) High resolution melt analysis (HRMA); a viable alternative to agarose gel electrophoresis for mouse genotyping. PLoS One 7(9):e45252

Thurtle-Schmidt DM, Lo T-W (2018) Molecular biology at the cutting edge: a review on CRISPR/CAS9 gene editing for undergraduates. Biochem Mol Biol Educ 46(2):195–205. https://doi.org/10.1002/bmb.21108

Torney F, Trewyn BG, Lin VSY, Wang K (2007) Mesoporous silica nanoparticles deliver DNA and chemicals into plants. Nat Nanotechnol 2(5):295–300. https://doi.org/10.1038/nnano.2007.108

Truong D-JJ, Kühner K, Kühn R, Werfel S, Engelhardt S, Wurst W, Ortiz O (2015) Development of an intein-mediated split–Cas9 system for gene therapy. Nucleic Acids Res 43(13):6450–6458

Tsai SQ, Zheng Z, Nguyen NT, Liebers M, Topkar VV, Thapar V, Wyvekens N, Khayter C, Iafrate AJ, Le LP, Aryee MJ, Joung JK (2015) GUIDE-seq enables genome-wide profiling of off-target cleavage by CRISPR-Cas nucleases. Nat Biotechnol 33(2):187–197. https://doi.org/10.1038/nbt.3117

Tsai SQ, Nguyen NT, Malagon-Lopez J, Topkar VV, Aryee MJ, Joung JK (2017) CIRCLE-seq: a highly sensitive in vitro screen for genome-wide CRISPR-Cas9 nuclease off-targets. Nat Methods 14(6):607–614. https://doi.org/10.1038/nmeth.4278

Tycko J, Myer VE, Hsu PD (2016) Methods for optimizing CRISPR-Cas9 genome editing specificity. Mol Cell 63(3):355–370

Upadhyay SK, Sharma S (2014) SSFinder: high throughput CRISPR-Cas target sites prediction tool. Biomed Res Int 2014:742482

Van Soolingen D, De Haas P, Hermans P, Groenen P, Van Embden J (1993) Comparison of various repetitive DNA elements as genetic markers for strain differentiation and epidemiology of Mycobacterium tuberculosis. J Clin Microbiol 31(8):1987–1995

Vannucci L, Lai M, Chiuppesi F, Ceccherini-Nelli L, Pistello M (2013) Viral vectors: a look back and ahead on gene transfer technology. New Microbiol 36(1):1–22

Walev I, Bhakdi SC, Hofmann F, Djonder N, Valeva A, Aktories K, Bhakdi S (2001) Delivery of proteins into living cells by reversible membrane permeabilization with streptolysin-O. Proc Natl Acad Sci 98(6):3185–3190. https://doi.org/10.1073/pnas.051429498

Wang D, Huang J, Wang X, Yu Y, Zhang H, Chen Y, Liu J, Sun Z, Zou H, Sun D (2013a) The eradication of breast cancer cells and stem cells by 8-hydroxyquinoline-loaded hyaluronan modified mesoporous silica nanoparticle-supported lipid bilayers containing docetaxel. Biomaterials 34(31):7662–7673

Wang H, Yang H, Shivalila CS, Dawlaty MM, Cheng AW, Zhang F, Jaenisch R (2013b) One-step generation of mice carrying mutations in multiple genes by CRISPR/Cas-mediated genome engineering. Cell 153(4):910–918

Wang T, Wei JJ, Sabatini DM, Lander ES (2014) Genetic screens in human cells using the CRISPR-Cas9 system. Science 343(6166):80–84

Wang D, Mou H, Li S, Li Y, Hough S, Tran K, Li J, Yin H, Anderson DG, Sontheimer EJ (2015) Adenovirus-mediated somatic genome editing of Pten by CRISPR/Cas9 in mouse liver in spite of Cas9-specific immune responses. Hum Gene Ther 26(7):432–442

Wang M, Zuris JA, Meng F, Rees H, Sun S, Deng P, Han Y, Gao X, Pouli D, Wu Q (2016) Efficient delivery of genome-editing proteins using bioreducible lipid nanoparticles. Proc Natl Acad Sci 113(11):2868–2873

Wang X, Tu M, Wang D, Liu J, Li Y, Li Z, Wang Y, Wang X (2018) CRISPR/Cas9-mediated efficient targeted mutagenesis in grape in the first generation. Plant Biotechnol J 16(4):844–855

Wanisch K, Yáñez-Muñoz RJ (2009) Integration-deficient lentiviral vectors: a slow coming of age. Mol Ther 17(8):1316–1332

Wiedenheft B, Lander GC, Zhou K, Jore MM, Brouns SJ, van der Oost J, Doudna JA, Nogales E (2011) Structures of the RNA-guided surveillance complex from a bacterial immune system. Nature 477(7365):486

Wightman L, Kircheis R, Rössler V, Carotta S, Ruzicka R, Kursa M, Wagner E (2001) Different behavior of branched and linear polyethylenimine for gene delivery in vitro and in vivo. J Gene Med 3(4):362–372

Wilson LOW, O'Brien AR, Bauer DC (2018) The current state and future of CRISPR-Cas9 gRNA design tools. Front Pharmacol 9:749. https://doi.org/10.3389/fphar.2018.00749

Wong N, Liu W, Wang X (2015) WU-CRISPR: characteristics of functional guide RNAs for the CRISPR/Cas9 system. Genome Biol 16(1):218

Woo JW, Kim J, Kwon SI, Corvalán C, Cho SW, Kim H, Kim S-G, Kim S-T, Choe S, Kim J-S (2015) DNA-free genome editing in plants with preassembled CRISPR-Cas9 ribonucleoproteins. Nat Biotechnol 33(11):1162

Wu Z, Yang H, Colosi P (2010) Effect of genome size on AAV vector packaging. Mol Ther 18 (1):80–86

Wu X, Kriz AJ, Sharp PA (2014a) Target specificity of the CRISPR-Cas9 system. Quant Biol 2 (2):59–70. https://doi.org/10.1007/s40484-014-0030-x

Wu X, Scott DA, Kriz AJ, Chiu AC, Hsu PD, Dadon DB, Cheng AW, Trevino AE, Konermann S, Chen S, Jaenisch R, Zhang F, Sharp PA (2014b) Genome-wide binding of the CRISPR endonuclease Cas9 in mammalian cells. Nat Biotechnol 32(7):670–676. https://doi.org/10.1038/nbt.2889

Xiao A, Cheng Z, Kong L, Zhu Z, Lin S, Gao G, Zhang B (2014) CasOT: a genome-wide Cas9/gRNA off-target searching tool. Bioinformatics 30(8):1180–1182. https://doi.org/10.1093/bioinformatics/btt764

Xie S, Shen B, Zhang C, Huang X, Zhang Y (2014) sgRNAcas9: a software package for designing CRISPR sgRNA and evaluating potential off-target cleavage sites. PLoS One 9(6):e100448. https://doi.org/10.1371/journal.pone.0100448

Xing H-L, Dong L, Wang Z-P, Zhang H-Y, Han C-Y, Liu B, Wang X-C, Chen Q-J (2014) A CRISPR/Cas9 toolkit for multiplex genome editing in plants. BMC Plant Biol 14(1):327

Xu H, Xiao T, Chen CH, Li W, Meyer CA, Wu Q, Wu D, Cong L, Zhang F, Liu JS, Brown M, Liu XS (2015) Sequence determinants of improved CRISPR sgRNA design. Genome Res 25(8):1147–1157. https://doi.org/10.1101/gr.191452.115

Yamada Y, Akita H, Harashima H (2012) Multifunctional envelope-type nano device (MEND) for organelle targeting via a stepwise membrane fusion process. Methods Enzymol 509:301–326. https://doi.org/10.1016/b978-0-12-391858-1.00015-0

Yamano T, Nishimasu H, Zetsche B, Hirano H, Slaymaker IM, Li Y, Fedorova I, Nakane T, Makarova KS, Koonin EV (2016) Crystal structure of Cpf1 in complex with guide RNA and target DNA. Cell 165(4):949–962

Yan F, Kuang Y, Ren B, Wang J, Zhang D, Lin H, Yang B, Zhou X, Zhou H (2018) Highly efficient A· T to G· C base editing by Cas9n-guided tRNA adenosine deaminase in rice. Mol Plant 11(4):631–634

Yang H, Wang H, Shivalila CS, Cheng AW, Shi L, Jaenisch R (2013) One-step generation of mice carrying reporter and conditional alleles by CRISPR/Cas-mediated genome engineering. Cell 154(6):1370–1379

Yang L, Grishin D, Wang G, Aach J, Zhang CZ, Chari R, Homsy J, Cai X, Zhao Y, Fan JB, Seidman C, Seidman J, Pu W, Church G (2014) Targeted and genome-wide sequencing reveal single nucleotide variations impacting specificity of Cas9 in human stem cells. Nat Commun 5:5507. https://doi.org/10.1038/ncomms6507

Yang H, Wu J-J, Tang T, Liu K-D, Dai C (2017) CRISPR/Cas9-mediated genome editing efficiently creates specific mutations at multiple loci using one sgRNA in Brassica napus. Sci Rep 7(1):7489

Yang L, Zhang X, Wang L, Yin S, Zhu B, Xie L, Duan Q, Hu H, Zheng R, Wei Y (2018) Increasing targeting scope of adenosine base editors in mouse and rat embryos through fusion of TadA deaminase with Cas9 variants. Protein Cell 9(9):814–819

Yin K, Han T, Liu G, Chen T, Wang Y, Yu AYL, Liu Y (2015) A geminivirus-based guide RNA delivery system for CRISPR/Cas9 mediated plant genome editing. Sci Rep 5:14926

Yin H, Song C Q, Dorkin JR, Zhu LJ, Li Y, Wu Q, Park A, Yang J, Suresh S, Bizhanova A (2016) Therapeutic genome editing by combined viral and non-viral delivery of CRISPR system components in vivo. Nat Biotechnol 34(3):328

Zaidi SS, Mansoor S (2017) Viral vectors for plant genome engineering. Front Plant Sci 8:539. https://doi.org/10.3389/fpls.2017.00539

Zaro JL, Shen W-C (2015) Cationic and amphipathic cell-penetrating peptides (CPPs): their structures and in vivo studies in drug delivery. Front Chem Sci Eng 9(4):407–427

Zhang S, Zhang R, Song G, Gao J, Li W, Han X, Chen M, Li Y, Li G (2018a) Targeted mutagenesis using the Agrobacterium tumefaciens-mediated CRISPR-Cas9 system in common wheat. BMC Plant Biol 18(1):302

Zhang Z, Ge X, Luo X, Wang P, Fan Q, Hu G, Xiao J, Li F, Wu J (2018b) Simultaneous editing of two copies of Gh14-3-3d confers enhanced transgene-clean plant defense against Verticillium dahliae in allotetraploid upland cotton. Front Plant Sci 9:842

Zhou J, Peng Z, Long J, Sosso D, Liu B, Eom JS, Huang S, Liu S, Vera Cruz C, Frommer WB (2015) Gene targeting by the TAL effector PthXo2 reveals cryptic resistance gene for bacterial blight of rice. Plant J 82(4):632–643

Zhu X, Xu Y, Yu S, Lu L, Ding M, Cheng J, Song G, Gao X, Yao L, Fan D (2014) An efficient genotyping method for genome-modified animals and human cells generated with CRISPR/Cas9 system. Sci Rep 4:6420

Zischewski J, Fischer R, Bortesi L (2017) Detection of on-target and off-target mutations generated by CRISPR/Cas9 and other sequence-specific nucleases. Biotechnol Adv 35(1):95–104

Zuris JA, Thompson DB, Shu Y, Guilinger JP, Bessen JL, Hu JH, Maeder ML, Joung JK, Chen Z-Y, Liu DR (2015) Cationic lipid-mediated delivery of proteins enables efficient protein-based genome editing in vitro and in vivo. Nat Biotechnol 33(1):73

CRISPR/Cas-Based Insect Resistance in Crops

4

Muhammad Kashif Zahoor, Aftab Ahmad, Muhammad Asif Zahoor, Humara Naz Majeed, Muhammad Zulhussnain, and Kanwal Ranian

Abstract

Insects are responsible for considerable crop losses worldwide through their direct damage and the transmission of various diseases as well. Recently, novel techniques would replace the most frequently used chemical insecticides and help facilitate the sustainability in crop production in near future. Different strategies to overcome crop resistance against insect, especially Clustered Regularly Interspaced Short Palindromic Repeats/CRISPR-associated system (CRISPR/Cas) based genome editing and gene drives are becoming available for insecticide resistance management. Recent advances and applications of CRISPR/Cas9 both in plants and insects offer promising mechanism of deterrence to insect pests through improving resistance of Bt, knockout, or insertion of new genes; in depth understanding of plant response against insect pests provides routes to optimize plant defenses against insects. In addition, directed evolution may play an important role to combat insect resistance against Bt crops. Although, various genome editing techniques have been developed, however; CRISPR-based

M. K. Zahoor (✉) · M. Zulhussnain · K. Ranian
Entomology Lab, Department of Zoology, Government College University Faisalabad, Faisalabad, Pakistan
e-mail: kashif.zahoor@gcuf.edu.pk

A. Ahmad
Center for Advanced Studies in Agriculture and Food Security (CASAFS), University of Agriculture, Faisalabad, Pakistan
e-mail: aftab.ahmad@uaf.edu.pk

M. A. Zahoor
Department of Microbiology, Government College University Faisalabad, Faisalabad, Pakistan

H. N. Majeed
Department of Biochemistry, Government College Women University Faisalabad, Faisalabad, Pakistan

A. Ahmad et al. (eds.), *CRISPR Crops*,
https://doi.org/10.1007/978-981-15-7142-8_4

approaches for insect management in crops are growing rapidly so far. Therefore, recently CRISPR-mediated gene drives are being established as potential insect management approaches in agriculture.

4.1 Introduction

Nutrient-rich and sustainable crop production stands as one of our greatest challenges to ensure the availability of affordable, healthier, and adequate nutrition to both farm as well as non-farm families. It is estimated that about 800 million people are malnourished due to consumption of inadequate number of calories per day. The essential micronutrients including minerals and vitamins being mandatory for human health are not easily accessible for more than half of the population in the world (Food and Agriculture Organization (FAO 2013). South Asia and sub-Saharan Africa are significantly concentrated with food-insecure populations. Although, with the onset of twenty-first century, the extreme poverty level has been decreased (below $1/day); however, cumulatively leading towards increased gross domestic product (GDP) have yet to be reached out to further decrease the poverty. Indeed, increase in population growth is sometimes happened to be faster than the real gross GDP (Bazuin et al. 2011). It is predicted that the population of the world within the next 30–40 years will be increased up to 9 or 10 billion exclusively much of increase is expected in poorer countries (McGuire 2015). The agricultural productivity must be doubled up to 2050 which has been suggested by Food and Agriculture Organization (FAO) to adequately feed the world. This target can be achieved by increasing the crop production through advancement in biotechnology without compromising environment (Crist et al. 2017; King 2017). Besides crop production, maximum agriculture productivity significantly reduced due to certain insect pests starting from field to the stores. Thus, it is obvious that world's food supply primarily depends upon effective crop production and its protection from insect pests. On an average 28% of annual crop productivity has been reduced due to insect pests. Of insecticides, annual global cost to reduce pest damage valued at US $10 billion. Additionally, application of numerous chemical pesticides has adverse effects towards environment. The development of insecticide resistance plus their non-degradable nature and biomagnification necessitates developing more selective and environment friendly techniques (Osman et al. 2015). Because of the wide use of synthetic pesticides during the last 50 years, resistance in insects against chemicals has been developed. Nevertheless, control strategies for insect pests generated high selection pressure which causes the evolution of pesticide resistance. Moreover, adverse effects of these chemical on environment and human health have been developed. The situation thus, necessitated to bringing new products to the market demanding novel strategies to control insect pests (Borel 2017). Hence, the initiation of recent advances in recombinant DNA technology has brought an increase in insect-resistant (IR) plant varieties and more recently the genome editing techniques in certain insect pests with huge economic benefits for the farming community. In nutshell, use of agro-chemicals will be reduced by the development of new techniques and improvement of existing techniques as well to counter the insect and weed resistance.

For the development of IR plants, most promising breakthrough in new technologies has been examined, which is CRISPR/Cas-based genome editing (Hall et al. 2018). This chapter addresses recent research advances in CRISPR/Cas being developed to manage the crop resistance against insect pests and other recent genome editing activities conducted in insects. The speed of discovery, both of essential insect traits and of reagents to disturb these traits, has increased dramatically through CRISPR/Cas genomics. The opportunities for the interventions are continuing to expand as technologies (e.g., sustained expression of multiple stacked transgenes in plants) are being optimized and new technologies (e.g., editing of plant genomes by CRISPR) are introduced. It is now a routine to address crop resistance to insect pests in molecular terms. Specifically, the insect trait is defined in terms of one or multiple gene targets, and plant resistance is defined as a gene or suite of genes with a product or products that inactivate or otherwise disable the product or products of the target insect gene or genes. A second consequence of the genomic revolution has been the development of molecular methods to inquire about the diversity and function of microbial communities. The sustained vigor and fitness of both plants and insects are, in fact, dependent on the interactions with this microbiome (Busby et al. 2017). It is becoming increasingly evident that members of the microbiome can influence insect–plant interactions and therefore, can contribute towards strategies for enhanced crop resistance to insects.

The main purpose of this chapter is to explore broad routes for the management of resistance through novel sources of genome editing tools in insects and to customize CRISPR/Cas against toxicity or deterrence to insect pests and to harness the immunity of plants against insect pests. Some aspects of these topics are supported by well-developed research disciplines, and others are more speculative.

4.2 Historical Perspective of Insect Control in Plants

Over the centuries, resistant varieties of plants are being selected by the farmers to control different kind of pests. The use of synthetic herbicides like 2,4-dichlorophenoxyacetic acid (2,4-D) began in 1946. Insect-resistant crops with high yields were developed as a result of breeding for resistance to insects. But unfortunately, it remained unsuccessful in contrast to successful breeding for disease resistance because of absence of strong efforts due to control of insect pests by use of insecticide. Use of chemicals to control insect pests has not only promoted the crop production but their extensive use also causes serious problems including insect resistance, food safety, and mostly ecological concerns (Sparks and Nauen 2015; Liu et al. 2016). These issues have brought a demand for safe and effective pest control technology. A gram-positive soil bacterium *Bacillus thuringiensis* (Bt) emerged with insecticidal potential as it produces poisonous parasporal crystals which causes the release of one or more insecticidal crystalline proteins (ICPs) by dissolving of this parasporal crystals in the alkaline environment of insect midgut (Bravo et al. 2007). Midgut proteases activate these ICPs, which then destroy membrane integrity of midgut epithelial cells of larvae by interacting with it and ultimately lead to insect death (Bravo et al. 2011).

More than 200 Bt genes are known, each active against a restricted range of insects and thus conferring exquisite specificity with minimal direct impact on non-target organisms, including natural enemies of the insect pest (Palma et al. 2014). The first of the crops introduced to express Bt toxins and grown commercially were corn and cotton approved in USA in 1995 and 2016, respectively. Bt varieties accounted for 79% and 84% of the US acreage for corn and cotton, respectively (USDA ERS 2016). Internationally, the use of Bt crops is increasing, although with much regional variation (ISAAA 2016). Although, resistance in Bt crops has been reported against various insect pests, however, the scale of the problem was reduced by proactive resistance management strategy, especially refuges technique for non-Bt host plants and pyramiding of multiple Bt genes against a single pest (Carriere et al. 2015). The first resistant variety of cotton having resistance against both insects as well as herbicide glyphosate was marketed by Monsanto in 1997. This resistant variety of cotton was resulted from crossing of Roundup Ready® Flex with Bollgard II® cotton. Since then, a constant improvement of the technique was performed, and thus, in 2010, Monsanto and DOW AgroSciences marketed SmartStaxTM maize with two heritable tolerance (HT) and six Bt genes.

4.2.1 RNAi as a Gene Silencing Tool

Although, significant environmental and economic benefits have been attained by decreasing usage of pesticides with Bt crops but however, several studies have reported the development of insect resistance against Bt toxins. Therefore, researchers were compelled to focus on other alternative techniques to control insect pests. Hence, RNAi technique was emerged and it was thought that it could serve as an alternate approach for the control of insect pests. RNAi was based on the introduction of long double-stranded RNA (dsRNA of ~500 bp) in cell and this dsRNA was then cleaved into small fragments of 21–23 nucleotides denoted as short interfering RNA (siRNA). Degradation of RNA is resulted by incorporation of siRNA into an RNAi-induced silencing complex (RISC). *Caenorhabditis elegans* a nematode was the first model organism wherein the cell to cell systemic movement of dsRNA was discovered (Fire et al. 1998). RNAi has been widely studied in *Drosophila melanogaster*, but not fully characterized in insects. To administer dsRNA, there are many demonstrated methods including feeding, injecting, and soaking. Henceforth, RNAi based genome editing has shown enormous potential for insect control in various model systems. However, the application of RNAi was limited by off-target effects, lack of reproducibility and difficult to target nuclear RNA. Recently, Cluster Interspaced Specific Palindromic Repeats CRISPR/Cas9 system emerged as a novel technology for genome editing (Cong et al. 2013). Therein precise genetic modifications can be introduced for any gene (Barrangou et al. 2007; Bassett et al. 2013; Wang et al. 2013; Heidenreich and Zhang 2016).

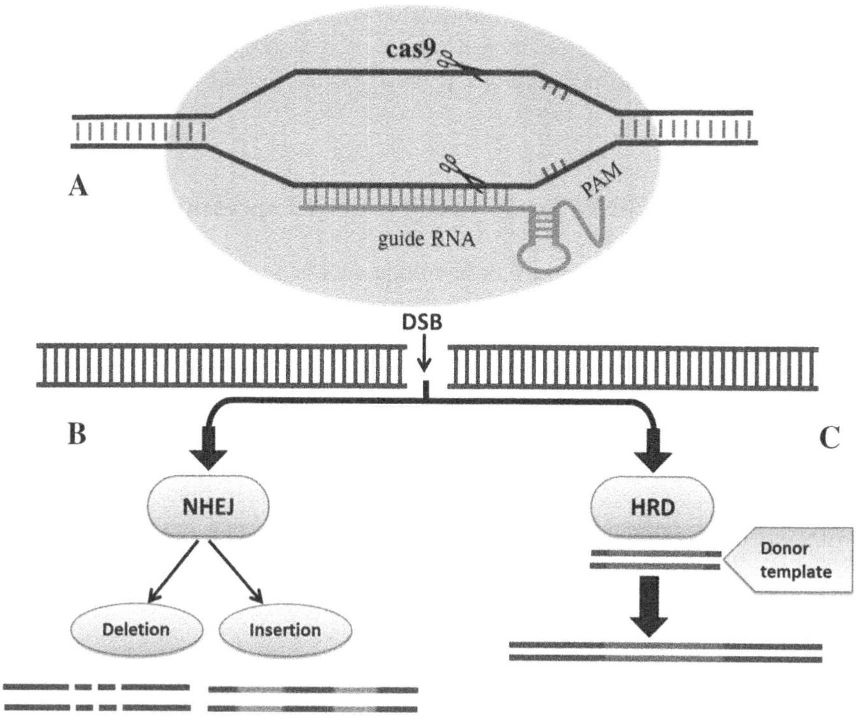

Fig. 4.1 Mechanism of CRISPR/Cas based genome editing: (**a**) Single guide RNA (sgRNA) along with Cas9 protein binds to target sequence of DNA and produce double-strand cut. (**b**) Repairing of double-strand breaks by non-homologous end joining (NHEJ) pathway causes either deletion or insertion. (**c**) On provision of donor template, DSB will br repaired by Homologous Direct Repair (HDR)

CRISPR/Cas9 is the newest synthetic endonucleases that facile genome editing in a short time.

CRISPR/Cas9 system, where the Cas9 is endonuclease CRISPR-associated protein originated from the bacterial immune system which not only disrupts gene expression unlike RNAi but also insert or alter coding sequences. Single guide RNA (sgRNA) forms base pairs with a DNA of the target sequence after finding the sequence of its complement which then is precisely cut by the Cas9 endonuclease (Fig. 4.1b) (Jinek et al. 2012). CRISPR-based genome editing is simple, cost effective, faster, has less off-target effects, precise, efficient, and easier to use in comparison to other techniques like zinc-finger nucleases (ZFNs) and transcription activator-like effector nucleases (TALENs). Gene editing using CRISPR has been demonstrated in insect cell lines, embryo or larval/immature stages of silkworms, Drosophila, mosquitoes, and others insects (Huang et al. 2016). A typical CRISPR/Cas9 gene drive with a deleterious traits are reduced fertility, distorted sex ratio and the chemical sensitivity which would be reared in laboratory and then released into the field to spread the said engineered mutation. Although, the potential and the applications of gene drive to control invasive species have been demonstrated in several studies (Deredec et al. 2008; Hodgins et al. 2009); however, the possible successes for its implementation have been mostly limited to the mosquitoes (Alphey 2014).

4.2.2 CRISPR vs RNAi for Controlling Gene Expression in Insects

CRISPR change the gene expression by editing the DNA of the cell and this change in expression may be permanent if the transformation is stable. While in case of RNAi, the continuous administration of dsRNA is required to achieve diminishing effects of RNAi. In fact, RNAi interferes with the existing gene expression, but with some exceptions wherein parental RNAi effect has been reported (Bucher et al. 2002). CRISPRi (CRISPR interference) is a modification of the CRISPR/Cas9 technology, but it has reversible effects which targets DNA instead of RNA by using dCas9 (deactivated Cas9) that inhibit gene expression by binding reversibly to target promoter region (Zhao et al. 2014). Permanent gene modification is an advantage of CRISPR; in addition unlike siRNAs or short hairpin RNAs (shRNA) it does not interfere with the endogenous cellular machinery which may lead to cell death (Doudna and Charpentier 2014).

4.3 What Is CRISPR/Cas Genome Editing?

CRISPR/Cas9 is a gene derive tool to alter the insect's genotype and potentially allows spreading the desired traits in a wild population of target species (Fig. 4.1) (Hsu et al. 2014). In nature, Bacteria and Archaea have some adaptive immune system to fend the invading phages and any other foreign genetic elements through employed defensive system therein via clustered regularly interspaced palindromic repeats (CRISPR)/CRISPR associated system9 (Cas9). It has been reported that ~40% of bacterial and most of the archaeal species have a defensive mechanism with a number of CRISPR/Cas systems capable of targeting invading DNA, RNA, or both for degradation (Jansen et al. 2002; Sorek et al. 2008).

When CRISPR equipped bacteria infected by phages, pieces of foreign DNA are acquired within the CRISPR arrays of infected bacterium and the phase is known as adaptation phase. Biogenesis phase is started with the transcription of CRISPR array into crRNAs which is then used as guides by Cas9 in immunity phase to target the phage genome during next attack of phage (Barrangou et al. 2007). The specificity to target the invading phages is ensured through recognizing the PAM sequence (NGG for *S. pyogenes* Cas9) by Cas9-gRNA complex which resulted in precise binding of crRNA with target DNA and led to cleavage of target DNA region by Cas9 endonuclease, 3–4 nucleotides proximal to the PAM (protospacer-adjacent motif) present in targeted DNA but not in bacterial genome (Jinek et al. 2012).

Double-stranded site-specific break occurs by nicking of crRNA-binding DNA strand and its opposite DNA strand through HNH and RuvC domains which are two

Cas9 subunits. It is reported that some important conformational changes occur in Cas9-gRNA complex to bind with target DNA and PAM recognition site ultimately Cas9 induces the DNA cuts (Fig. 4.1, Jiang et al. 2014). Normally, the PAM sequences happen to occur at every 16 nucleotides in a random DNA sequence and hence, raise the possibility of multiple targeting by designing multiples gRNAs and then applying simultaneously within one or several genes (Xue et al. 2014). Of cellular DNA repair system, NHEJ resulted error due to indels, while HDR can be used to insert desired DNA fragments into the target sites for knockout (KO) or knock-in (KI) the gene. This would help introduce the modification of precisely engineering the exogenous repair template in-vivo (Fig. 4.1). Cas9 can be deactivated by introducing the mutation in Cas9 subunits HNH and RuvC domains, which resulted in a dead Cas9 (dCas9), having no or reduced nuclease activity but still capable to bind DNA as sgRNA guide (Doudna and Charpentier 2014). Inactivation of only one domain either of HNH or RuvC led to dCas9 to induce cut in one DNA strand only. The use of pairs of such dCas9 mutant and sgRNAs resulted in targeted and precise indel formation in both strands (Hsu et al. 2014; Ran et al. 2013; Sander and Joung 2014). The specificity of target can be improved by multiplexed nick strategy up to 1500x as compared to wild-type Cas9 due to precise repairing of off-target nick sites (Ran et al. 2013).

Therefore, modified versions of Cas9 have been engineered to achieve epigenetic modifications, in-vivo DNA/RNA labeling, and transcriptional activation/ repression of targeted genes. dCas9 with both mutated domains cannot make cut or DSB in the DNA strands, and hence, the fusion of a modular domain to dCas9 led to the dCas9-gRNA complex which resulted in the activation or suppression of gene expression after interaction with the promoter (Gilbert et al. 2013; Sander and Joung 2014). Similarly, fluorescently labeled dCas9 can be used for in-situ labeling of genomic loci in fixed cells as well as in live imaging (Deng et al. 2015; Ochiai et al. 2015).

4.3.1 Emerging CRISPR/Cas Systems with New Cas Proteins

4.3.1.1 CRISPR/Cas9

Limitations faced by the CRISPR/Cas9 system such as bigger size of Cas9 and NGGPAM have been overcome through the varieties of CRISPR systems for purpose of efficient genome editing (GE). For example, *Campylobacter jejuni*, derived Cas9 variant CjCas9 (984 amino acid residues; 2.95 kbp) has been used successfully in the genomes of mouse or humans. Various targeted mutations have been reported via adeno-associated virus (AAV) using this system (Kim et al. 2017). In the following section, emerging CRISPR/Cas systems are discussed.

4.3.1.2 CRISPR/Cas12

CRISPR/Cas12a (Cpf1) belongs to class II type V CRISPR system that is also an RNA-guided system. Although, *Francisella novicida* U112 (FnCpf1) is analogous to CRISPR/Cas9, but this system showed distinct characteristics including its reliability on 5′ T-rich PAM sequence (5′-TTTN-3′ or 5′-TTTV-3′; V = A, C, or G) instead of G-rich, NGG as in case of Cas9 (Zetschi et al. 2015). Cas9 produces blunt-ended DNA breaks, in contrast Cpf1 produces DSBs with staggered ends at the distal position of a PAM and improves the knock-in strategies and efficiency of (NHEJ)-based gene insertion (Moon et al. 2018). Additionally, the efficiency of HDR to insert DNA fragment using complementary DNA ends is enhanced due to cohesive DNA ends. Cpf1 is more cost effective than Cas9 due to its requirement of a 42-nt CRISPR RNA (crRNA) instead of 100-nt scaffold crRNA required by Cas9 (Zaidi et al. 2017).

4.3.1.3 CRISPR/Cas13

CRISPR/Cas13a is entirely a new type of CRISPR belonging to class II type VI which can more efficiently target endogenous RNAs in plants and viral RNAs with RNA target specificity (Kim et al. 2017; Wolter and Puchta 2018). It is associated with RNase activity due to the presence of higher eukaryotes and prokaryotes nucleotide-binding (HEPN) domains (Kim et al. 2017).

4.3.1.4 CRISPR/CasX

An SpCas9 variant, xCas9 was developed by phage-assisted continuous evolution and a broader range of protospacer adjacent motifs (PAMs) is recognized by this variant (Rees and Liu 2018). The broader compatibility to PAM (including NG, GAT, and GAA), lower off-target activity, higher editing efficiency, and specificity is demonstrated by xCas9 as compared to SpCas9 (Hu et al. 2018). The replacement of Cas9 with xCas9 in the BE3 vector generates CBE which can edit *loci* containing various PAMs such as NGN, GAA, and GAT (Hu et al. 2018).

4.3.1.5 CRISPR/Cas14

The compact CRISPR/Cas14a system is used as a guided GE tool to cut single-stranded DNA (ssDNA) (Khan et al. 2010). Cas14a, Cas14b, and Cas14c are three subgroups of 24 different Cas14 gene variants recognized on the bases of comparative sequence analysis (Harrington et al. 2018). As compared to Cas9 protein the size of CRISPR/Cas14a is half and it does not require a flanking sequence (PAM) near to the target site (Harrington et al. 2018; Khan et al. 2010).

4.3.2 Design Tools for CRISPR/Cas System

Specificity of CRISPR/Cas9 and selection of a target site is the most important to edit or disrupt a gene to achieve the desired modification. Mismatches between target DNA and gRNA can lead to the reduction of CRISPR/Cas9 specificity and three mismatches of such type are not well tolerated in generating DSB in *Drosophila* (Ren et al. 2014). Of higher specificity, different online tools such as CRISPR flydesign and flyCRISPR have been developed (Gokcezade et al. 2014; Prykhozhij et al. 2015). Similarly, focusing on reference genome of *D. melanogaster* and the on-target scoring algorithm, DESKGEN has been widely used for sgRNA design with additional features of target specificity and low off-target effects. Users of these tools should select sgRNAs with high on-target activity score to make sure either they bind specifically to target region or not (Doench et al. 2014). In addition, to minimize the possibility of guide somewhere else in the genomc, it is needed to look for the off-target activity. The availability of many more such platforms and resources are summarized in Chap. 3.

4.3.3 Practical Considerations for Construct Designing in CRISPR/Cas

A large number of sgRNA and Cas9 plasmid constructs are available in the plasmid repository Addgene (https://www.addgene.org/) and have been successfully exploited in genome editing of certain insects. In 2013, two plasmids pHsp70-Cas9 and pU6-BbsI-chiRNA coding for Cas9 and sgRNA expression, respectively, were constructed and co-injected in embryos which successfully produced germline transmitted mutagenesis into the next generation. pHsp70-Cas9 uses a *Drosophila* Hsp70 promoter because it highly induces the expression of transgenes in insects; whereas pU6-BbsI-chiRNA is driven by pol III promoter from the U6 gene, commonly used to express small RNAs in various mammalian systems (Gratz et al. 2013a). After subsequent injection of a single bi-cistronic pDCC6 plasmid construct expressing both *S. pyogenes* Cas9 nuclease under Hsp70Bb and a sgRNA cassette under U6:96Ab (U6:2) improved the efficiency of targeted mutagenesis in *Drosophila*. Thus, the percentage of the mutant phenotype of injected flies was found as 66% (Gokcezade et al. 2014) which is much higher than two plasmid method which was resulted only 6% mutant phenotype among all of injected flies (Gratz et al. 2013b). More than one sgRNA in single construct makes possibility of the production of multiplex gRNAs (Nakanishi et al. 2014). In addition, the incorporation of both sgRNA and Cas9 into one construct showed highly efficient mutagenesis (Gantz and Bier 2015). A tripartite construct including three components; i) a Cas9 gene, ii) a synthetic guide-RNA (sgRNA) and iii) homology arms flanking the Cas9/gRNA cassettes, was constructed against a recessive mutation; thereby, converted heterozygous *Drosophila* to homozygous line and produced a phenotype (Gantz and Bier 2015).

4.3.4 Delivery Tools for CRISPR/Cas System

Efficient CRISPR-based genetic engineering requires safe transfer and expression of Cas9 protein and sgRNA in the target cell lines or intact organism.

4.3.4.1 Direct Methods

Direct delivery methods include microinjection (MI), transfection and virus mediated delivery.

1. Microinjection is frequently used in insects. DNA, RNA, or proteins are injected into embryos usually at preblastoderm stage to reach into both somatic and germline cells.
2. Transfection mostly is used in insect cells (Liu et al. 2014; Mabashi-Asazuma et al. 2015).
3. A microfluidic membrane deformation method can be applied to deliver Cas9-sgRNA in case if hard-to-transfect cells (Han et al. 2015).
4. Electroporation is used mostly in mammalian systems (ovary) but not reported yet in insects (Hashimoto and Takemoto 2015).
5. Virus-mediated Cas9-sgRNA delivery methods are reported feasible mainly in mammals and zebrafish, but also attempted to adapt in insects. In case of Cas9 system, the efficiency or the efficacy of genome editing is directly dependent on the delivery pathway following the type of species involved (Reid and O'Brochta 2016). Baculoviruses are reported to be in use in mosquitoes thanks to their mammalian safety (Naik et al. 2018).

4.3.4.2 Indirect Methods

The transgenic lines expressing sgRNA and Cas9 can be generated keeping in view the life cycle of insects. Subsequently, two separate *Drosophila* strains were stably transformed (Kondo and Ueda 2013). In one line, nos-Cas9 was expressed in germline cells using specific nos driver, while in the second strain, sgRNA for a target gene was driven by U6 promoter. Both U6 promoter and guide were incorporated into a fixed landing site using phiC31 integrase system. These two lines were crossed to generate a founder generation with both nos-Cas9 and sgRNA for target gene in the germline stage; after obtaining this established offspring line, the incorporated transgene could be removed by crossing out using classical genetics. Another technique has been reported using injection of sgRNA expression plasmids under the control of U6 driver into transgenic Cas9 fly strains against germline-specific vasa promoter in order to avoid any somatic mutations having deleterious effect on the survival of embryos (Sebo et al. 2014). Using this technique, a highly efficient mutagenesis was observed in the offspring derived from injected flies, with a transmission rate of up to 100%. Detailed direct and indirect delivery tools have been discussed in Chap. 3.

4.3.5 Detection Tools in CRISPR/Cas

The presence of mutated cells with reference to the control cells and also the individuals without any visible phenotypes after being employed targeted mutagenesis can be detected using various methods including enzymatic detection via presence of heteroduplex PCR products, high resolution-melt-curve analysis, sequencing and loss of a restriction site. These methods are effective with varied sensitivity to produce quantitative results. The development of such method for the genotype of individual insects without disrupting their reproduction abilities and killing them is an ongoing challenge (Merlin et al. 2013; Basu et al. 2015; Kistler et al. 2015). Detailed description of detection methods have been discussed in Chapter 3.

4.4 CRISPR/Cas Technology: A Pragmatic Approach to Control Insects

Despite being routinely used, gene disruption approaches viz. gene insertion, gene deletion, or frame-shift in coding sequences in fact are labor-intensive and expensive because of inefficient homologous recombination-based protocols (Cong et al. 2013). Moreover, due to the complex mechanism of genome editing, limited alternative protocols are available using meganuclease enzymes. Thus, by far, CRISPR/Cascade 9 system emerged for genome editing for a broad range of applications. CRISPR/Cas9 system has been developed to modify the DNA sequence of an insect's genome mainly insect functional genomics studies. In insect models, mutants as well as null alleles provide a wide opportunities to look insight for the gene function in question. Gene editing tools using engineered endonucleases provide new prospects to create desired mutations to generate null alleles but also of any alleles using sequence-specific endonuclease activity to induce DNA repair mechanism that results in indel at the targeted *loci*. CRISPR/Cas systems have been found fruitful to create null mutations commonly named as knockouts in insects (Zhang et al. 2015). Though other techniques like ZFNs, TALENs have been previously used as mutagens to generate different variants of genes, but also, other Cas (dCas9, dCas13) variants conferred without endonuclease activity were used as tunable site-specific DNA binding protein upon which other functions were attributed such as transcription activation or repression or chromatin modifications. The chromatin modification has paved new ways for researchers to exert control over gene expression and ultimately insect's phenotype.

Ever since, development in genome editing using CRISPR/Cas9, two landmarks are noticed in *Drosophila* (Table 4.1). A successful gene knock-out was performed both with euchromatic and heterochromatic genes such as yellow loci and RpL15 loci on autosomes chromosomes, respectively; as well as the kl-3 fertility factor on Y-chromosome (Yu et al. 2014). Second, conditional CRISPR/Cas9-induced mutagenesis was performed for germline-specific expression of Cas9 (Gratz et al. 2013a). Most significantly, mutagenesis specifically for both temporal and spatial genes was also attained, wherein Gal4 line was crossed with UASCas9/sgRNA transgenic line to induce Cas9/sgRNA activity for tissue-specific expression (Xue et al. 2014). The target gene can be disrupted and modified using CRISPR/Cas9-mediated conditional mutagenesis method. This method also facilitates functional studies of genes throughout development in different tissues. Resultantly, to replace other conventional controlling measures including chemical control and genetic control like radiation mutagenesis, genome editing has been successfully performed in a range of insects including their genetic control perspective through knockout of target sites, impaired sex ratio as well as fecundity of obstructing insects. Alternatively, host plant immunity has been successfully improved and/or developed for S-genes (susceptibility genes), R-genes (resistance genes), and the genes controlling interaction between plant and the insect effectors through CRISPR system.

Table 4.1 Summary of various insects incorporated in genetic manipulations

Insect	Target gene	References	Type of modification	Results/objectives
Drosophila melanogaster	*Sxl*	Li and Scott (2016)	Mutation	Female *D. melanogaster* with abnormal reproductive tissues and genitalia
Drosophila melanogaster	*G275E*	Zimmer et al. (2016)	Mutation	Level of resistance against spinosad in G275E mutant flies was 66-fold higher than non-mutated flies
Drosophila melanogaster	*Scsa*	Quan et al. (2017)	Mutation	Scsa-deficient individuals showed delays in development, increased mortality under starvation conditions, and impaired loco motor activity
Bombyx mori	*BmBLOS2*	Wang et al. (2013)	Mutation	Integument was changed from opaque to translucent in larvae
Bombyx mori	*BmWnt1*	Zhang et al. (2015)	Mutation	Defects in Pigmentation and body segmentation was observed in mutant *B. mori*
Bombyx mori	*BmOrco*	Liu et al. (2016)	Mutation/ deletion	Severely disrupts the olfactory system, moreover, the homozygous mutant was unable to respond to sex pheromones
Bombyx mori	*BmJHE*	Zhang et al. (2017)	Mutation	Extended life span was observed in mutant individuals
Spodoptera litura	*Slabd-A*	Bi et al. (2016)	Mutation	Anomalous pigmentation and abnormal body segmentation occurred in *Slabd-A*-deficient individuals
Plutella xylostella	*Pxabd-A*	Huang et al. (2016)	Mutation	Defects in gonad development and segmentation
Spodoptera litura	*SlitPBP3*	Zhu et al. (2015)	Mutation	Response to sex pheromone components was significantly decreased in mutant males
Spodoptera littoralis	*Orco*	Koutroumpa et al. (2016)	Mutation	Response to plant odors and sex pheromones was blocked in mutant *S. littoralis*
Helicoverpa armigera	*HaCad*	Wang et al. (2016)	Mutation	549-fold higher resistance against Cry1Ac was observed in mutant *H. armigera* as compared to control strain
Locusta migratoria	*Orco*	Li and Scott (2016)	Mutation	Problem in olfactory functions
Aedes aegypti	*Nix*	Hall et al. (2015)		To convert female into harmless male mosquitoes

4.4.1 Precise Knockout (KO) Approach in Insects

CRISPR/Cas9-mediated knockout technique has been exploited in various insects, *Caenorhabditis elegans* and humans as well (Tsai et al. 2014; Liu et al. 2014; Xue et al. 2014). Mutation of yellow gene was successfully generated through sgRNA injection in *Drosophila* embryos (Bassett et al. 2013). In *Arabidopsis*, the success of the powerful tool CRISPR/Cas9 system was reported without detection of any off-target events (Peterson et al. 2016).

Recently, wildtype CRISPR/Cas9-mediated gene knockout and CRISPR-STOP approaches have been developed as second generation genome editing tools. The conversion of single specific base C to T or A to G using the nickase activity of Cascade 9 by WT Cascade9 method is reported (Komor et al. 2016; Gaudelli et al. 2017). The insertion of early STOP codons by using CRISPR-STOP approach in genes with the codons CAA (Gln), CAG (Gln), and CGA (Arg) resulted in conversion of these effectively into stop codons (Kuscu et al. 2017). It has been reported that the gene silencing through knockout CRISPR-STOP approach is more efficient than the WT-Cas9 (Billon et al. 2017).

The NHEJ is employed wherein the broken DNA ends are ligated directly without the need for homologous template. NHEJ is actually an error prone approach to generate mutations through small indels surrounding the cut site guided by gRNAs target. When multiple sgRNAs are used, NHEJ repair system can be induced resulting in the deletion of larger DNA fragment in-between the two sgRNAs. NHEJ produce mutations in coding or regulatory region of the gene using single sgRNA and generate the functional knockouts (KOs). Mostly, this mechanism is used in the investigations of gene function. CRISPR/Cas9 mediated NHEJ pathway is involved in the knockout of certain factors like Ku70, Ku80, LigIV, XRCCA, and XLF in *B. mori* cells (Zhu et al. 2015).

4.4.2 Targeted Knock-in with HDR

HDR genome editing tool works when homologous ends of DNA are present in the cell to repair double-stranded DNA. Once CRISPR/Cas9 produces a cut with the help of gRNAs complex, knock-in is resulted by the action of HDR machinery between homologous sequences. Generally, this pathway requires integral sister chromatids as a template for the correction of dsDNA breaks.

In comparison to NHEJ, the HDR is a more precise genome editing tool (Gratz et al. 2013a, b). In addition, Kistler et al. (2015) also reported that if a donor template is present, HDR gives successful results in both germ-line and somatic tissues. The first successful knocked-out gene by using the CRISPR/Cas9 mediated HDR was BmBIOS2 in *B. mori* (Wang et al. 2013). Subsequent multiplex mutagenesis has also been conducted using CRISPR/Cas9 mediated HDR for six genes BmNs (Liu et al. 2014). In *Drosophila*, a 4.6 kb fragment in yellow locus as already mentioned earlier has been replaced with an *attp* docking very successfully. This HDR based gene knock-in became highly efficient with 98–99% accuracy when DNA ligase4 is mutated (Beumer et al. 2013; Yu et al. 2014). Similarly, insertion via CRISPR/Cas9 based HDR of two other *Drosophila* endogenous genes were performed successfully when tagged with Myc and eGFP (Yu et al. 2014).

4.4.3 Gene Drive for Insect/Pest Management

Gene editing systems particularly, the CRISPR/Cas9 system due to its relative simplicity provides new opportunities to use genetics for insect control. The sterile insect technique (SIT), chromosome breakage is induced by radiation to produce such adult males; those can produce inviable offspring after mating with wild females (Klassen and Curtis 2005). But large number of irradiated males and their continual release is required for the significant reduction in pest population by using this (SIT) technique. Hence, if the pest population is low, then SIT is effective. For instance, Medfly, *Ceratitis capitata*, having a temperature-sensitive lethal mutation and a reciprocal translocation is used throughout the world resulting in female-specific embryonic lethality (Franz 2005).

In parallel, Klassen and Curtis (2005) suggested that desired gene could be driven into a wild population by translocation mechanism. Genetic drive elements, transposons are naturally able to replicate themselves in a genome. Thus, the existed phenomenon of inheritance of these transposons in a super-Mendelian inheritance manner can be borrowed to develop a system that can kill the offspring which are then unable to inherit the element. The mechanisms of natural gene drives are not easily adapted or well understood which makes it technically difficult. The artificial drives have been developed based on the principles of natural elements. Among these artificial drives, CRISPR/Cas9 based gene drive is most effective and simplest (Bhaya et al. 2011). Using an appropriate sgRNA, a double-stranded break is introduced in insects by Cas9 which can be then repaired through either HDR and/or NHEJ (Jinek et al. 2012; Mali et al. 2013). HDR based editing causes a self-replicating insertion and mutates a target gene in every generation continuously, which results in spread of the introduced gene in the entire population in a given area (Esvelt et al. 2014). The Cas9 based gene drive can be exploited in agriculture to control certain insect pests.

Through gene drive, selective gene can be introduced from parent to offspring so that generation after generation most of the member of given wild population would express the same gene, i.e. female-sterility phenotype controlling genes were disrupted in mosquitoes. Each sterility gene and its homologue were targeted and edited using CRISPR/Cas9 based gene drive constructs and a massive increase of sterile females was resulted due to insertion of these gene drive constructs into sterility gene locus in female. Because only female mosquitoes feed on blood, so use of gene drive technology is suggested to target female reproduction effectively and to edit genome of mosquitoes to make them unable to bite in order to combat dengue, malaria, and Zika (Reid and O'Brochta 2016; Singer and Frischknecht 2016). Gene drive has been also performed in Plasmodium carrying Anopheles mosquito (Esvelt et al. 2014; Gabrieli et al. 2014). Interestingly, gene drive can be used to cope with

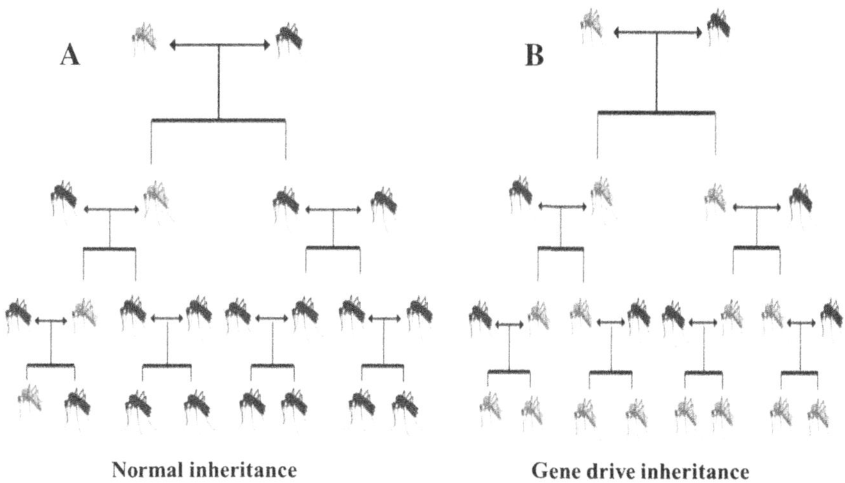

Fig. 4.2 A typical inheritance pattern of gene drive. (**a**) A decreasing population ratio of a mutant as compared to the wild type in normal inheritance pattern (**b**) An increasing population ratio of a mutant as compared to the wild type in gene drive inheritance. Gene drive inheritance pattern gives more penetrance of mutation under given conditions

biodiversity concern by eliminating the species threatening the local biodiversity via altering the pathogen of plants or saving the endangered species of plants by introducing the new genes into existing population of that species and control of invasive species of insects and their subsequent vector-borne diseases (Languin 2014; Li and Scott 2016).

Natural phenomenon of gene drive is exhibited by endonuclease genes which cut the corresponding locus of chromosomes and then through homologous recombination, nuclease gene is copied onto the damaged chromosome to induce the cell to repair the break (Burt and Koufopanou 2004). The process of copying is termed as "homing", while "gene drive" is used for endonuclease-containing cassette. A gene drive is spread within an insect population using a small number of individuals via self-sustaining process which resulted in a spread after many generations (Fig. 4.2) (Esvelt et al. 2014). If a natural sequence in organisms is cut efficiently and reliably to generate an endonuclease gene drive by insertion of endonuclease transgene in place of a natural sequence, the transgene is copied and ultimately spread into the wild populations. The genes for lethality and infertility can be disrupted by using gene drive (Burt 2003). Feasibility of CRISPR/Cas 9 based gene drive for the control of agriculture pest has been investigated in several insect pest including drosophila, beetle, moth, grasshopper, and others (Scott et al. 2014; Shukla and Palli 2013). The target population of insect can be controlled by a typical CRISPR/Cas9 gene drive, which would be based on rearing and release of a lab strain having a certain deleterious trait (e.g., reduced fertility, chemical sensitivity, and distorted sex ratio, etc.) into the field (Fig. 4.2) (Deredec et al. 2008; Oye et al. 2014). Disease transmission can be blocked by using two main strategies to explore gene drives:

1. editing the gene in female mosquito to combat the disease carrying load (Gantz and Bier 2015); and
2. sterility introduced in male mosquitoes to limit the lifespan and to prevent reproduction (Hammond et al. 2016).

In screwworm *transfer* gene is considered possible responsible for female development, which switches the *fruitless* and *doublesex* genes and results in development of female characteristics (Scott et al. 2014). So the *transfer* gene is recognized as a target gene to develop female having male phenotype. Various studies have been reported in model organisms such as *Drosophila, Lucilia cuprina, Aedes aegypti* and *Tribolium castaneum* to develop male characteristics in female using CRSPR/Cas9 gene drive system (Scott et al. 2014; Shukla and Palli 2013; Hall et al. 2015). Similarly, Nix (an *Aedes aegypti* male determining gene) has been knocked out using CRISPR/Cas9 resulted in feminized male while the expression of *Nix* gene resulted in development of female having nearly complete male reproductive organ (Hall et al. 2015). In addition, pathogen resistance in *Ae. Aegypti* has been developed by using the CRISPR/Cas9 system (Basu et al. 2015). Subsequently, deletions induced by multiplex mutagenesis were inherited by 18–21% of the offspring in *Ae. aegypti* (Dong et al. 2015; Kistler et al. 2015). Similarly, HDR was found successful both in somatic as well as germline tissues when a donor template was introduced (Kistler et al. 2015). Here are some other examples of insects wherein gene drive system has been successfully employed or potentially can be exploited.

4.4.3.1 Gene Drive in Spotted-Wing Drosophila

Spotted-wing Drosophila, *Drosophila suzukii* is considered as the most significant pest of cherry, plum, peach, and berries. This pest is endemic to Asia but from 2008 onward when first reported in California and Europe, now has been established in temperate regions in Europe and North America (Lee et al. 2011; Cini et al. 2014; Tochen et al. 2014; Asplen et al. 2015). The development of resistance in *D. suzukii* against insecticides is expected; therefore, non-chemical control measures for *D. suzukii* are needed (Diepenbrock et al. 2016). A Cas9-mediated gene drive can suppress the population of *D. suzukii* very effectively by targeting an essential gene. Thanks to Gal4-UAS system widely studied in *D. melanogaster*; germline-specific *nanos*, *vasa*, and U6:3 gene promoters can also be exploited in *D. suzukii* embryos (Chiu et al. 2013; Li and Scott 2016).

4.4.3.2 Diamond Back Moth

The diamond back moth (DBM), *Plutella xylostella* feeds on Brassica plants including cabbage and other vegetable crops, e.g., cauliflower, broccoli, collards, canola, wild mustards, and Brussels sprouts with an annual economic loss of 4–5 billion dollars (Zalucki et al. 2012). Recently a new species diamond back was identified in

Australia which further aggravates towards the economic loss (Landry and Hebert 2013). A gene drive system would then seem to be developed to its suppression mainly between the harvesting and planting season, when host plants are not available.

4.4.3.3 The Red Flour Beetle

The red flour beetle, *Tribolium castaneum* is a global pest of cereals and stored grain products. Interestingly, *T. castaneum* have *Medea* elements which are selfish genetic elements and are driven in non-*Medea* populations (Beeman et al. 1992). The mode of action of Medea elements is under study which can provide a most reliable and stable mechanism of gene drive as compared to Cas9-based systems. The improvements in using Cas9-mediated gene drive systems would also be important for *T. castaneum*.

4.4.3.4 Whitefly A Vector for Diseases in Plants

The whitefly *Bemisia tabaci* have been reported to not only feed on over 500 host plants but also transmit many viruses of different genera i.e. Torradovirus, Crinivirus, and Ipomovirus. The extended embryonic development of *B. tabaci* (required approximately 9 days) is suitable for microinjection protocols in preblastoderm embryos and to introduce gene drive into the germline. It is worth mentioning here that *B. tabaci* is excellent subject for genetic studies due to haplo/diploid sex determination; unfertilized eggs hatch haploid males while fertilized eggs hatch females. Because the gene drive involves copying the Cas9 gene to homologous chromosome, Cas9-mediated gene drive would be expressed only in heterozygous females (Esvelt et al. 2014).

4.4.3.5 Precision-Guided Sterile Insect Technology (PGSIT)

Despite being sophisticated system, some flaws are demonstrated by the current gene drive designs including the rapid evolution of resistance that can be overcome and limited by necessitated future research to develop gene drives. The precision-guided sterile insect technique (pgSIT) is newly developed technique that disrupts genes responsible for male fertility and female viability by using the accuracy and precision of CRISPR which ensures the emergence of only sterile adult males from the genetically modified eggs. A novel CRISPR-based system has been developed across species by Akbari et al. (2013) and initially employed in *Drosophila* and resulted in survival of only sterile males. In fact, the biallelic knockouts were generated simultaneously wherein recessive phenotype was converted to a dominant one. The pgSIT depends on DNA cleavage by CRISPR/Cas and then NHEJ-based repair system. pgSIT in fruit fly is 100% efficient that results in the production of fit sterile males capable to successfully compete for mates (Kandul et al. 2019). Critically, the effective suppression of insect population through pgSIT is predicted by mathematical modeling (Sanchez et al. 2018).

4.4.4 Directed Evolution for Improving Bt Protein

Directed evolution is a powerful tool that allows simulating natural selection to produce a desirable trait under selective pressure by generating diverse sequence variants of a gene of interest. The programming of CRISPR/Cas9 systems can result in targeted directed evolution via targeting any genomic locus. The beginning of CRISPR/Cas technologies for site-specific mutagenesis and editing has entirely transformed towards site-specific changes, including single-stranded nicks, DSBs, and specific base changes via nucleotide base editing and can be summed up as CRISPR-mediated directed evolution (CDE). This platform has been used to design all potential sgRNAs of Rice *Splicing Factor 3b subunit 1* (*OsSF3B1*) which is essential for RNA splicing and conserved among eukaryotes. In human, pladienolide B (PB) and herboxidiene (GEX1A) splicing inhibitors have been used to target SF3B1. GEX1A was found as new variants of *OsSF3B1* that confer resistance to one of the drugs (Butt et al. 2019).

Bt toxins, the insecticidal proteins expressed by *Bacillus thuringiensis* proved as successful control measure against various agricultural pests. Bt crops were mainly accepted thanks to some substantial economic increase of $78 billion from 1996 to 2013 and various other numerous environmental benefits. Unfortunately, the continued success of this approach to control pest is threatened by resistance development in insect pests against Bt toxin. The interaction of Bt toxins with protein receptors located in insect midgut causes pore formation in the cell membrane which results in cell death. The resistance against Bt toxin is linked with downregulation, the mutation, or deletion of these receptors. A phage-assisted continuous evolution (PACE) selection method has been developed that evolves high-affinity protein–protein interactions. Cry1Ac is a Bt toxin that binds a cadherin-like receptor on insect midgut, whereas *Trichoplusia ni* (TnCAD) is reported to not bound. The variants of Cry1Ac are thus, generated through PACE which can bind with high affinity to TnCAD and results in the death of TnCAD-expressing insect cells which are not susceptible to wild-type Cry1Ac. Insects that are Cry1Ac-resistant are killed more potently as compared to wild-type Cry1Ac. Thus, insect Bt toxin resistance can be overcome through PACE with novel insect cell receptor affinity and lead to the lethality of the wild-type Bt toxin against nonresistant insects (Badran et al. 2016). CRISPR/Cas9 based direct evolution could also help in the development of new Bt proteins with improved characteristics.

4.4.5 Gene Regulation with CRISPRi and CRISPRa

The CRISPR/Cas9 system has the potential to activate or repress gene expression in vivo. CRISPR inference (CRISPRi) is a modified CRISPR/Cas9 system used in bacteria for systematic phenotypic analysis in which dCas9 paired with sgRNA can hamper transcription at the sgRNA locus (Qi et al. 2013; Peters et al. 2015). Subsequently, CRISPRa/CRISPRi system can be used to induce activation or repression of target gene through fusion of dCas9 to the activator or repressor modules (Gilbert et al. 2013; Mali et al. 2013). It is also reported that multiple genes can be regulated in different directions (activation or repression) simultaneously. Using CRISPRi, the coexpression of sgRNAs and dCas9 targeted an endogenous *roX* locus (long non-coding RNAs) in *Drosophila* cells which resulted in a specific knockdown of *roX2* RNAs and *roX1* (Ghosh et al. 2016). Whereas, CRISPRa system involves the use of different activator modules to induce gene expression in fly, mouse cell lines, and humans. Therefore, a tripartite activator, VP64-p65-rta (VPr) has been developed with an improved transcriptional regulator (Chavez et al. 2015). The sgRNA was converted to scaffold RNA (scRNA) by extending the sgRNA sequence with modular RNA domains (MS2, PP7, and com), thus enabling flexible and parallel programmable locus-specific regulation (Zalatan et al. 2015).

4.4.6 Conditional Knockout

Embryonic lethality and sometimes lethality at the certain developmental stages through gene silencing or knockout models are circumvented by conditional knockout (cKO) models which are mostly exploited for tissue-specific gene deletion, i.e. in mouse and zebrafish models (Kuhn et al. 1995). The limitations occurred through CRISPR/Cascade 9 system can be overcome by conditional knockout strategies. For detailed study in embryogenesis, cKO can be best used in order to decipher the role of certain genes expressed during various developmental stages. cKO models also help favor for understanding and investigation of a number of human diseases. In the biomedical field, conditional mutant models are becoming essential and conditional knockouts for *C. elegans*, mice, and rats have been generated successfully through the CRISPR/Cas9 system. In addition, in developmental biology CRISPR/Cas9 is a powerful tool to investigate the relationship between phenotype and genotype (Ma et al. 2014).

4.4.7 RNA Editing with CRISPR/Cas

The field of functional genomics and molecular genetics is revolutionized by CRISPR/Cascade 9 technology. With the use of CRISPR/Cas9 system, the efficiency of gene knockdown is varied in relation to location of the particular gene, length, and complementary pairing to the crRNA sequence (Gilbert et al. 2013). Recently, it was reported that Cas13a from *Leptotrichia wadei* (LwCas13a) enables targeted knockdown of RNA transcripts in mammalian cells, bacteria, and plant cells with higher specificity and efficiency than RNA interference (Abudayyeh et al. 2017). The activity of LwCas13a depends on the stabilization domain i.e. msGFP, which concern with cellular localization and fusion of nuclear localization. LwCas13a also mediated RNA knockdown in plants and rapidly reduces the cytoplasmic RNA pool (Abudayyeh et al. 2016). Cas13 can target the specific transcript for downregulation of gene expression, has very efficient RNA cleavage activity, and does not require msGFP for stabilization. The RNA targeting ability of Cas13b ortholog in eukaryotic cells is reported to be more efficient than LwaCas13a (Cox et al. 2017). Cas13 has an ability to cleave the long non coding RNA (lncRNA) in-vivo and provide flexibility in manipulation. In addition, Cas13 is not limited to cytoplasmic transcript but also target the nuclear transcript by fusing with proteins of nuclear localization signals (NLS) (Wang et al. 2017).

4.5 Bt Crops

There is a great threat to food safety and environment due to extensive use of synthetic pesticides. A new era of pest control was opened when the efficient insecticidal activity of *Bacillus thuringiensis* (*Bt*) toxins was discovered. Later on, the introduction of Bt gene was made in various other plants in order to develop transgenic insect-resistant crops. Genetically modified (GM) crops producing Bt toxin were approved to be commercialized including cotton, maize, potato, soya bean, tomato, brinjal, sugarcane, and rice. Currently, Bt crops are being grown against many insect pests of lepidoptera, coleoptera, and some hemiptera. In 1986, first Bt-transgenic maize resistant to insects was developed in USA but commercialized in 1996. Three lines of Bt-maize were then commercialized in USA, and meanwhile, among 14 countries, 59.7 mha (million hectares) were planted with transgenic maize in 2017 (ISAAA 2017). For more than 20 years, genetically modified (GM) cotton was commercially grown and raised up to 80% among all of the cotton grown countries in 2017. The top four producers of Bt cotton are India, USA, Pakistan, and China with planting area of 11.40 million, 4.58 million, 3.00 million, and 2.78 million hectares, respectively. Since the introduction of Bt soya beans in Brazil in 2013, it was grown in other seven countries. In Bangladesh, the planting area reached to 2400 ha for Bt aubergine, while in Brazil Bt sugarcane (expressing Cry1Ab protein) was grown for the first time for commercial purpose in 2018. The insect resistance against Bt crops has been evolved by different mechanisms including:

a. Variations in toxin activation,
b. Mutation in genes for toxin receptor, and
c. Regulation of the immune system.

The hydrolysis of Bt protoxin occurs in alkaline intestine and activation takes place during degradation in midgut by protease and released as toxin. The activation of insecticidal proteins is affected by the changes in proteases present in midgut thus affecting the activation of insecticidal proteins. The protoxin is not activated in the absence of trypsin in midgut of *Plodia interpunctella*, which thus results in Bt resistance (Herrero et al. 2001). In the lepidopteran larvae, ATP binding cassette (ABC) transporters, midgut membrane-bound cadherin (CAD), alkaline phosphatase (ALP), amino-peptidase N (APN), and perhaps unknown receptors have important roles in the insecticidal activity of Bt toxins. Important reasons for Bt resistance are regulation of gene expression of receptors and mutations in insects. Cadherin, because of its important role in toxin oligomerization is one of the most important

Bt toxin receptors. Mutation in the CAD gene resulted in *Bt* resistance in many pests. Resistant in *H. virescens* against Cry1Ac toxin is related with early stop codon and disruption of the gene caused by retrotransposon-mediated insertion (Gahan et al. 2001).

4.6 CRISPR to Study Mechanism of Resistance Development

The mechanism of resistant against spinosad is investigated recently by using CRISPR/Cas9. The resistance in multiple insect pests to spinosad is associated with (Alpha6) subunit of the nicotinic acetylcholine receptor (nAChRs). High levels of resistance to spinosad in *D. melanogaster* conferred by the mutation P146S in DmAlpha6 was identified by generating the CRISPR/Cas9-induced (P146S) fly strain resistant to spinosad (Somers et al. 2015).

Lepidopteran pests are controlled by using diamide insecticides that are potent activators of insect Ryanodine Receptors (RyRs) but the resistance against diamide has been developed in tomato leafminer, (*Tuta absoluta*), and diamondback moth (*Plutella xylostella*) due to mutations in RyR gene in these insects (Guo et al. 2014; Steinbach et al. 2015; Roditakis et al. 2017).

Three candidate resistance mutations, I4790M, G4946V and G4946E were introduced in RyR of *D. melanogaster* using CRISPR/Cas9 and the high resistance level was presented by G4946V and G4946E against mutant strains of *D. melanogaster* and beet armyworm, *Spodoptera exigua*, respectively. Similarly, high resistance level was also showed by G4946E mutant strains of beet armyworm, *Spodoptera exigua* introduced by CRISPR/Cas9 against diamides (Zuo et al. 2017). When the I1043M and I1043L mutations in BPU diflubenzuron resistance *Culex pipiens* were introduced in *D. melanogaster* in their CHS gene using CRISPR/Cas9 system, a high level of resistance was found in these *D. melanogaster* mutant strains against BPU (Grigoraki et al. 2017). The relationship between pyrethroid resistance and detoxifying enzymes in *Culex quinquefasciatus* has also been examined by using CRISPR/Cas9. The knockout of P450 gene CYP9M10 in a resistant *C. quinquefasciatus* when the cytochrome P450 gene CYP9M10 was targeted in a resistant *Culex quinquefasciatus* results in reduced permethrin resistance (Itokawa et al. 2016).

To control the damage caused by Lepidoptera, the Bt expressing transgenic crops produce *Bacillus thuringiensis* derived insecticidal toxins. The mutations in receptors for Cry1 proteins, cadherin (CAD) like protein and/or the ABCC2, ATP dependent binding cassette transporter causes the development of resistance against Cry1-type toxins in Lepidopteran. Susceptible *H. armigera* were converted into resistant strains after confirming CAD receptors for Cry1 protein and disrupting the CAD gene using a reverse genetic (CRISPR/Cas9) technique (Wang et al. 2016), while the identification of ABCC2 as a Cry1 receptor was performed by using GAL4/UAS method in *D. melanogaster* (Stevens et al. 2017). The knockout of ABCA2, an ABC transporter in *H. armigera* resulted in not only the resistant strain of Cry2Ab but also against Cry2Aa (Wang et al. 2017).

Recently, many of the tetraspanin genes have been identified which play an important role in cell to cell communication, annotated with dominant resistance. It has been reported that a point mutation in a tetraspanin gene confers dominant resistance to Cry1Ac in *H. armigera* (Jin et al. 2018). In addition, CRISPR/Cas9 system based complete knockout of HaTSPAN1 gene resulted in susceptibility of insects to Bt cotton. As already mentioned, the mutation in *cadherin* gene is a receptor of *Bt* toxin Cry1Ac and linked to resistance in *Bt* cotton. Therefore, CRISPR/Cas9 can be used for the modification of inheritable *Cadherin* resistant gene (single Cry1Ac gene as well as multiplex-gene editing for Cry1Ac and Cry2Ab) into inheritable nonresistant gene in pink bollworm, *Pectinophora gossypiella*. The disruption of *Cadherin* gene in cotton to nonresistant gene would cause non-feeding on *Bt* cotton or corn and ultimately lead to the safety of the crop.

4.7 Application of CRISPR/Cas in Insects

4.7.1 CRISPR-Based Genome Editing in Insect Pests

Plutella xylostella, *Pectinophora gossypiella*, *Helicoverpa armigera*, *Halymorpha halys*, *Ostrinia nubilalis*, *Diabtotica virgifera virgifera*, and many others are considered as major damaging insect pests of cash crops pests worldwide. *Plutella xylostella* diamondback moth (DMB) is found to be most destructive and cosmopolitan pests of cruciferous, oilseed rape, and cabbage crops. Furthermore, resistance has developed in this pest to the conventional chemicals used against it frequently (Huang et al. 2016). Pink bollworm (*Pectinophora gossypiella*) is the most damaging pest of cotton. The larvae mostly stay inside the fruiting bodies and causes rotting, malformation, premature or sometimes partial opening of bolls, reduced fiber length, and staining of lint which ultimately leads to severe loss in quality of cotton. Pink bollworm has been reported to cause 10.70–59.20% loss in normal opening of bolls, 2.8–61.9% loss in seed cotton yield, and 2.1–47.10% loss in oil content (Patil 2003). Besides, various techniques including mating disruption with synthetic pheromones, transgenic cotton which expresses insecticidal proteins from *Bacillus thuringiensis* (Bt cotton), and recently many other genetic control methods are also in use (Tabashnik et al. 2010). Recently, CRISPR/Cas9 technology has been applied in various insects and few examples are as follows (see Fig. 4.3):

Drosophila melanogaster as a model organism, the first insect which was used for the application of CRISPR/Cas9 system in arthropods followed by *Bombyx mori* and other non-model insects including different types of flies, moths, mosquitoes, butterflies, and other non-insect arthropods (Bassett and Liu 2014). The first mutation into the *D. melanogaster* by using CRISPR/Cas9 system was the deletion of 4.6 kb in the yellow locus with a single-stranded oligonucleotide donor (ssODN) and two target sgRNAs (Gratz et al. 2013a, b). The introduction of site-specific mutations in sex lethal genes (*Sxl*) of *D. melanogaster* by CRISPR/Cas9 technology resulted in the abnormality of reproductive tissues and genitalia in female (Li and Scott 2016). Another site-specific mutation G275E developed by CRISPR/Cas9

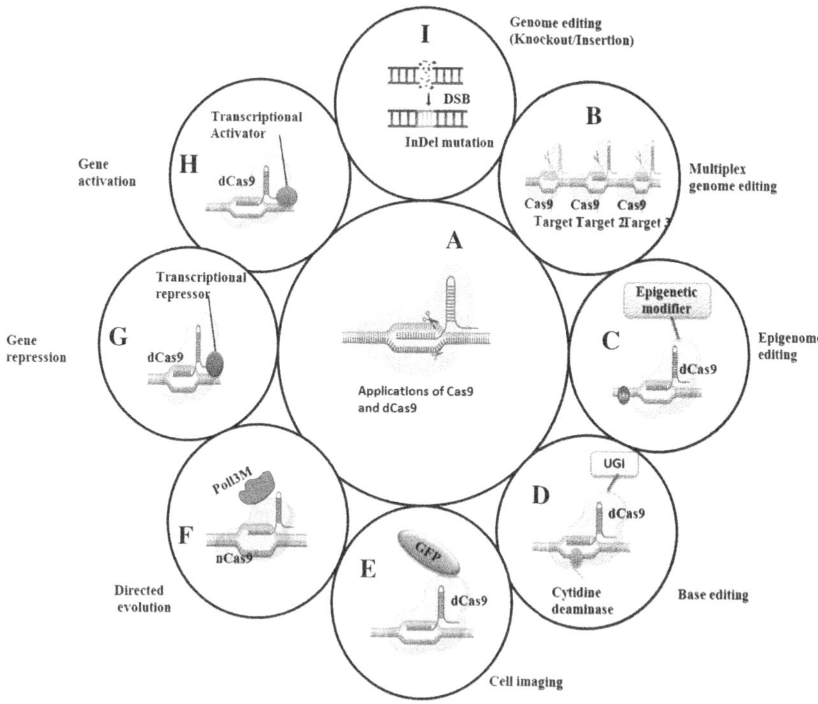

Fig. 4.3 A brief summary of applications of CRISPR/Cas system: (**a**) Applications of Cas9 to create indels through NHEJ repair pathway. (**b**) Multiplex genome editing; Modification of multiple *loci* can be performed. (**c**) Epigenome editing; *Epigenome* is modified at specific sites. (**d**) Base editing; To install point mutations in DNA without making double-stranded DNA breaks. (**e**) Cell imaging; To study cellular processes, including DNA replication, DNA damage repair, and gene expression. (**f**) Directed evolution; A method used in protein engineering that mimics the process of natural selection to steer proteins or nucleic acids toward a user-defined goal. (**g**) Gene expression; To repress or express the gene of our interest. (**h**) Gene activation; To activate the gene which we need. (**i**) Genome editing (knockout/insertion); to knock out the target gene or insert a DNA sequence

system in *Drosophila* demonstrated the direct relation of this mutation and spinosad resistance (Zimmer et al. 2016). The use of CRISPR/Cas9 system in *B. mori* verified the mutation of BmBLOS2 gene which resulted in the change of opaque to translucent integument *B. mori* larvae (Wang et al. 2013; Liu et al. 2014). The Wnt1 signaling pathway in insects is important in their embryonic development which was studied in *B. mori* resulted due to mutation in BmWnt1 gene which showed the defects in pigmentation and segment of the body (Zhang et al. 2015).

Spodoptera litura is an omnivorous and very destructive pest worldwide. A gene abdominal-A (abd-A) is important during insect development in 3rd–6th body segments (Pan et al. 2009). CRISPR/Cas9 system was used to target the Slabd-A gene in *S. litura* and the deficient individuals showed abnormality in the body segmentation as well as irregular pigmentation (Bi et al. 2016). Olfactory response of *Spodoptera littoralis* to sex pheromones and plant odors was blocked by targeting the olfactory receptor co-receptor (Orco) gene using CRISPR/Cas9 system (Koutroumpa et al. 2016). To study the functional gene studies in Lepidopterans, CRISPR/Cas9 system is a stable method as compare to RNAi. *B. thuringiensis* engineered crops and Bt insecticides are extensively used for insect pest control (Bravo et al. 2011). In several insects belonging to Lepidoptera a cadherin-like receptor is identified as a receptor of Cry1A toxin in Bt crops. CRISPR/Cas9 system was used to ensure the involvement of cadherin in Cry1Ac resistance by targeting the exon 9 of HaCad gene of *Helicoverpa*. Mutant individuals showed higher resistance than that of control strain against Cry1Ac (Wang et al. 2016).

Antennal nerve tissues are described to process the chemical signals and pheromone received by peripheral tissues and finally integrated by brain to direct insect behaviors including feeding, foraging, spawning, and mating (Leal 2013). Odorant receptor co-receptor (Orco) gene of *Locusta migratoria* was targeted by injecting Orco-sgRNA and Cas9-mRNA into eggs to produce homozygous and heterozygous mutant lines of *Locusta migratoria* (Li and Scott 2016). These results gave new idea to control locust and also helped to provide guidelines to control others orthopteran using CRISPR/Cas9 system.

Mosquitoes are vectors of several diseases such as malaria, Zika, dengue, filariasis, and chikungunya (Gabrieli et al. 2014; Reegan et al. 2016). The use of synthetic pesticides is very common for several decades to control mosquitoes. These pesticides are not only hazardous to environment and non-target organism including human but also cause resistance in mosquitoes (Tikar et al. 2009; Bayen 2012). In order to curb the vector mosquitoes, safe and environmental friendly techniques are needed. Genetic control is thus considered as safe controlling measure. TALENs and ZFNs have been used successfully to target the genes of interest of mosquitoes in vivo (Aryan et al. 2013; Smidler et al. 2013).

CRISPR/Cas9 technology was used to target the ECFP gene in *Ae. aegypti* mosquitoes having both ECFP and DsRed marker genes but the individuals obtaining have expressed only DsRed but not ECFP (Dong et al. 2015). CRISPR/Cas9 is identified as a new genetic technique to induce mutations in mosquito (Kistler et al. 2015). *Ae. aegypti* mosquitoes with decreased follicle numbers, reduced egg hatching rate, loss of ovarian function, and growth retardation have been produced by targeting microRNA-309 with the CRISPR/Cas9 system (Zhang et al. 2016). Because female mosquitoes have to feed on blood for development of eggs only female mosquito bite their host so the new vector management technique is to convert female into harmless male mosquitoes targeting the Nix gene of male determining factor in *A. aegypti* using CRISPR/Cas9 system (Hall et al. 2015). Three genes in *Anopheles gambiae* related to a female-sterility phenotype are targeted by using CRISPR/Cas9 system to control malaria vector (Hammond et al. 2016). The relationship between insecticide resistance and detoxifying enzymes in *Culex quinquefasciatus* was studied by using CRISPR/Cas9 method. It was reported that cytochrome P450 gene CYP9M10 is a key factor related with pyrethroid resistance (Itokawa et al. 2016). The biosafety risks of CRISPR/Cas9 technology should be considered during the release of CRISPR/Cas9-edited insects into the field (Taning et al. 2017).

4.8 Biosafety Issues of CRISPR Applications in Insects

There are some major concerns with CRISPR/Cas system with respect to its efficiency and affordability which could lead to the release of such genetically edited insects carrying gene drives with potential to replace wild-type insects to mutants and/or to alter the sex ratio of certain pests or even disturbing the ecosystem in longer run (Champer et al. 2016; Esvelt et al. 2014; Oye et al. 2014)

Thus, the stakeholders should carefully take into account the concerns and focus on CRISPR work accordingly. Having observed with other gene editing techniques, off-target mutations in the target genome must be worked out properly when designing the guide RNA. Any risk of gene flow from target species to other species should be observed. In addition, non-target effects should also be considered during biocontrol or invasive species and insect vectors for various human diseases carrying insect vectors (Briese 2004; Bae et al. 2014; Sander and Joung 2014). It is suggested that genetically edited through CRISPR/Cas system may be tagged in order to track the gene flow within a given population and to trace the future damages at species level or at ecosystem level (Esvelt et al. 2014).

Despite of being safe the CRISPR technique as declared by EPA; however, currently both national and international regulations do not provide legal provisions and adequate guidance for the introduction and applications regarding CRISPR/Cas based genetically edited insects (Caplan et al. 2015; Cotter and Steinbrecher 2016). A framework would be needed in future for the regularization of CRISPR-edited insects which could work in coordination both at national and international level as well (Holdren et al. 2015). It has been reported that about 30 genetically modified organisms (GMOs) have by-passed the USDA regulatory system inclusively herein already developed using gene editing techniques such as the zinc-finger nuclease (ZFN) and transcription activator-like effector nuclease (TALEN). In addition, even the European Food Safety Agency (EFSA) in EU did not address the regulations for CRISPR-edited insects (Taning et al. 2017). Overall, CRISPR has become a very popular technology worldwide, however, it is requisite to remove biosafety concerns regarding its true application. It also necessitates proper regulations to set within the framework of national and international standards (Oye et al. 2014).

4.9 Outlook

The evolutionary pace of genome editing technology from the first generation ZFNs to the second generation TALENs and now the third generation CRISPR/Cas9 is regarded as marvelous steps in genome editing history. The CRISPR/Cas9 system encompasses gene editing from cellular to intact organisms including wide areas to promote cancer immunotherapy, edited cattle, pigs, insects, human eye disease treatment, etc. The list may be enhanced if counting individually; nonetheless, the most important landmark development has been proposed with new CRISPR/Cpf1 system which is even more convenient than CRISPR/Cas9 and requires only a single crRNA in an activated complex with Cpf1 (Cas12a) to perform multiplexed genome editing (Zetsche et al. 2017). Another improved system uses Cas13a, which enable the editing of ssRNA target sequences (Abudayyeh et al. 2016).

Additionally, CRISPR interference (CRISPRi) technology provides a precise and efficient expression of target DNA (Choudhary et al. 2015). These emerging systems suggest that in the future, gene editing would become even more efficient and convenient and more impressive achievements would be awaited in agriculture and more particularly disease carrying insect vectors.

References

Abudayyeh OO, Gootenberg JS, Konermann S, Joung J, Slaymaker IM, Cox DB, Severinov K (2016) C2c2 is a single-component programmable RNA-guided RNA-targeting CRISPR effector. Science 353(6299):5573

Abudayyeh OO, Gootenberg JS, Essletzbichler P, Han S, Joung J, Belanto JJ, Lander ES (2017) RNA targeting with CRISPR–Cas13. Nature 550(7675):280

Akbari OS et al (2013) A synthetic gene drive system for local, reversible modification and suppression of insect populations. Curr Biol 23:671–677

Alphey L (2014) Genetic control of mosquitoes. Annu Rev Entomol 59:205–224

Aryan A, Anderson MA, Myles KM, Adelman ZN (2013) TALEN-based gene disruption in the dengue vector Aedes aegypti. PLoS One 8(3):60082

Asplen MK, Anfora G, Biondi A, Choi DS, Chu D, Daane KM, Isaacs R (2015) Invasion biology of spotted wing Drosophila (Drosophila suzukii): a global perspective and future priorities. J Pest Sci 88(3):469–494

Badran AH, Guzov VM, Huai Q, Kemp MM, Vishwanath P, Kain W, Wang P (2016) Continuous evolution of Bacillus thuringiensis toxins overcomes insect resistance. Nature 533(7601):58

Bae S, Park J, Kim JS (2014) Cas-OFFinder: a fast and versatile algorithm that searches for potential off-target sites of Cas9 RNA-guided endonucleases. Bioinformatics 30:1473

Barrangou R, Fremaux C, Deveau H, Richards M, Boyaval P, Moineau S, Romero DA, Horvath P (2007) CRISPR provides acquired resistance against viruses in prokaryotes. Science 315:1709–1712

Bassett AR, Liu JL (2014) CRISPR/Cas9 and genome editing in Drosophila. J Genet Genomics 41 (1):7–19

Bassett AR, Tibbit C, Ponting CP, Liu JL (2013) Highly efficient targeted mutagenesis of Drosophila with the CRISPR/Cas9 system. Cell Rep 4(1):220–228

Basu S, Aryan A, Overcash JM, Samuel GH, Anderson MA, Dahlem TJ, Adelman ZN (2015) Silencing of end-joining repair for efficient site-specific gene insertion after TALEN/CRISPR mutagenesis in *Aedes aegypti*. Proc Natl Acad Sci 112(13):4038–4043

Bayen S (2012) Occurrence, bioavailability and toxic effects of trace metals and organic contaminants in mangrove ecosystems. Environ Int 48:84–101

Bazuin S, Azadi H, Witlox F (2011) Application of GM crops in Sub-Saharan Africa: lessons learned from green revolution. Biotechnol Adv 29(6):908–912

Beeman RW, Friesen KS, Denell RE (1992) Maternal-effect selfish genes in flour beetles. Science 256(5053):89–92

Beumer KJ, Trautman JK, Mukherjee K, Carroll D (2013) Donor DNA utilization during gene targeting with zinc-finger nucleases. Genes Genom Genet 3(4):657–664

Bhaya D, Davison M, Barrangou R (2011) CRISPR-Cas systems in bacteria and archaea: versatile small RNAs for adaptive defense and regulation. Annu Rev Genet 45:273–297
Bi HL, Xu J, Tan AJ, Huang YP (2016) CRISPR/Cas9-mediated targeted gene mutagenesis in *Spodoptera litura*. Insect Sci 23(3):469–477
Billon P, Bryant EE, Joseph SA, Nambiar TS, Hayward SB, Rothstein R (2017) CRISPR-mediated base editing enables efficient disruption of eukaryotic genes through induction of STOP codons. Mol Cell 67:1068–1079
Borel B (2017) When the pesticides run out. Nature 543:302–304
Bravo A, Gill SS, Soberon M (2007) Mode of action of *Bacillus thuringiensis* Cry and Cyt toxins and their potential for insect control. Toxicon 49(4):423–435
Bravo A, Likitvivatanavong S, Gill SS, Soberón M (2011) *Bacillus thuringiensis*: a story of a successful bioinsecticide. Insect Biochem Mol Biol 41(7):423–431
Briese DT (2004) Weed biological control: applying science to solve seemingly intractable problems. Aust J Entomol 43:304–317
Bucher G, Scholten J, Klingler M (2002) Parental RNAi in *Tribolium* (Coleoptera). Curr Biol 12:85–86
Burt A (2003) Site-specific selfish genes as tools for the control and genetic engineering of natural populations. Proc R Soc London Ser B 270(1518):921–928
Burt A, Koufopanou V (2004) Homing endonuclease genes: the rise and fall and rise again of a selfish element. Curr Opin Genet Dev 14(6):609–615
Busby PE, Soman C, Wagner MR, Friesen ML, Kremer J et al (2017) Research priorities for harnessing plant microbiomes in sustainable agriculture. PLoS Biol 15:2001793
Butt H, Eid A, Momin AA, Bazin J, Crespi M, Arold ST, Mahfouz MM (2019) CRISPR directed evolution of the spliceosome for resistance to splicing inhibitors. Genome Biol 20(1):73
Caplan AL, Parent B, Shen M, Plunkett C (2015) No time to waste-the ethical challenges created by CRISPR. EMBO Rep 16:1421–1426
Carriere Y, Crickmore N, Tabashnik BE (2015) Optimizing pyramided transgenic Bt crops for sustainable pest management. Nat Biotechnol 33:161–168
Champer J, Buchman A, Akbari OS (2016) Cheating evolution: engineering gene drives to manipulate the fate of wild populations. Nat Rev Genet 17:146–159
Chavez A, Scheiman J, Vora S, Pruitt BW, Tuttle M, Iyer EP, Ter-Ovanesyan D (2015) Highly efficient Cas9-mediated transcriptional programming. Nat Methods 12(4):326
Chiu JC, Jiang X, Zhao L, Hamm CA, Cridland JM, Saelao P, Hamby KA, Lee EK, Kwok RS, Zhang G, Zalom FG (2013) Genome of Drosophila suzukii, the spotted wing drosophila. G3: Genes Genome Genet 3(12):2257–2271
Choudhary E, Thakur P, Pareek M, Agarwal N (2015) Gene silencing by CRISPR interference in mycobacteria. Nat Commun 6:6267
Cini A, Anfora G, Escudero-Colomar LA, Grassi A, Santosuosso U, Seljak G, Papini A (2014) Tracking the invasion of the alien fruit pest Drosophila suzukii in Europe. J Pest Sci 87 (4):559–566
Cong L, Ran FA, Cox D, Lin SL, Barretto R, Habib N, Hsu PD, Wu XB, Jiang WY, Marraffini LA, Zhang F (2013) Multiplex genome engineering using CRISPR/Cas systems. Science 339:819e823
Cotter J, Steinbrecher R (2016) Gene-editing. In: Urgent need of regulation. Elsevier, London
Cox DB, Gootenberg JS, Abudayyeh OO, Franklin B, Kellner MJ, Joung J, Zhang F (2017) RNA editing with CRISPR-Cas13. Science 358(6366):1019–1027
Crist E, Mora C, Engelman R (2017) The interaction of human population, food production, and biodiversity protection. Science 356:260–264
Deng W, Shi X, Tjian R, Lionnet T, Singer RH (2015) CASFISH: CRISPR/Cas9-mediated in situ labeling of genomic loci in fixed cells. Proc Natl Acad Sci 112(38):11870–11875
Deredec A, Burt A, Godfray HCJ (2008) The population genetics of using homing endonuclease genes in vector and pest management. Genetics 179(4):2013–2026

Diepenbrock LM, Rosensteel DO, Hardin JA, Sial AA, Burrack HJ (2016) Season-long programs for control of *Drosophila suzukii* in southeastern US blueberries. Crop Prot 81:76–84
Doench JG, Hartenian E, Graham DB, Tothova Z, Hegde M, Smith I, Root DE (2014) Rational design of highly active sgRNAs for CRISPR-Cas9–mediated gene inactivation. Nat Biotechnol 32(12):1262
Dong S, Lin J, Held NL, Clem RJ, Passarelli AL, Franz AW (2015) Heritable CRISPR/Cas9-mediated genome editing in the yellow fever mosquito, *Aedes aegypti*. PLoS One 10(3): e0122353
Doudna JA, Charpentier E (2014) The new frontier of genome engineering with CRISPR-Cas9. Science 346(6213):1258096
Esvelt KM, Smidler AL, Catteruccia F, Church GM (2014) Emerging technology: concerning RNA-guided gene drives for the alteration of wild populations. elife 3:e03401
Fire A, Xu S, Montgomery MK, Kostas SA, Driver SE, Mello CC (1998) Potent and specific genetic interference by double-stranded RNA in Caenorhabditis elegans. Nature 391 (6669):806–811
Food and Agriculture Organisation (FAO) (2013) The state of food insecurity in the world, executive summary
Franz G (2005) Genetic sexing strains in mediterranean fruit fly, an example for other species amenable to large-scale rearing for the sterile insect technique. In: SIT. Springer, Dordrecht, pp 427–451
Gabrieli P, Smidler A, Catteruccia F (2014) Engineering the control of mosquito-borne infectious diseases. Genome Biol 15:535
Gahan LJ, Gould F, Heckel DG (2001) Identification of a gene associated with Bt resistance in *Heliothis virescens*. Science 293(5531):857–860
Gantz VM, Bier E (2015) The mutagenic chain reaction: a method for converting heterozygous to homozygous mutations. Science 348(6233):442–444
Gaudelli NM, Komor AC, Rees HA, Packer MS, Badran AH, Bryson DI et al (2017) Programmable base editing of A • T to G • C in genomic DNA without DNA cleavage. Nat Publ Gr 551:464–471
Ghosh S, Tibbit C, Liu JL (2016) Effective knockdown of Drosophila long non-coding RNAs by CRISPR interference. Nucleic Acids Res 44(9):84–84
Gilbert LA, Larson MH, Morsut L, Liu Z, Brar GA, Torres SE et al (2013) CRISPR-mediated modular RNA-guided regulation of transcription in eukaryotes. Cell 154(2):442–451
Gokcezade J, Sienski G, Duchek P (2014) Efficient CRISPR/Cas9 plasmids for rapid and versatile genome editing in Drosophila. Genes Genom Genet 4(11):2279–2282
Gratz SJ, Cummings AM, Nguyen JN, Hamm DC, Donohue LK, Harrison MM et al (2013a) Genome engineering of Drosophila with the CRISPR RNA-guided Cas9 nuclease. Genetics 194 (4):1029–1035
Gratz SJ, Wildonger J, Harrison MM, O'Connor-Giles KM (2013b) CRISPR/Cas9-mediated genome engineering and the promise of designer flies on demand. Fly 7(4):249–255
Grigoraki L, Puggioli A, Mavridis K, Douris V, Montanari M, Bellini R, Vontas J (2017) Striking diflubenzuron resistance in *Culex pipiens*, the prime vector of West Nile Virus. Sci Rep 7 (1):11699
Guo L, Liang P, Zhou X, Gao X (2014) Novel mutations and mutation combinations of ryanodine receptor in a chlorantraniliprole resistant population of *Plutella xylostella* (L.). Sci Rep 4:6924
Hall AB, Basu S, Jiang X, Qi Y, Timoshevskiy VA, Biedler JK, Sharakhov IV (2015) A male-determining factor in the mosquito *Aedes aegypti*. Science 348(6240):1268–1270
Hall JC, Donnelly-Vanderloo MJ, Hume DJ (2018) Triazine-resistant crops: the agronomic impact and physiological consequences of chloroplast mutation. In: Herbicide-resistant crops. Springer, New York, pp 107–126
Hammond A, Galizi R, Kyrou K, Simoni A, Siniscalchi C, Katsanos D et al (2016) A CRISPR-Cas9 gene drive system targeting female reproduction in the malaria mosquito vector *Anopheles gambiae*. Nat Biotechnol 34(1):78

Han X, Liu Z, Chan JM, Zhang K, Li Y, Zeng Z et al (2015) CRISPR-Cas9 delivery to hard-to-transfect cells via membrane deformation. Sci Adv 1(7):1500454

Harrington LB et al (2018) Programmed DNA destruction by miniature CRISPR-Cas14 enzymes. Science 362:839–842

Hashimoto M, Takemoto T (2015) Electroporation enables the efficient mRNA delivery into the mouse zygotes and facilitates CRISPR/Cas9-based genome editing. Sci Rep 5:11315

Heidenreich M, Zhang F (2016) Applications of CRISPR–Cas systems in neuroscience. Nat Rev Neurosci 17(1):36

Herrero S, Ferré J, Escriche B (2001) Mannose phosphate isomerase isoenzymes in *Plutella xylostella* Support common genetic bases of resistance *to Bacillus thuringiensis* toxins in lepidopteran species. Appl Environ Microbiol 67(2):979–981

Hodgins KA, Rieseberg L, Otto SP (2009) Genetic control of invasive plants species using selfish genetic elements. Evol Appl 2(4):555–569

Holdren JP, Shelanski H, Vetter D, Goldfuss C (2015) Improving transparency and ensuring continued safety in biotechnology. Office of Science and Technology Policy

Hsu PD, Lander ES, Zhang F (2014) Development and applications of CRISPR-Cas9 for genome engineering. Cell 157:1262–1278

Hu JH et al (2018) Evolved Cas9 variants with broad PAM compatibility and high DNA specificity. Nature 556(7699):57–63

Huang Y, Chen Y, Zeng B, Wang Y, James AA, Gurr GM et al (2016) CRISPR/Cas9 mediated knockout of the abdominal-A homeotic gene in the global pest, diamondback moth (*Plutella xylostella*). Insect Biochem Mol Biol 75:98–106

ISAAA (2016) ISAA Brief 52-2016: executive summary. http://www.isaaa.org/resources/publications/briefs/52/executivesummary/default.asp

ISAAA (2017) Global status of commercialized biotech/GM crops in 2017: Biotech crop adoption surges as economic benefits accumulate in 22 years

Itokawa K, Komagata O, Kasai S, Ogawa K, Tomita T (2016) Testing the causality between CYP9M10 and pyrethroid resistance using the TALEN and CRISPR/Cas9 technologies. Sci Rep 6:24652

Jansen R, van Embden JD, Gaastra W, Schouls LM (2002) Identification of a novel family of sequence repeats among prokaryotes. OMICS 6:23–33

Jiang W, Brueggeman AJ, Horken KM, Plucinak TM, Weeks DP (2014) Successful transient expression of Cas9 and single guide RNA genes in *Chlamydomonas reinhardtii*. Eukaryot Cell 13(11):1465–1469

Jin MH, Xiao YT, Cheng Y, Hu J, Xue CB, Wu KM (2018) Chromosomal deletions mediated by CRISPR/Cas9 in *Helicoverpa armigera*. Insect Sci 26(6):1029–1036

Jinek M, Chylinski K, Fonfara I, Haur M, Doudna JA, Carpentier EA (2012) Programmable dual-RNA-guided DNA endonuclease in adaptive bacterial immunity. Science 337:816–821

Kandul NP, Liu J, Wu SL, Marshall JM, Akbari OS (2019) Transforming insect population control with precision guided sterile males with demonstration in flies. Nat Commun 10(1):84

Khan MZ et al (2010) Targeting plant ssDNA viruses with engineered miniature CRISPR-Cas14a. Trends Biotechnol 37:800–804

Kim E et al (2017) In vivo genome editing with a small Cas9 orthologue derived from *Campylobacter jejuni*. Nat Commun 8:14500

King A (2017) Technology: the future of agriculture. Nature 544:21–23

Kistler KE, Vosshall LB, Matthews BJ (2015) Genome engineering with CRISPR-Cas9 in the mosquito *Aedes aegypti*. Cell Rep 11(1):51–60

Klassen W, Curtis CF (2005) History of the sterile insect technique. In: SIT. Springer, Dordrecht, pp 3–36

Komor AC, Kim YB, Packer MS, Zuris JA, Liu DR (2016) Programmable editing of a target base in genomic DNA without double-stranded DNA cleavage. Nature 533:420–424

Kondo S, Ueda R (2013) Highly improved gene targeting by germline-specific Cas9 expression in Drosophila. Genetics 195(3):715–721

Koutroumpa FA, Monsempes C, François MC, De Cian A, Royer C, Concordet JP, Jacquin-Joly E (2016) Heritable genome editing with CRISPR/Cas9 induces anosmia in a crop pest moth. Sci Rep 6:29620

Kuhn R, Schwenk F, Aguet M, Rajewsky K (1995) Inducible gene targeting in mice. Science 269 (5229):1427–1429

Kuscu C, Parlak M, Tufan T, Yang J, Szlachta K, Wei X et al (2017) CRISPR-STOP: gene silencing through base-editing-induced nonsense mutations. Nat Methods 14:710–712

Landry JF, Hebert PD (2013) Plutella australiana (Lepidoptera, Plutellidae), an overlooked diamondback moth revealed by DNA barcodes. ZooKeys 327:43

Languin K (2014) Genetic engineering to the rescue against invasive species? National Geographic

Leal WS (2013) Odorant reception in insects: roles of receptors, binding proteins, and degrading enzymes. Annu Rev Entomol 58:373–391

Lee JC, Bruck DJ, Dreves AJ, Ioriatti C, Vogt H, Baufeld P (2011) In focus: spotted wing drosophila, *Drosophila suzukii*, across perspectives. Pest Manag Sci 67(11):1349–1351

Li F, Scott MJ (2016) CRISPR/Cas9-mediated mutagenesis of the white and Sex lethal loci in the invasive pest, *Drosophila suzukii*. Biochem Biophys Res Commun 469(4):911–916

Liu Y, Ma S, Wang X, Chang J, Gao J, Shi R, Xia Q (2014) Highly efficient multiplex targeted mutagenesis and genomic structure variation in *Bombyx mori* cells using CRISPR/Cas9. Insect Biochem Mol Biol 49:35–42

Liu Q, Hallerman E, Peng Y, Li Y (2016) Development of Bt rice and Bt maize in China and their efficacy in target pest control. Int J Mol Sci 17(10):1561

Ma Y, Zhang X, Shen B, Lu Y, Chen W, Ma J et al (2014) Generating rats with conditional alleles using CRISPR/Cas9. Cell Res 24:122–125

Mabashi-Asazuma H, Kuo CW, Khoo KH, Jarvis DL (2015) Modifying an insect cell *N*-Glycan processing pathway using CRISPR-Cas technology. ACS Chem Biol 10:2199–2208

Mali P, Esvelt KM, Church GM (2013) Cas9 as a versatile tool for engineering biology. Nat Methods 10(10):957–963

Merlin C, Beaver LE, Taylor OR, Wolfe SA, Reppert SM (2013) Efficient targeted mutagenesis in the monarch butterfly using zinc finger nucleases. Genome Res 23:159–168

McGuire S (2015) International Food Policy Research Institute. 2014. Washington, DC: Global Nutrition Report 2014: actions and accountability to accelerate the world's progress on nutrition. Adv Nutr 6(3):278–279

Moon SB et al (2018) Highly efficient genome editing by CRISPR-Cpf1 using CRISPR RNA with a uridinylate-rich 3′-overhang. Nat Commun 9:3651

Naik NG, Lo YW, Wu TY, Lin CC, Kuo SC, Chao YC (2018) Baculovirus as an efficient vector for gene delivery into mosquitoes. Sci Rep 8(1):17778

Nakanishi T, Kato Y, Matsuura T, Watanabe H (2014) CRISPR/Cas-mediated targeted mutagenesis in *Daphnia magna*. PLoS One 9(5):98363

Ochiai H, Sugawara T, Yamamoto T (2015) Simultaneous live imaging of the transcription and nuclear position of specific genes. Nucleic Acids Res 43(19):e127–e127

Osman GH, Assem SK, Alreedy RM, El-Ghareeb DK, Basry MA, Rastogi A, Kalaji HM (2015) Development of insect resistant maize plants expressing a chitinase gene from the cotton leaf worm, *Spodoptera littoralis*. Sci Rep 5:18067

Oye KA, Esvelt K, Appleton E, Catteruccia F, Church G, Kuiken T et al (2014) Regulating gene drives. Science 345(6197):626–628

Palma L, Munoz D, Berry C, Murillo J, Caballero P (2014) *Bacillus thuringiensis* toxins: an overview of their biocidal activity. Toxins 6:3296–3325

Pan MH, Wang XY, Chai CL, Zhang CD, Lu C, Xiang ZH (2009) Identification and function of Abdominal-A in the silkworm, *Bombyx mori*. Insect Mol Biol 18(2):155–160

Patil SB (2003) Studies on the Management of Cotton Pink Bollworm *Pectinophora Gossypiella* (saunders)(Lepidoptera: Gelechiidae) (Doctoral dissertation, UAS)

Peters JM, Silvis MR, Zhao D, Hawkins JS, Gross CA, Qi LS (2015) Bacterial CRISPR: accomplishments and prospects. Curr Opin Microbiol 27:121–126

Peterson BA, Haak DC, Nishimura MT, Teixeira PJ, James SR, Dangl JL et al (2016) Genome-wide assessment of efficiency and specificity in CRISPR/Cas9 mediated multiple site targeting in Arabidopsis. PLoS ONE 11:0162169

Prykhozhij SV, Rajan V, Gaston D, Berman JN (2015) Correction: CRISPR MultiTargeter: a web tool to find common and unique CRISPR single guide RNA targets in a set of similar sequences. PLoS One 10(9):0138634

Qi Y, Li X, Zhang Y, Starker CG, Baltes NJ, Zhang F et al (2013) Targeted deletion and inversion of tandemly arrayed genes in *Arabidopsis thaliana* using zinc finger nucleases. Genes Genom Genet 3(10):1707–1715

Quan X (2017) Genetic analysis of desiccation resistance in Drosophila (Doctoral dissertation, 首都大学東京)

Ran FA, Hsu PD, Lin CY, Gootenberg JS, Konermann S, Trevino AE et al (2013) Double nicking by RNA-guided CRISPR Cas9 for enhanced genome editing specificity. Cell 154(6):1380–1389

Reegan AD, Ceasar SA, Paulraj MG, Ignacimuthu S, Al-Dhabi NA (2016) Current status of genome editing in vector mosquitoes. Rev Biosci Trends 20:34–38

Rees HA, Liu DR (2018) Base editing: precision chemistry on the genome and transcriptome of living cells. Nat Rev Genet 19:770–788

Reid W, O'Brochta DA (2016) Applications of genome editing in insects. Curr Opin Insect Sci 13:43–54

Ren XL, Ma Y, Cui JJ, Li GQ (2014) RNA interference-mediated knockdown of three putative aminopeptidases N affects susceptibility of *Spodoptera exigua* larvae to *Bacillus thuringiensis* Cry1Ca. J Insect Physiol 67:28–36

Roditakis E, Steinbach D, Moritz G, Vasakis E, Stavrakaki M, Ilias A et al (2017) Ryanodine receptor point mutations confer diamide insecticide resistance in tomato leafminer, *Tuta absoluta* (Lepidoptera: Gelechiidae). Insect Biochem Mol Biol 80:11–20

Sanchez CHM, Wu SL, Bennett JB, Marshall JM (2018) A modular simulation framework for the spread of gene drives through spatially-explicit mosquito populations. Preprint at https://www.biorxiv.org/content/early/2018/06/19/350488

Sander JD, Joung JK (2014) CRISPR-Cas systems for editing, regulating and targeting genomes. Nat Biotechnol 32(4):347

Scott MJ, Pimsler ML, Tarone AM (2014) Sex determination mechanisms in the Calliphoridae (blow flies). Sex Dev 8(1-3):29–37

Sebo ZL, Lee HB, Peng Y, Guo Y (2014) A simplified and efficient germline-specific CRISPR/Cas9 system for Drosophila genomic engineering. Fly 8(1):52–57

Shukla JN, Palli SR (2013) Tribolium castaneum Transformer-2 regulates sex determination and development in both males and females. Insect Biochem Mol Biol 43(12):1125–1132

Singer M, Frischknecht F (2016) Time for genome editing: next-generation attenuated malaria parasites. Trends Parasitol 33(3):202–213

Smidler AL, Terenzi O, Soichot J, Levashina EA, Marois E (2013) Targeted mutagenesis in the malaria mosquito using TALE nucleases. PLoS One 8(8):74511

Somers J, Nguyen J, Lumb C, Batterham P, Perry T (2015) In vivo functional analysis of the Drosophila melanogaster nicotinic acetylcholine receptor Dα6 using the insecticide spinosad. Insect Biochem Mol Biol 64:116–127

Sorek R, Kunin V, Hugenholtz P (2008) CRISPR--a widespread system that provides acquired resistance against phages in bacteria and archaea. Nat Rev Microbiol 6:181–186

Sparks TC, Nauen R (2015) IRAC: mode of action classification and insecticide resistance management. Pestic Biochem Physiol 121:122–128

Steinbach D, Gutbrod O, Lümmen P, Matthiesen S, Schorn C, Nauen R (2015) Geographic spread, genetics and functional characteristics of ryanodine receptor based target-site resistance to diamide insecticides in diamondback moth, *Plutella xylostella*. Insect Biochem Mol Biol 63:14–22

Stevens T, Song S, Bruning JB, Choo A, Baxter SW (2017) Expressing a moth abcc2 gene in transgenic Drosophila causes susceptibility to Bt Cry1Ac without requiring a cadherin-like protein receptor. Insect Biochem Mol Biol 80:61–70

Tabashnik BE, Sisterson MS, Ellsworth PC, Dennehy TJ, Antilla L, Liesner L et al (2010) Suppressing resistance to Bt cotton with sterile insect releases. Nat Biotechnol 28(12):1304

Taning CNT, Van Eynde B, Yu N, Ma S, Smagghe G (2017) CRISPR/Cas9 in insects: applications, best practices and biosafety concerns. J Insect Physiol 98:245–257

Tikar SN, Kumar A, Prasad GBKS, Prakash S (2009) Temephos-induced resistance in *Aedes aegypti* and its cross-resistance studies to certain insecticides from India. Parasitol Res 105 (1):57–63

Tochen S, Dalton DT, Wiman N, Hamm C, Shearer PW, Walton VM (2014) Temperature-related development and population parameters for Drosophila suzukii (Diptera: Drosophilidae) on cherry and blueberry. Environ Entomol 43(2):501–510

Tsai SQ, Wyvekens N, Khayter C, Foden JA, Thapar V, Reyon D et al (2014) Dimeric CRISPR RNA-guided FokI nucleases for highly specific genome editing. Nat Biotechnol 32:569–576

USDA ERS (2016) Adoption of genetically engineered crops in the US. https://www.ers.usda.gov/dataproducts/adoption-of-genetically-engineered-crops-in-the-us.aspx

Wang Y, Li Z, Xu J, Zeng B, Ling L, You L et al (2013) The CRISPR/Cas system mediates efficient genome engineering in *Bombyx mori*. Cell Res 23(12):1414

Wang J, Zhang H, Wang H, Zhao S, Zuo Y, Yang Y, Wu Y (2016) Functional validation of cadherin as a receptor of Bt toxin Cry1Ac in *Helicoverpa armigera* utilizing the CRISPR/Cas9 system. Insect Biochem Mol Biol 76:11–17

Wang J, Wang H, Liu S, Liu L, Tay WT, Walsh TK et al (2017) CRISPR/Cas9 mediated genome editing of *Helicoverpa armigera* with mutations of an ABC transporter gene HaABCA2 confers resistance to *Bacillus thuringiensis* Cry2A toxins. Insect Biochem Mol Biol 87:147–153

Wolter F, Puchta H (2018) The CRISPR/Cas revolution reaches the RNA world: Cas13, a new Swiss Army knife for plant biologists. Plant J 94:767–775

Xue Z, Wu M, Wen K, Ren M, Long L, Zhang X, Gao G (2014) CRISPR/Cas9 mediates efficient conditional mutagenesis in Drosophila. Genes Genom Genet 4(11):2167–2173

Yu Z, Chen H, Liu J, Zhang H, Yan Y, Zhu N, Liang X (2014) Various applications of TALEN-and CRISPR/Cas9-mediated homologous recombination to modify the Drosophila genome. Biol Open 3(4):271–280

Zaidi SS et al (2017) CRISPR-Cpf1: a new tool for plant genome editing. Trends Plant Sci 22:550–553

Zalatan JG, Lee ME, Almeida R, Gilbert LA, Whitehead EH, La Russa M et al (2015) Engineering complex synthetic transcriptional programs with CRISPR RNA scaffolds. Cell 160 (1-2):339–350

Zalucki MP, Shabbir A, Silva R, Adamson D, Shu-Sheng L, Furlong MJ (2012) Estimating the economic cost of one of the world's major insect pests, Plutella xylostella (Lepidoptera: Plutellidae): just how long is a piece of string? J Econ Entomol 105(4):1115–1129

Zetsche B, Heidenreich M, Mohanraju P, Fedorova I, Kneppers J, Degennaro EM et al (2017) Multiplex gene editing by CRISPR-Cpf1using a single crRNA array. Nat Biotechnol 35:31–34

Zetschi A et al (2015) Cpf1 is a single RNA-guided endonuclease of a class 2 CRISPR-Cas system. Cell 163:759–771

Zhang Z, Aslam AF, Liu X, Li M, Huang Y, Tan A (2015) Functional analysis of *Bombyx* Wnt1 during embryogenesis using the CRISPR/Cas9 system. J Insect Physiol 79:73–79

Zhang Y, Zhao B, Roy S, Saha TT, Kokoza VA, Li M, Raikhel AS (2016) MmicroRNA-309 targets the Homeobox gene SIX4 and controls ovarian development in the mosquito *Aedes aegypti*. Proc Natl Acad Sci 113(33):4828–4836

Zhang Z, Liu X, Shiotsuki T, Wang Z, Xu X, Huang Y et al (2017) Depletion of juvenile hormone esterase extends larval growth in Bombyx mori. Insect Biochem Mol Biol 81:72–79

Zhao Y, Dai Z, Liang Y, Yin M, Ma K, He M, Ouyang H, Teng C (2014) Sequence-specific inhibition of microRNA via CRISPR/CRISPRi system. Sci Rep 4:3943

Zhu L, Mon H, Xu J, Lee JM, Kusakabe T (2015) CRISPR/Cas9-mediated knockout of factors in non-homologous end joining pathway enhances gene targeting in silkworm cells. Sci Rep 5:18103

Zimmer CT, Garrood WT, Puinean AM, Eckel-Zimmer M, Williamson MS, Davies TE, Bass C (2016) A CRISPR/Cas9 mediated point mutation in the alpha 6 subunit of the nicotinic acetylcholine receptor confers resistance to spinosad in *Drosophila melanogaster*. Insect Biochem Mol Biol 73:62–69

Zuo Y, Wang H, Xu Y, Huang J, Wu S, Wu Y, Yang Y (2017) CRISPR/Cas9 mediated G4946E substitution in the ryanodine receptor of *Spodoptera exigua* confers high levels of resistance to diamide insecticides. Insect Biochem Mol Biol 89:79–85

Disease Resistance in Crops Through CRISPR/Cas

5

Zulqurnain Khan, Tahira Saboor, Muhammad Ashfaq, Abubakar Saddique, and Plosha Khanum

Abstract

Plants, being sessile, face a lot of abiotic and biotic stresses. Owing to these stresses, a big loss, up to 40%, has been observed in crop growth and production. Diseases are caused by malicious biotic factors result in huge economic losses. Disease loss may depend upon the type of disease such as bacterial, fungal, and viral disease. Integrated disease management (IDM) program has been adopted by plant protection departments to cope with the diseases and protect plants. In IDM, a number of physical, chemical, and biological methods are being applied to protect crops against diseases. It has been observed that some diseases are still not fully controlled and due to the climate change, new diseases have been emerged in the field causing huge economic losses. In this scenario, there is a need to mount new programs focusing on using novel approaches for plant protection and disease resistance in plants. Genome editing has been revolutionizing all fields of life sciences due to its marvelous applications. CRISPR/Cas, latest genome editing tool, has been used widely in plants, animals, bacteria, and other organisms. Due to its easy designing and effortless cloning, CRISPR/Cas has become a method of choice for researchers. Apart from its applications in crops for better growth, more production and high quality, CRISPR/Cas has also been used for disease resistance in plants. In this chapter, a brief introduction of CRISPR/Cas technology and its applications for diseases resistance in plants will be discussed in details.

Z. Khan (✉) · T. Saboor · A. Saddique · P. Khanum
Institute of Plant Breeding and Biotechnology, MNS University of Agriculture, Multan, Pakistan
e-mail: zulqurnain.khan@mansuam.edu.pk; abubakar.siddique@mnsuam.edu.pk; plosha.khanum@mnsuam.edu.pk

M. Ashfaq
Institute of Plant Protection, MNS University of Agriculture, Multan, Pakistan
e-mail: mashfaq@mnsuam.edu.pk

A. Ahmad et al. (eds.), *CRISPR Crops*,
https://doi.org/10.1007/978-981-15-7142-8_5

5.1 Introduction

After domestication, the plant breeding has been the best approach for making significant advances in world nutrition and food security by developing new crop varieties. Crops are vulnerable to fungal, bacterial and viral diseases causing significant economical loss (FAO 2017). Increasing plant resistance plays an important part in changing crop production to counter rise in the world's population. Disease control methods that rely on resistant species and agrochemicals are usually successful. Nonetheless, because many plant pathogens have a high evolutionary potential, novel genotypes that quickly develop by mutation or recombination which show no longer response to the resistance. This can easily make specific disease prevention strategies ineffective. The advanced and high-performance molecular technologies have made possible a more detailed inventory of the pathogens linked to certain cultures and offered an insight into how environmental conditions and the plant genotype can impact these populations. Disease is a complex interaction between the host plant and the pathogen, which can include many components in the resistance/susceptibility response. The relationship between natural and induced mutations and the mechanism of infection could alter resistance/susceptibility response (Dracatos et al. 2018; Boyd and O'Toole 2012). Prior to pre-genomic era, the traditional breeding programs based their activities on the detection of natural and induced mutant alleles for resistance and their integration into elite genotypes through breeding techniques. Such methods have been unclear and unreliable, for example, leading to transfer of broad genome regions rather than single gene insertions. The breeding methods for mutation have been effective in improving resistance against diseases and conventional breeding has been used for decades to produce or develop new plant varieties. Numerous mutants were produced by induction of mutations that showed increased resistance to different diseases. MLO mutants are very useful against fungal diseases like powdery mildew in barley (Miklis et al. 2007).

The homozygous mutant *MLO* alleles have shown sustainability for resistance against powdery mildew in Barley (Panstruga and Schulze-Lefert 2002). The gene knockout is responsible for this longevity. In addition, mutagenesis should be used to provide more reliable single nucleotide mutations in the target gene sequences in cases where immunity to different pathotypes is provided by a certain host gene allele. A new start in plant breeding programs has been the revolution driven by the genomic, transcriptome, and other omics techniques. Major quantitative trait loci (QTLs) are used for the quantitative resistance against pathogens in plants apart from resistance genes (R) in varieties that are highly agronomic with disease resistance. New breeding techniques (NBTs) are of great interest because of their wide applications in a variety of fields such as developmental biology, resistance against biotic factors, and abiotic stress tolerance (Nelson et al. 2018). The new and best molecular approaches, NBT, have been used to precisely and genetically modify one or more target genes. NBT such as genome editing is based on site-driven nucleases to introduce double stranded breaks (DSBs) at default DNA sites (Sovova et al. 2016). Such breaks are repaired by a variety of host cell repair mechanisms

(Khandagale and Nadaf 2016; Qi et al. 2013). The breakdowns are repaired with repair mechanism of the host cell either by Non-homologous end joining (NHEJ) with small insertions/deletions or by homologous recombination (HR) which results in gene modifications using a donor sequence. The genome editing (GE) have four types of nucleases which may be used to improve plant resistance like Mega nucleases (MNs), Zinc Finger Nucleases (ZFNs), Transcription Activator-Like Effector Nucleases (TALENs) and CRISPR/CRISPR-associated protein 9 (Cas9). The exponential increase in CRISPR/Cas9 publications shows that the CRISPR technology needs less financial resources and that the rate of success of gene modification is higher than that of other available technologies. The CRISPR/Cas9 editor is a powerful way of improving agronomic features in plants (Mohanta et al. 2017). The CRISPR/Cas9 of *Streptococcus pyogenes* (SpCas9) has quickly played an essential role in different fields of plant research applications as well as in many other areas of research. A single guide RNA (sgRNA) in the CRISPR/Cas9 system can bind to Cas9 and aim at certain DNA sequences (Ding et al. 2018). In a gene of interest, the protospacer-adjacent motif (PAM) element restricts the potential target sequences. This restriction is not a big deal in case of targeted mutagenesis (Zaidi et al. 2016; Fu et al. 2013). CRISPR/Cas is a very efficient bacterial immune system against bacteriophages (Fig. 5.1) which may be harnessed for targeted gene modifications in different organisms. Diseases have been a major threat to our crops and production. A number of strategies may be employed to cope with the plant diseases (Fig. 5.2).

5.2 Major Diseases in Crops

5.2.1 Bacteria as a Sources of Plant Diseases

Rod shape bacteria cause some different types of diseases in plants. Bacteria enter the plant through natural openings, such as leaf stomata or plant tissue wounds (Strange and Scott 2005; Kannan et al. 2015). Once inside the vascular system (the water vessels and nutrients) of the plant, the bacteria will plug and wilt the plant. Rotting and swollen plant tissues are also common symptoms of bacterial disease. Water, insects, infected soil or contaminated tools may spread bacteria. In crops, pear and apple fire- blight (caused by Erwinia amylovora) was the first bacterial disease ever discovered, by T. J. Burrill (1877–1885) from the University of Illinois. *Erwinia*, *Pectobacterium*, *Pantea*, *Pseudomonas*, *Ralstonia*, *Burkholderia*, *Acidovorax*, *Xanthomonas*, *Clavibacter*, *Streptomyces*, *Xylella*, *Spiroplasma*, and *Phytoplasma* (Yabuuchi et al. 1995) are the most common herb's pathogenic bacteria. Plant pathogenic bacteria cause many forms of symptoms including galls and overgrowths, wilts, leaf stains, soft rots, scabs, and cankers. In contrast to viruses, which are inside host cells, walled bacteria grow in the spaces between cells and do not invade them (Mansfield et al. 2012).

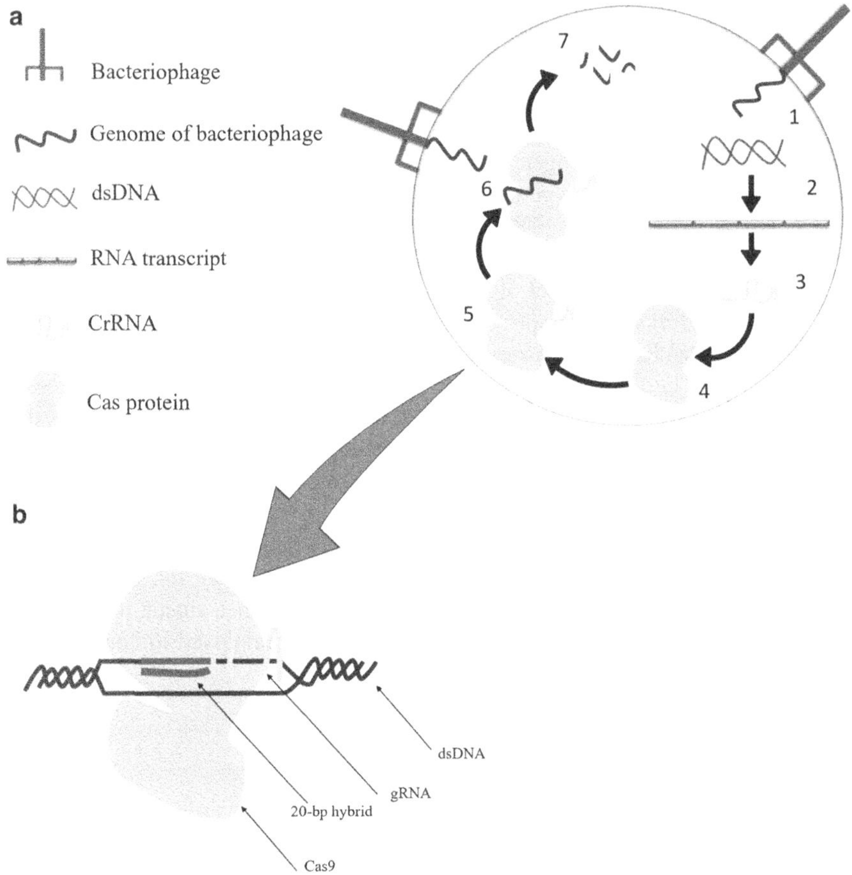

Fig. 5.1 CRISPR-based immune system in bacteria against Bacteriophages. Figure a and b showing mechanism of action of the CRISPR system against bacteria invading viruses and the derived artificial CRISPR system which may be used to target any sequence of choice. (**a**) Bacterial cell is attacked by bacteriophages. After entrance into the host, the genetic material is replicated (1) and transcribed (2). The CRISPR RNA (CrRNA) of short length (3) recruit the CRISPR-associated 9 (Cas9) protein (4) to make CRISPR-CrRNA complex (5). CrRNA guides the Cas protein to target the new entering viruses having complementary base pairs (6). Targeting of the viral genome resulted in viral interference (7). (**b**) Artificially designed CRISPR/Cas system. In artificial system, a single guided RNA (gRNA) of 20bp complementary sequence to the target DNA is used for the site specific mutation with the help of Cas protein. The gRNA can be redesigned according to the target by replacing 20 bp subsequently

5.2.2 Fungal Diseases in Plants

Roughly, about 80% of plant diseases are caused by fungi which can reproduce both sexually and asexually (Hibbett et al. 2007, 2003). They can penetrate the tissue of the plant or grow on the surface of the plants. Air, water, soil, and animals disperse fungal spores to other plants which process acting like seeds. Warm and wet

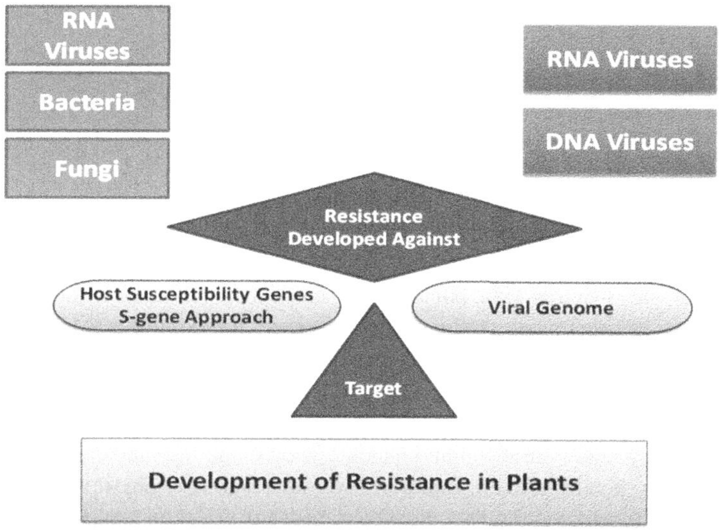

Fig. 5.2 A schematic work-flow shows the CRISPR/Cas9-based resistance strategy against pathogens (Mushtaq et al. 2019)

conditions promote fungal growth (Purvis and Hector 2000). Although, many fungi have useful plant growth functions, particularly by forming mycorrhizal ties and associated with the roots of the plants, but they also cause significant diseases such as anthracnose, late blight, apple scab, club root, black spot, damping off, and powdery mildew. Fungus-like organisms (FLOs) collectively are responsible for more plant diseases than any other class of plant-like pests of over 8000 species. FLO organisms, such as *Phytophthora* and *Pythium*, are now known to belong to another taxonomic group (Boerjan et al. 2003) for causing downy mildew disease in plants.

5.2.3 Viruses and Viroids

The most difficult pathogens to control are viruses and viroids that attack plants. Destroying the infected plants is usually the best control method, since chemicals to inactivate plant viruses (Mysore and Ryu 2004) and viroids have not proven effective. While there have been more than 300 plant viruses, new strains continue to develop because these organisms can mutate. Either rod-formed or isometric (polyhedra) are the most common plant viruses. A short rod-formed rigid, a long rod-formed flexion, and an isometric virus are examples of TMV, potato virus Y (PVY), and a Cucumber mosaic virus (CMV), (Verlaan et al. 2013; Ishibashi et al. 2007). Yellowing, the slow growth of some part of the plant and plant malformations such as leaf rolls and untypical narrow leaf growth are signs of viral infection. Many plants can be infected with mosaic viruses. Plants that have been infected with this virus are mottled or streaked; fruit trees infected produce poor fruit and low yields (Yamaji et al. 2012; Brotman et al. 2013).

5.2.4 Nematodes Causing Plant Diseases

Nematodes are small, worm-like animals, having long needlelike structures called stylets, suck plant cell nutrients (Decraemer and Hunt 2006; Davis et al. 2004). They lay eggs like larvae and go through four stages before they become adults. Nematodes are 30 days long but can live more than 30 years in dormant state. Nematicides are chemical products that are used for controlling infestations of nematode. Marigolds are nematode-resistant and are often planted to prevent infected soil from nematodes (Hussey et al. 1989). Nematodes attack mainly plant roots, but they can destroy other plant parts internally or externally as well. They live in warm, sandy, and humid soil and attack a wide range of plant species, including maize, lettuce, potato, tomato, liquorice, rye, and onions (Habash et al. 2017; Ali et al. 2017). All nematodes do not, however, have any effect on plants. Some of them are in fact used for controlling other pests like cutworms, armyworms, and beetle grubs.

5.2.5 Other Sources

Mycoplasmas are single-cell organisms without rigid cell walls and contain the cell membranes in layers. They are the the simplest and smallest self-replicating bacteria. Mycoplasmas can be saprotrophic or parasitic. The diseases of plant materials, called yellow diseases, are transmitted by insects like the leaf hopper. Parasitic plants such as mistletoe cannot receive soil nutrients, but must adhere to other plants and use host plant nutrients to survive. The woods of their host trees are weakened and branches deformed.

5.3 Using Genome Editing for Disease Resistance in Plants

CRISPR/Cas9 is a toolkit that is versatile and uses nuclease-engineered sgRNA to make accurate modification in the genome at certain locations (Ran et al. 2013; Langner et al. 2018). Resistance to disease may be achieved in plants either through the editing of the pathogen genomes or the S genes. In case of viruses, CRISPR/Cas has been successfully used to target the host S gene for creating broad spectrum disease resistance (Chandrasekaran et al. 2016). The CRISPR/Cas9 method is being used by Kis et al. (2019) to inhibit a phloem controlled, insect transmission virus, an economically important wheat dwarf virus (WDV) that is in the Barley Geminiviridae family. Four WDV genome sites have been targeted and WDV tolerant transgenic lines have been produced. Genome editing (GE) was suggested, targeting satellite molecule (a and b satellite) of the virus, as a key tool to control the

CLCuD (complex of begomoviruses) (Iqbal et al. 2016). CLCuD caused by begomoviruses is a complex of viruses including different types such as CLCuRaV (cotton leaf curl Rajasthan virus), CLCuAlV (cotton leaf curl Alabad virus), CLCuKoV (cotton leaf curl kokhran virus), CLCuKov-Bu (Burewala Strain), CLCuMV (cotton leaf curl Multan virus), CLCuBaV (cotton leaf curl Banglor virus) and many others. GE techniques have successfully been circumvented to curtail the proliferation of these begomoviruses (Uniyal et al. 2019; Iqbal et al. 2016). GE technology can be adopted to provide wide range of natural resistance targeting the intergenic (IR) area with the CRISPR/Cas9 system to induce mutation or inhibiting replicating-associated (Rep) protein of CLCuD associated Begomoviruses (CABs). The Rice tungro disease (RTD) is caused by the interaction between the Rice tungro spherical virus (RTSV), and Rice tungro bacilliform virus (RTSV). RTD is a serious constraint in the production of rice. Resistance against RTSV may be achieved by mutating a recessive trait that governed by the factor (eIF4G) which is known as a translation initiation factor 4 gamma gene. In eIF4G of the RTSV susceptible variety, numerous mutations were introduced using CRISPR/Cas9 system which resulted in the resistance againt RTD in rice (IR-64) (Macovei et al. 2018). Multiplex genome editing technology has added new insights to unparalleled opportunities to overcome emerging diseases, where different disease-developing pathogens complement one another. The array of CRISPR against different crops are listed below in Table 5.1.

Ji et al. (2015) have succeeded in blocking up to 70% of the beet severe curly top virus (BSCTV) in *A. thaliana* and *N. benthamian* by actively targeting the specific regions within viral genome through CRISPR/Cas system. Similarly, CRISPR/Cas9 technique has also been used by Ali et al. (2016) to induce mutation in the non-coding (IR) region of the three different types of geminiviruses; begomovirus, curtovirus and mastrevirus (Baltes et al. 2015). Three motifs were considered indispensable in the replication of virus (motifs I, II and III) through rolling circle replication mechanism which may be targeted using CRISPR/Cas to limit virus infection of bean yellow dwarf virus (BeYDV) (Baltes et al. 2015). CRISPR/Cas9 was used to inhibit endogenous banana streak virus (eBSV) in the genome B. The stress and in-vitro culture activates the latent eBSV in the B genome and the transmission into active viral proteins in targeted site mutations escape proper transcription (Tripathi et al. 2019). *Francisella novicida* derived CRISPR/Cas9 (FnCas9) was used to develop immunity in Arabidopsis and *N. benthamiana* against RNA viruses (CMV and TMV) (Zhang et al. 2018). The plants transformed with FnCas9 showed considerably reduced symptoms of virus infection and reduced accumulation of viral RNA. CRISPR/Cas system has been used for disease resistance against a number of pathogens such as bacteria, fungi and viruses.

Table 5.1 Successful report of CRISPR/Cas for disease resistance in plants

S. No crop	Gene function	Resistance to pathogen	Target gene/ gene mutated	Strategies	Mutation type	Effect on growth and development	References
Cotton	Replication of virus	Cotton leaf curl disease associated Begomoviruses (CABs) and helper a and b satellites begomoviruses	The replication associated protein like (Rep), CABs, and (IR), the non-coding intergenic regions, a-Satellite (Rep) and b-Sat (IR)	Agroinfiltration or stable agro-mediated transformation	InDel	N.A	Uniyal et al. (2019), Iqbal et al. (2016)
Rice	Translation of viral RNA growth Regulation of genomes, opening stomatal	Rice tungro spherical virus (RTSV)	eIF4G, translation initiation factor	Agrobacterium mediated transformation of immature rice embryos	InDel	Insignificant	Macovei et al. (2018)
Banana	Viral growth and development	Endogenous banana streak virus (eBSV)	The integrated genome region of eBSV in the banana genome	Agrobacterium Mediated transformation	InDel	Insignificant	Tripathi et al. (2019)
Nicotiana benthamiana	Required for replication and viral growth	CLCuKoV and merremia mosaic virus (MeMV) and bean yellow dwarf virus	(CP) coat protein, RCRII The domain of replication associated with (Rep) protein	(RNA2) genome having, SgRNA into (NB-Cas9OE) plants, Agroinfection of tobacco rattle virus	InDel	NA	Ali et al. (2016), Baltes et al. (2015)
Barley	Required for replication and viral growth	Wheat dwarf virus (WDV)	Rep, MP, LIR	Agrobacterium Mediated transformation	InDel	Insignificant	Kis et al. (2019)

Wheat (*Triticum aestivum* L.,)	Immunity repress against powdery mildew	*Graminis f.* sp. tritici (Bgt), *Blumeria graminis f.* sp. tritici	Mildew resistance, (MLO) locus)-A1 allele, TaMLO-A1 allele	Biolistic transformation of wheat embryos	InDel	NA	Wang et al. (2016)
Citrus (grape fruit Duncan)	The pathogenicity factor PthA4, canker susceptibility induced by the pathogen	*Xanthomonas citri* subsp. citri	Binding elements (EBEs) in the type-1, PthA4 effector-(CsLOBP) of the (CsLOB1), (*Citrus sinensis* lateral organ Bounda, CsLOB1 promoter (EBEPthA4)	Agrobacterium mediated transfromation	InDel	NA	Jia et al. (2016)
Tomato	Regulates stomatal opening	*Pseudomonas syringae* pv tomato (Pto) DC3000	SlJAZ2	Agrobacterium mediated transformation	InDel	NA	Ortigosa et al. (2019), Thomazella et al. (2016)
Grapes	Susceptibility confers to powdery mildew	*Erysiphe necator*	MLO-7	CRISPR/Cas9 purified delivery into ribonucleoproteins (RNPs) to the protoplast	InDel	NA	Malnoy et al. (2016)
Grapefruit			CsLOB1			Insignificant	Jia et al. (2016)

(continued)

Table 5.1 (continued)

S. No crop	Gene function	Resistance to pathogen	Target gene/ gene mutated	Strategies	Mutation type	Effect on growth and development	References
	(LBD), the Factor (CsLOB1), is the member of lateral organ, boundaries domain (gene family) by the factors of plant transcription	*Xanthomonas citri* subsp. citri (Xcc)		Agrobacterium mediated	InDel Frame shift		
Rice	Blast resistance in rice negative resistance	*Oryza magnaphorthe*	OsERF922 factor	Agrobacterium mediated	InDel	Insignificant	Wang et al. (2016)
Arabidopsis	Factor susceptibility	Clover yellow vein virus	Gene eIF4E1	Agrobacterium Mediated transformation	Substitution single base	Insignificant	Bastet et al. (2019)
Cucumber	For (Potyviridae life cycle), Eukaryotic translational initiation factor	Cucumber vein yellowing virus	eIF4E gene	Agrobacterium mediated	InDel	Insignificant	Chandrasekaran et al. (2016)
Tomato	Powdery mildew confers	*Oidium neolycopersici*	slmlo1	Agrobacterium-mediated	InDel	Insignificant	Nekrasov et al. (2017)

5.4 CRISPR/Cas Against Viral Diseases in Plants

The CRISPR/Cas method has now been used to provide virus resistance to several plants (Chaparro-Garcia et al. 2015) in an effective manner. Various studies demonstrated the CRISPR/Cas approach to plant defense against geminivirus and the widespread usefulness of this approach has also been reported previously (Ji et al. 2015). This technology has increased resistance of the plants to geminivirus species including BCTV, TYLCV, and MeMV. Initially the CRISPR/Cas system recognizes and targets the particular genetic material site through three step-by-step processes: selection, expression, and intervention. The acquisition phase is the leading or original event involving the recognition and incorporation of a foreign DNA as a spacer on the CRISPR locus leading side followed by repeated replication. In general, a small fragment of exogenous DNA is included in the (Fang and Tyler 2016; Shanmugam et al. 2019) bacterial genome via CRISPR array and acts as a new spacer sequence in the first step. During the CRISPR-Case expression stage, both Cas protein (Specific Cas) and transactivating RNA (tracrRNA) first transcribe the long pre-CRISPR RNA (pre-crRNA) and then transform it actively into crRNAs. A certain sequence of foreign genomic elements is targeted and then break into small fragments during the third or final intervention activity. The crRNA guides the cleavage Cas9 protein to the additional target area of viral/plasmid DNA and leads thereby induction of resistance (Kumar and Jain 2014). The bacterial immune system has been harnessed for developing disease resistance against viruses in plants. So far, a number of researchers have reported successful disease resistance in a number of plants species using GE tools (Ji et al. 2015; Khan et al. 2017; Khan et al. 2019, Ali et al. 2016).

5.4.1 Plant Resistance Against RNA Viruses

No earlier evidence has been found that RNA viruses were directly targeted and removed. The device sgRNA of CRISPR/Cas9 was commonly used to inactivate certain genes efficiently and later, with CRISPR/Cas9, these target regions of the genome mutated to generate virus resistant plants (Chandrasekaran et al. 2016). Cas9-guide RNA can be used to target DNA viruses and has a drawback of the direct targeting of RNA viral genomes. Nonetheless, it might be adjusted in future because the Cas9 nuclease can be programmed and modified to break RNA. Type III-B and other CRISPR/Cas systems have the ability to target RNA as well (Romay and Bragard 2017). Currently, CRISPR/Cas9 technology has been highly demonstrated to generate novel genetic resistance to the TuMV potyvirus in DNA-free *A. thaliana* plants (Pyott et al. 2016), targeting the host *factor eIF(iso)4E*. Heritable and homozygous mutations were observed in transgene free Arabidopsis plants. Virus resistance was also obtained in *Cucumis sativus* L. by the disruption of *eIF4E* and by the development of *eIF4E* mutant plants, which were non-transgenic. The non-transgenic plants showed partial resistance to *ipomovirus (Cucumber vein yellowing virus)* and two essential *potyvirus (Zucchini yellow mosaic virus and Papaya ring spot mosaic virus-W)* (Chandrasekaran et al. 2016). With the advent of time, new addition in the CRISPR toolbox has been made which enabled the system for further diverse applications from DNA to RNA targeting for editing and modifications through insertions, deletions, corrections and replacement. Moreover, modifications at the SNPs level may be possible. Addition of prime editing and base editing has further broadened the spectrum and opened new horizons for future applications of the marvelous CRISPR tools especially for disease resistance.

5.4.2 Plant Resistance Against DNA Viruses

The geminiviruses constitute the main group of single stranded DNA viruses and most past research has been taken place on these ssDNA geminiviruses with respect to CRISPR/mediated viral interference. Recently, the use of CRISPR/Cas systems targeting geminiviruses in *Nicotiana benthamian*, beet severe curly top virus (BSCTV, genus *curtovirus*), has been reported. The resistance has also been reported against *tomato yellow leaf curl viruses* (TYLCV) and *bean yellow dwarf viruses* (BeYDV, genus *Mastrevirus*) in *Nicotiana benthamiana* and *Arabidopsis* (Ali et al. 2015). SgRNA-Cas9 inhibits DNA replication of the virus at various levels. Over-expressing the Cas9 in the genome has led to plants resistant to the virus. Symptoms of the disease were also reduced to various levels ranging from severe to mild. Ali et al. (2015) excellent work on the CRISPR/Cas9 machinery expression *N. benthamian* plants has provided strong resistance against TYLCV, BCTV, and *Merremia mosaic virus* (MeMV). Ali et al. (2015) engineered sgRNA target open reading frames encrypting viral genes, such as Rep, coat proteins, and preserved non-coding intergenic area (IR). It has been stated that the most successful target to reduce TYLCV is the preserved intergenic area of the gene. In addition, the use of sgRNA aiming a conserved IR region could achieve concurrent resistance to TYLCV, BCTV, and MeMV viruses. One of the major and damaging diseases of cotton is the cotton leaf curl disease (Cheng et al. 2015) caused by *begomoviruses*. A CRISPR multiplex editing technique may be developed as a wide-ranging approach to the management of cotton leaf curl disease (Iqbal et al. 2016). Baltes et al. (2015) have shown that a sgRNA targeting the BeYDV genome could confer plant resistance without cleavage activity that suggests catalytic inactive Cas9 (dCas9) can be used to prevent interference viruses through double stand breaks and thus to avoid concerns over plant genome off-target activities. Ali et al. (2016) studied CRISPR/Cas system to target the cotton leaf curl kokhran virus (*CLCuKoV)* and reported that it not only aimed at the conserved nucleotide sequence but could simultaneously target multiple begomoviruses and consequently confers a wide-spectrum activity of *geminivirus* resistance. All these experiments have shown that in *N. Benthamiana*, the CRISPR/Cas9 can significantly lower viral titers, which have eliminated or decreased the symptoms of the disease substantially. The capacity of multiple geminivirus systems to evade CRISPR/Cas9 focusing the coding and non-coding sequences was systemically identified by Ali et al. (2016). This study revealed that coding sequences resulted in the generation of viral versions that can evade the machine CRISPR/Cas9. Focusing on intergenic non-coding sequences has provided high virus interference rates, no viral detectable movement from CRISPR/Cas9 was

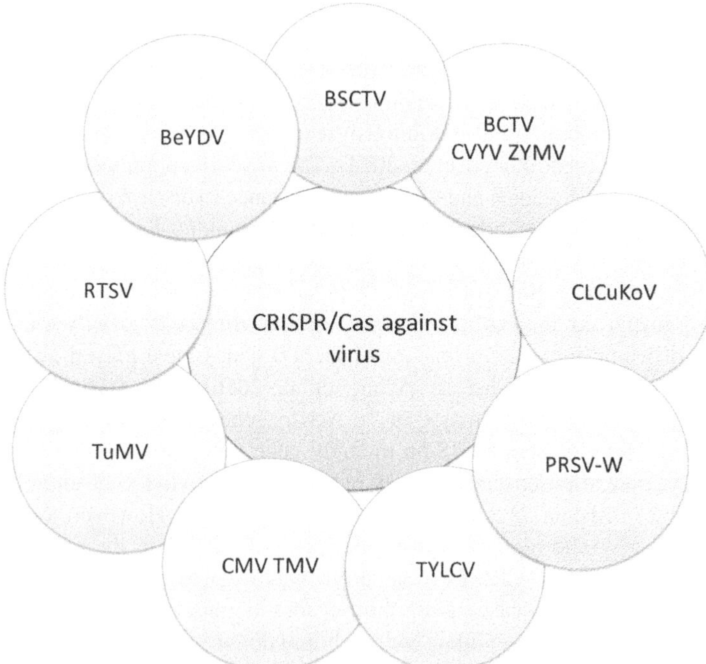

Fig. 5.3 Application of CRISPR/Cas against viruses represented in diagram: papaya ring spot mosaic virus-W(PRSV-W), Potato virus X (PVX), Turnip crinkle virus (TCV), Rice tungro spherical virus (RTSV) (Macovei et al. 2018), Bean yellow dwarf virus (BeYDV), Beet severe curly top virus (BSCTV) (Ji et al. 2015), Merremia mosaic virus (MeMV), Tobacco rattle virus (TRV), Tomato yellow leaf curl virus (TYLCV) (Ali et al. 2015; Pyott et al. 2016), Beet curly top virus (BCTV) (Baltes et al. 2015),Cotton leaf curl Kokhran virus (CLCuKoV), Turnip mosaic virus (TuMV) (Aman et al. 2018), Cucumber mosaic virus (CMV) (Zhang et al. 2018), Tobacco mosaic virus (TMV), Cucumber vein yellowing virus (CVYV), Zucchini yellow mosaic virus (ZYMV)

found, and it may be used as an effective strategy to impart possible long-lasting resistance in plants against multiple DNA viruses (Fig. 5.3).

5.5 Resistance to Fungal Diseases Through CRISPR/Cas

Many diseases in plants such as powdery mildew, smut, red rot, rust, etc. are caused by fungal pathogens. These diseases cause dramatic worldwide losses in yield every year, and also jeopardize the quality of the products harvested. Moreover, the development of mycotoxins known as secondary metabolites, cause severe health problems in animals and as well as in humans following exposure to contaminated food. Moreover, feed habits are also seriously affected by mycotoxigenic fungi. Various strategies to increase fungal resistance have been developed in the species of plant based on the understanding of current techniques and molecular mechanisms involved in the interaction between plant and pathogens. The main aim of CRISPR/

Cas9 editing today is to target genes and the gene products involved in fungal plant resistance. In the three different plants; tomato, grapevine, and bread wheat, (Nekrasov et al. 2017), *MLO* loci are tackled in a number of ways by RNA-guided Cas9 endonuclease. The *MLO* is the protein containing seven domains located in the plants plasma membrane and ubiquitously expressed in the dicots and monocots plants species (Acevedo-Garcia et al. 2014). The *MLOs* were previously reported as susceptiblity genes (S genes) and to increase resistance to disease (powdery mildew) in barley, Tomato and Arabidopsis, loss of function mutants have been used significantly (Piffanelli et al. 2004; Consonni et al. 2006; Bai et al. 2008). In one (*TaMLO-A1)* of the three homoalleles of the *MLO*, bread wheat plants mutated by CRISPR/Cas9 and displayed increased resistance against *Blumeria graminis* f. sp. The infection of triotic has shown that the (*TaMLO)* genes have played an important role in powdery mildew disease (Wang et al. 2016). The *SlMlo1*, in tomato, identified previously as the major of 16 *SlMlo* genes, has been targeted at two sites, and has been deleted by 48 bp to study gene function for disease resistance. "Tomelo," a new non-transgenic variety of tomato, was found to be entirely *oidium neolycopersici* resistant after targeting with CRISPR/Cas9. However, there was no effect on the genomic regions beyond the locus of *SlMlo1* in the off-target study (Nekrasov et al. 2017). CRISPR/Cas9 RNP was used to target *MLO7, DIPM-1, DIPM-2, and DIPM-4* disease susceptibility genes in grapevine and apple to develop resistance against powdery mildew and fire blight disease respectively (Malnoy et al. 2016). It was concluded that DNA-free non-transgegnic plants of apple and grapevine may be developed by direct delivery of RNPs into protoplast. In parallel studies, the knockout of *VvMLO7*, through RNAi have shown decreased the sensitivity to *Erysiphe necator* (Pessina et al. 2016). In order to avoid fire blight disease, RNP approaches were also applied to modify *DIPM-2, DIPM-1*and also *DIPM-4* genes in apple protoplasts (Malnoy et al. 2016). The transient application of genome editing tools in protoplast is particularly interesting for economically important crops such as apple and grapevine, which take several years to bloom, as the separation of CRISPR/Cas9 stably integrated cassettes through backcrossing will take much longer time to be achieved. Secondly, the Cas9/sgRNA complex is a quick approach as RNP, enabling transformed protoplasts to be achieved in the space of one to two days and an assessment of sgRNA is easy with effectiveness. Thirdly, the foreign DNA is not integrated into the genome, and during the cell culture regeneration process, the Cas9/sgRNA complexes can quickly be degraded. Moreover, in transient approaches, plasmid can sometimes cause unwanted integration into the host genome with prolonged presence of CRISPR/Cas9 editing tools increasing the risk of off-target mutations which can be improved by target specificities with CRISPR/Cas9 RNP. To increase the efficiency of this technology, it is important to refine regeneration protocols. The transcription factor *(VvWRKY52)* was targeted by the four gRNAs (Wang et al. 2018) to enhance grapevine's effective defense through the CRISPR/Cas9. CRISPR/Cas9 has developed rice blast disease resistant plants by disrupting the rice genes, *OsERF922* and *OsSEC3A* (Ma et al. 2018a, b; Wang et al. 2016). A multifaceted phenotype, with enhanced *Magnaporthe oryzae* resistance, increased salicylic acid levels (SA, high levels) and up-regulation of pathogenesis, was revealed by *Osséc3*a mutant plant. In addition, in non-transgenic plants mutated

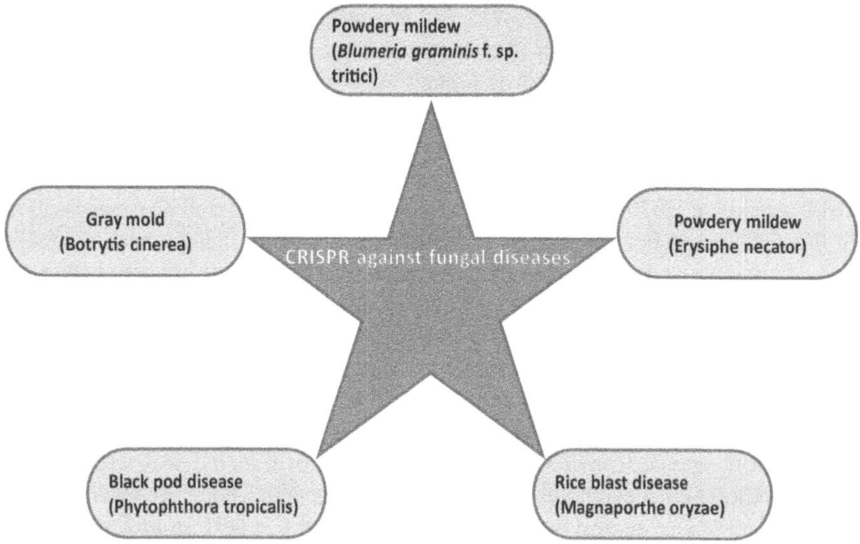

Fig. 5.4 CRISPR/Cas against fungal diseases represented in diagram such as powdery mildew disease (Wang et al. 2016), powdery mildew (Erysiphe necator) (Nekrasov et al. 2017; Malnoy et al. 2016), rice blast disease (Wang et al. 2016), black pod disease (Fister et al. 2018) and Gray mold disease (Ma et al. 2018a, b)

into the ethylene response transcription factor (ERF) 922, involved in several stress responses, there was no alteration in the variety of agronomic characteristics was observed. In both seedling and tillering stages, the plants that are mutant have reduced number of blast lesions (Wang et al. 2016). In general, the results showed that the CRISPR/Cas9 system was powerful and beneficial for improving the crop resistance to fungal diseases (Fig. 5.4).

5.6 Resistance to Bacterial Diseases Through CRISPR/Cas

Hundreds of bacterial species exist in the soil, which often show some signs of disease, involved in plant damage (Schloss and Handelsman 2004). Phytopathogens are not easily controlled at large due to the lack of suitable agrochemical products and unrevealed asymptomatic infections. In general, bacteriological control by means of genetic resistance, agronomic practices, and biocontrol agent is based upon the prevention and exclusion of pathogen (Kerr 2016). Phytopathogenic bacteria can be categorized as plant pathogens, for instance, the *Clavibacter michiganensis* that is the tomato bacterial ring rot causative agent and *Ralstonia solanacearum*, that causes multiple diseases in the monocot and dicot species. There have been very few studies on the application of the CRISPR/Cas system to combat plant bacterial diseases. The *OsSWEET13* mutagenesis was performed in rice for the purpose of achieving resilience to Probacterium, *Xanthomonas oryzae pv*, causing bacterial blight disease. The *OsSWEET13* is a susceptibility (S) gene (Zhou et al. 2015) that encodes a transporter of sucrose which interacts with plant pathogens. The *X. Oryzae* produces a factor that is protein effector (*PthXo2*), which induces host expression and a corresponding susceptibility condition leads to the expression of *OsSWEET13*. The disruption of this gene makes the effector incapable of binding, *OsSWEET11/OsSWEET14*, eventually resulting in immunity to infection (Li et al. 2012; Zeng et al. 2020). The next step towards achieving bacterial resistance will be further GE approaches, for recessive multiplexed resistance using a main combination of effectors and also some other (R) genes. CRISPR/Cas9 has also been used successfully for developing resistance against other Xanthomonas bacteria such as *Xanthomonas axonopodis* pv. citri cause citrus bacterial canker (CBC). Jia et al. (2016) developed canker-resistant mutants in the Duncan grapefruit by targeting the PthA4 effector binding elements in *Lateral Organ Boundaries 1* (*CsLOB1*) gene.

Peng et al. (2017) verified the relation between CBC and *CsLOB1* in Wanjincheng orange (*Citrus sinensis Osbeck*). EBE of *CsLOB1* was targeted with CRISPR/Cas9 for devevloping resistance against *Xcc*. Four mutants showed enhanced resistance against the CBC. In addition, after *CsLOB1* promoter modification, there was no alteration in plant growth. So far, TALEs, TALENs, CRISPR/Cas9 and multiplexed Cas9 have been used for disrupting regulatory elements (effector binding elements) in the promoter region of the S-genes to develop resistance against bacterial blight. More insight into plant-microbe interaction is needed to explore further options for creating disease resistance using precise genome editing tools. Additional efforts to develop non-transgenic canker-resistant citrus varieties will be needed to promote their agricultural applications in CBC prevention (Fig. 5.5).

5.7 CRISPR/Cas Based Directed Evolution

Butt and his colleagues developed a CRISPR/Cas platform of directed evolution where a library for a target gene was used with all possible sgRNAs. The *3B subunit 1 (OsSF3B1)* of the rice splicing factor has been selected as a target gene as in eukaryotes it is essential to the splicing of the RNA. Splicing inhibitors such as herboxidiene (GEX1A) and pladienolide B (PB) (Kotake et al. 2007) were targeted (SF3B1) in human, which were used in plants and their widespread mode of action has also been demonstrated in different species. Butt and his associates decided to demonstrate the use of their CRISPR/Cas directing evolutionary platform by developing new variants of *OsSF3B1*, that provide resistance to one of its medications GEX1A, after having confirmed the harm of genome-wide splicing repression by drugs (Butt et al. 2019).

A total of 119 sgRNAs were developed based on the NGG protospacer-adjacent motif pattern for *Streptococcus pyogenes* Cas9 (SpCas9). They aimed at the entire *OsSF3B1* coding sequence. A maximum of 15,000 transformed calli have been subcultured on GEX1A selection medium at concentrations high enough to inhibit the growth of wild callus. It was observed that 21 SF3B1–GEX1A-resistant (SGR)

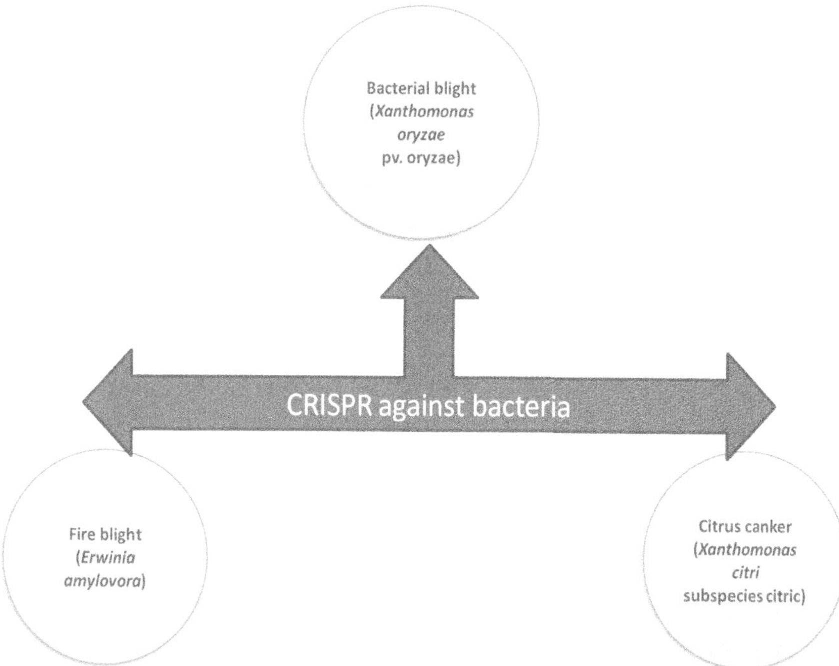

Fig. 5.5 CRISPR/Cas against different types of bacterial diseases. Bacterial blight disease (Li et al. 2012; Zhou et al. 2015), Citrus canker disease (Jia et al. 2016; Peng et al. 2017), Fire blight disease (Malnoy et al. 2016) are shown in diagram

rice shoots were recovered on selection medium containing 0.4 μM GEX1A. With each sgRNA's protospacer sequence as a barcode, the resulting mutations at *OsSF3B1* were clearly identified in these lines. The majority of mutations were in frame deletions, causing the loss of 1–10 amino acids at different protein positions without any active knockout variants. The knockout mutant was not heritable becuase the gene was essential for seed viability. Study of evolution and heredity are very important to improve crops and develop durable and sustainable resistance against diseases.

5.8 Domain-Focused Directed Evolution

This type of directed evolution is based on the targeted mutagenesis of the active protein domains responsible for stability, activity and function of the protein. Butt et al. (2019) also tested their CRISPR/Cas directed evolution platform for domain focused directed evolution. They targeted protein domain of SF3B1 which is responsible for protein-drug interaction to generate mutants resistant to GEX1A. All mutations recovered were possibly affecting the interaction between SF3B1 and the GEX1A inhibitor, based on structural protein analysis. The domain-based directed evolution has great applications for further genetic and molecular investigations. Domain-focused directed evolution platform may be efficiently used for understanding protein functions and interactions. Studies may be conducted on model organisms and cells to understand the mechanism and its further applications for basic and applied research. Host-pathogen interactions and cross-talk using molecular signals and proteins may be unraveled which would be helpful for creating resistance against pathogens.

5.8.1 Crop Improvement with CRISPR/Cas-Directed Evolution

Butt and colleagues developed a CRISPR/Cas platform for directed evolution in plants. Resistance to splicing inhibitors was observed in SGR (SF3B1-GEX1A-resistant) mutants. Phenotypically similar to wild plants, homozygous mutants showed that SF3B1 exhibit the full splicing behavior in rice. The SGR4 mutant showed the strongest GEX1A resistance. The seeds of this mutant may well grow on a medium with GEX1A up to 10 μM; other SGR mutants have not germinated under the same conditions. Although, SGR4 has transmitted three mutations, it is probable that the K1050E missense mutation has greatly helped weaken the relationship between SF3B1 and GEX1A. This study has shown that directed evolution in plants, which has important implications, is feasible (Butt et al. 2019).

Plants evolve in a typically long process, adapting to their growth environments. In the era of global warming and climate change, rapid technologies may provide an efficient way to achieving high productivity of agriculture for food security. In fact, saturation mutagenesis in vivo is impossible given the enormous sizes of crop genomes. As demonstrated in a study, CRISPR system may be used to achieve saturation mutagenesis. Using saturation mutagenesis a number of crop improvements may be made such as herbicide resistance, improved photosynthesis and enhanced tolerance or resistance to abiotic or biotic stresses, the strategy may also be effective in (Engqvist and Rabe 2019) improving or gaining useful traits in crops. The mapping and the discovery of important genes and alleles related to stress tolerance in major crops may be achieved using tools like genome-wide associations studies (GWASs). CRISPR is expected to contribute to improving agronomic traits or to creating novel alleles directly into elite cultivars by using directed evolution of these target genes. Although, directional design is widely used in protein technology, it can be applied by concentrating on cis-regulatory elements (Rodriguez-Leal et al. 2017) in terms of quantitative characteristics. Furthermore, similar directed developmental approaches can be implemented in the animals such as worm, fruit fly and zebrafish. A schematic flow of CRISPR/Cas-based directed evolution is shown below (Fig. 5.6). Directed evolutions have great utility for functional genomics interrogating gene functions, evolution and traits for crop improvement.

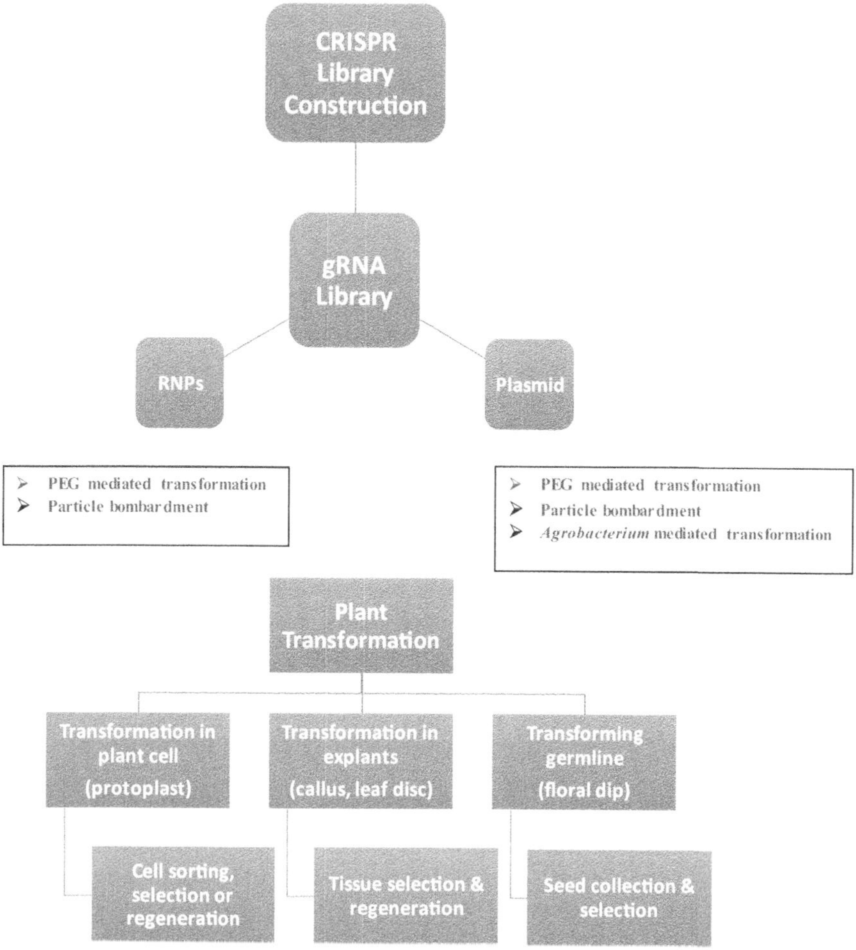

Fig. 5.6 CRISPR-based directed evolution system. CRISPR library constructed through cloning gRNA library into plasmid or RNP assembly. Then this CRISPR library delivered into plants through agrobacterium transformation or PEG mediated transformation. Plant cell or whole plant can be used as evolution platform

5.9 CRISPR/Cas Based Gene Derive and Resistance

The potential of using gene editing technology is extensive and the biotechnology industry has attracted massive interest in this field. But the use of CRISPR for gene drive is more complicated and controversial, involving scientific, social, and political debate (Pennisi 2015). The theory of gene drive (GD) is based on selfish, gene elements or self-replicating DNA pieces that guarantee their own transit into the next generation. As established by the National Academy of Sciences, Engineering and

Medicine, gene drive increases the ability of alleles to pass on to the next generation. Gene drive technologies are a set of molecular tools that promote ecological engineering by raising the likelihood that a gene component will move from generation to generation and spread through a population. The first concept of gene drive dates back to the sixties when a group of scientists proposed to control the Mendelian rule of a 50 percent chance of inheritance, which would free the "Y" chromosome from the inhibitory control of the rest of the genome (Hayirli and Martelli 2019).

In a few years' time, the concept was published. Austin Burt from Imperial College of London used genes, homing endonuclease genes, in order to lead to gene alterations into the population. It was necessary to suggest that transportations could be used to drive the desired gene into the genetic pool of a population. The CRISPR gene drive use the CRISPR revision method by copying modified nuclear sequences to homologous chromosomes, thus ensuring the inheritance of genes edited in all offspring. A total of 19 studies have demonstrated use of the genetic drive theory for yeast, 20 for fruit fly, 21 mosquitoes, since the initial Burt and Esvelt laboratoires proposal. Likewise, the evolutionary stability of homing GD can be enhanced with the use of many gRNAs that are closely distanced from the target area (Esvelt et al. 2014). Prowse (2017) used in silico modeling to test this hypothesis to show that a multitude of gRNA are required to prevent GD from driving and successfully repress populations of vertebrate plagues.

5.10 CRISPR/Cas for Third Generation Disease Resistance

GE methods are used for the control of harmful phytopathogens basically focused on the strategies to prevent their attacks. The genes conferring pathogen resistance providing molecular immune against a wide range of phytopathogens play an important role for GE engineering processes. In specific areas, the threats of plant viruses may cause more damage to plants (Mushtaq et al. 2019). CRISPR multiplexed system may be used to combat with many DNA viruses. Resistance against bacterial diseases and also some fungal diseases has been developed using GE in different types of crops to increase crop production. It also gives an insight into a virus or bacterial cellular pathogenesis process by knocking down or knocking off various genes involved in pathogenesis. GE can also be used to describe an overview of key players detecting plant pathogens, the role of various signal molecules, receptor proteins, and potential target genes. Plant epigenome structure and phytopathogens may also be affected by the genetic tolerance mehcanisms. Applicability of plasmid vectors in the host genome is limited by the random inclusion and off-target effects of DNA (Koo et al. 2015). Furthermore, stable integration in the host genome could be adverse with cytotoxicity by plasmid-mediated RGENs (Woo et al. 2015). Advances in the design of GE approaches for direct transmission of Cas9, RNAs and DNA (donor) has made gene editing successful in the protoplast without loading the genome with sequences of vectors. Because the CRISPR Cpf1 GE model is useful through the CRISPR-Cas9 framework, it can be an excellent tool for plant growth enhancement and sustainable crop resistance. CRISPR-Cas13a (formerly called C2c2) is a powerful and more accurate tool for RNA directed DNA editing derived from prokaryotes which could be recruited into plants to withstand and control gene expression. The use of CRISPR/Cas13 (Aman et al. 2018; Khan et al. 2018) to resist the plant RNA virus indicates that the system may be used to protect future crops.

5.11 Future Aspects

The CRISPR/Cas9 tool offers groundbreaking molecular approach for engineering biological systems and has recently become one of the most powerful and dominant platforms. An important aspect of this technology is the ability to target multiple genes simultaneously for gene addition, deletion, replacement, correction, and mutation (Rodriguez-Moreno et al. 2017). Although, CRISPR/Cas9 system is an excellent powerful tool for genome editing, it is necessary to explore the extent of off-target mutations between different but perfectly matched targets with high cleavage efficiency. However, to combat deadly phytopathogenic diseases, improvement in this technique could be deployed in the future. Moreover, different variants of CRISPR can be used for different purposes such as Cas9 to target dsDNA and Cas13 to target ssDNA and RNA. Modified Cas9 can be used for higher specificity, binding efficiency, and precision. CRISPR system has been becoming a promising reagent (Latorre et al. 2016) and favorite tool for development of 3rd generation disease resistance against fungal, bacterial, and viral diseases in changing climate.

References

Acevedo-Garcia J, Kusch S, Panstruga R (2014) Magical mystery tour: MLO proteins in plant immunity and beyond. New Phytol 204(2):273–281

Ali Z, Abulfaraj A, Idris A, Ali S, Tashkandi M, Mahfouz MM (2015) CRISPR/Cas9-mediated viral interference in plants. Genome Biol 16(1):238

Ali Z, Ali S, Tashkandi M, Zaidi SSA, Mahfouz MM (2016) CRISPR/Cas9-mediated immunity to geminiviruses: differential interference and evasion. Sci Rep 6:26912

Ali MA, Azeem F, Abbas A, Joyia FA, Li H, Dababat AA (2017) Transgenic strategies for enhancement of nematode resistance in plants. Front Plant Sci 8:750

Aman R, Ali Z, Butt H, Mahas A, Aljedaani F, Khan MZ, Ding S, Mahfouz M (2018) RNA virus interference via CRISPR/Cas13a system in plants. Genome Biol 19(1):1

Bai Y, Pavan S, Zheng Z, Zappel NF, Reinstädler A, Lotti C, De Giovanni C, Ricciardi L, Lindhout P, Visser R (2008) Naturally occurring broad-spectrum powdery mildew resistance in a Central American tomato accession is caused by loss of Mlo function. MPMI 21(1):30–39

Baltes NJ, Hummel AW, Konecna E, Cegan R, Bruns AN, Bisaro DM, Voytas DF (2015) Conferring resistance to geminiviruses with the CRISPR–Cas prokaryotic immune system. Nat Plants 1(10):15145

Bastet A, Zafirov D, Giovinazzo N, Guyon-Debast A, Nogué F, Robaglia C, Gallois JL (2019) Mimicking natural polymorphism in eIF 4E by CRISPR-Cas9 base editing is associated with resistance to potyviruses. Plant Biotechnol J 17(9):1736–1750

Boerjan W, Ralph J, Baucher M (2003) Lignin biosynthesis. Annu Rev Plant Biol 54(1):519–546

Boyd CD, O'Toole GA (2012) Second messenger regulation of biofilm formation: breakthroughs in understanding c-di-GMP effector systems. Annu Rev Cell Dev Biol 28:439–462

Brotman Y, Normantovich M, Goldenberg Z, Zvirin Z, Kovalski I, Stovbun N, Doniger T, Bolger AM, Troadec C, Bendahmane A (2013) Dual resistance of melon to Fusarium oxysporum races 0 and 2 and to Papaya ring-spot virus is controlled by a pair of head-to-head-oriented NB-LRR genes of unusual architecture. Mol Plant 6(1):235–238

Butt H, Eid A, Momin AA et al (2019) CRISPR directed evolution of the spliceosome for resistance to splicing inhibitors. Genome Biol 20:73. https://doi.org/10.1186/s13059-019-1680-9

Chandrasekaran J, Brumin M, Wolf D, Leibman D, Klap C, Pearlsman M, Sherman A, Arazi T, Gal-On A (2016) Development of broad virus resistance in non-transgenic cucumber using CRISPR/Cas9 technology. Mol Plant Pathol 17(7):1140–1153

Chaparro-Garcia A, Kamoun S, Nekrasov V (2015) Boosting plant immunity with CRISPR/Cas. Genome Biol 16(1):254

Cheng X, Li F, Cai J, Chen W, Zhao N, Sun Y, Guo Y, Yang X, Wu X (2015) Artificial TALE as a convenient protein platform for engineering broad-spectrum resistance to begomoviruses. Viruses 7(8):4772–4782

Christopoulou M, Wo SR-C, Kozik A, McHale LK, Truco M-J, Wroblewski T, Michelmore RW (2015) Genome-wide architecture of disease resistance genes in lettuce. Genes Genom Genet 5 (12):2655–2669

Consonni C, Humphry ME, Hartmann HA, Livaja M, Durner J, Westphal L, Vogel J, Lipka V, Kemmerling B, Schulze-Lefert P (2006) Conserved requirement for a plant host cell protein in powdery mildew pathogenesis. Nat Genet 38(6):716

Davis EL, Hussey RS, Baum TJ (2004) Getting to the roots of parasitism by nematodes. Trends Parasitol 20(3):134–141

Decraemer W, Hunt D (2006) Structure and classification. In: Perry RN, Moens M (eds) Plant nematol. CABI Publishing, Wallingford, pp 3–32

Ding D, Chen Y, Li H, Xie K (2018) Engineering introns to express RNA guides for Cas9- and Cpf1-mediated multiplex genome editing. Mol Plant 11:542–552. https://doi.org/10.1016/j.molp.2018.02.005

Dracatos PM, Haghdoust R, Singh D, Park RF (2018) Exploring and exploiting the boundaries of host specificity using the cereal rust and mildew models. New Phytol 218(2):453–462

Engqvist MK, Rabe KS (2019) Applications of protein engineering and directed evolution in plant research. Plant Physiol 179(3):907–917

Esvelt KM et al (2014) Concerning RNA-guided gene drives for the alteration of wild populations. elife 3:e03401

Fang Y, Tyler BM (2016) Efficient disruption and replacement of an effector gene in the oomycete Phytophthora sojae using CRISPR/Cas9. Mol Plant Pathol 17(1):127–139

FAO (2017) The future of food and agriculture – trends and challenges. FAO, Rome

Fister AS, Landherr L, Maximova SN, Guiltinan MJ (2018) Transient expression of CRISPR/Cas9 machinery targeting TcNPR3 enhances defense response in *Theobroma cacao*. Front Plant Sci 9:268

Fu Y, Foden JA, Khayter C, Maeder ML, Reyon D, Joung JK, Sander JD (2013) High-frequency off-target mutagenesis induced by CRISPR-Cas nucleases in human cells. Nat Biotechnol 31 (9):822–826

Habash SS, Radakovic ZS, Vankova R, Siddique S, Dobrev P, Gleason C, Grundler FM, Elashry A (2017) Heterodera schachtii Tyrosinase-like protein-a novel nematode effector modulating plant hormone homeostasis. Sci Rep 7(1):1–10

Hayirli TC, Martelli PF (2019) Gene drives as a response to infection and resistance. Infect Drug Resist 12:229

Hibbett DS, Binder M, Wang Z, Goldman Y (2003) Another fossil agaric from Dominican amber. Mycologia 95(4):685–687

Hibbett DS, Binder M, Bischoff JF, Blackwell M, Cannon PF, Eriksson OE, Huhndorf S, James T, Kirk PM, Lücking R (2007) A higher-level phylogenetic classification of the Fungi. Mycol Res 111(5):509–547

Hussey RS (1989) Disease-inducing secretions of plant-parasitic nematodes. Annu Rev Phytopathol 27(1):123–141

Iqbal Z, Sattar MN, Shafiq M (2016) CRISPR/Cas9: a tool to circumscribe cotton leaf curl disease. Front Plant Sci 7:475

Ishibashi K, Masuda K, Naito S, Meshi T, Ishikawa M (2007) An inhibitor of viral RNA replication is encoded by a plant resistance gene. Proc Natl Acad Sci 104(34):13833–13838

Ji X, Zhang H, Zhang Y, Wang Y, Gao C (2015) Establishing a CRISPR–Cas-like immune system conferring DNA virus resistance in plants. Nat Plants 1(10):15144

Jia H, Orbovic V, Jones JB, Wang N (2016) Modification of the PthA4 effector binding elements in Type I Cs LOB 1 promoter using Cas9/sg RNA to produce transgenic Duncan grapefruit alleviating XccΔpthA4: dCs LOB 1.3 infection. Plant Biotechnol J 14(5):1291–1301

Kannan VR, Bastas KK, Devi RS (2015) Scientific and economic impact of plant pathogenic bacteria. In: Sustainable approaches to controlling plant pathogenic bacteria. CRC Press, Boca Raton, p 369

Kerr A (2016) Biological control of crown gall. Australas Plant Pathol 45(1):15–18

Khan Z, Khan SH, Sadia B, Jamil A, Mansoor S (2017) TALE-mediated inhibition of replication of begomoviruses. Int J Agric Biol 20:109–118

Khan MZ, Amin I, Hameed A, Mansoor S (2018) CRISPR–Cas13a: prospects for plant virus resistance. Trends Biotechnol 36(12):1207–1210

Khan Z, Khan SH, Ahmad A, Aslam S, Mubarik MS, Khan S (2019) CRISPR/dCas9-mediated inhibition of replication of begomoviruses. Int J Agric Biol 21:711–718

Khandagale K, Nadaf A (2016) Genome editing for targeted improvement of plants. Plant Biotechnol Rep 10(6):327–343

Kis A, Hamar E, Tholt G, Bán R, Havelda Z (2019) Creating highly efficient resistance against wheat dwarf virus in barley by employing CRISPR/Cas9 system. Plant Biotechnol J. https://doi.org/10.1111/pbi.13077

Koo T, Lee J, Kim J-S (2015) Measuring and reducing off-target activities of programmable nucleases including CRISPR-Cas9. Mol Cells 38(6):475

Kotake Y, Sagane K, Owa T, Mimori-Kiyosue Y, Shimizu H, Uesugi M (2007) Splicing factor SF3b as a target of the antitumor natural product pladienolide. Nat Chem Biol 3:570–575

Kumar V, Jain M (2014) The CRISPR–Cas system for plant genome editing: advances and opportunities. J Exp Bot 66(1):47–57

Langner T, Kamoun S, Belhaj K (2018) CRISPR crops: plant genome editing toward disease resistance. Annu Rev Phytopathol 56:479–512

Latorre A, Latorre A, Somoza Á (2016) Modified RNAs in CRISPR/Cas9: an old trick works again. Angew Chem Int Ed 55(11):3548–3550

Li T, Liu B, Spalding MH, Weeks DP, Yang B (2012) High-efficiency TALEN-based gene editing produces disease-resistant rice. Nat Biotechnol 30(5):390

Ma J, Chen J, Wang M, Ren Y, Wang S, Lei C, Cheng Z (2018a) Corrigendum: disruption of OsSEC3A increases the content of salicylic acid and induces plant defense responses in rice. J Exp Bot 69(7):1817

Ma J, Chen J, Wang M, Ren Y, Wang S, Lei C, Cheng Z (2018b) Disruption of OsSEC3A increases the content of salicylic acid and induces plant defense responses in rice. J Exp Bot 69 (5):1051–1064

Macovei A, Sevilla NR, Cantos C, Jonson GB, Slamet-Loedin I, Čermák T, Voytas DF, Choi IR, Chadha-Mohanty P (2018) Novel alleles of rice eIF4G generated by CRISPR/Cas9-targeted mutagenesis confer resistance to Rice tungro spherical virus. Plant Biotechnol J 16 (11):1918–1927

Malnoy M, Viola R, Jung M-H, Koo O-J, Kim S, Kim J-S, Velasco R, Nagamangala Kanchiswamy C (2016) DNA-free genetically edited grapevine and apple protoplast using CRISPR/Cas9 ribonucleoproteins. Front Plant Sci 7:1904

Mansfield J, Genin S, Magori S, Citovsky V, Sriariyanum M, Ronald P, Dow M, Verdier V, Beer SV, Machado MA (2012) Top 10 plant pathogenic bacteria in molecular plant pathology. Mol Plant Pathol 13(6):614–629

Miklis M, Consonni C, Bhat RA, Lipka V, Schulze-Lefert P, Panstruga R (2007) Barley MLO modulates actin-dependent and actin-independent antifungal defense pathways at the cell periphery. Plant Physiol 144(2):1132–1143

Mohanta TK, Bashir T, Hashem A, Bae H (2017) Genome editing tools in plants. Gene 8(12):399

Mushtaq M, Sakina A, Wani SH, Shikari AB, Tripathi P, Zaid A, Galla A, Abdelrahman M, Sharma M, Singh AK (2019) Harnessing genome editing techniques to engineer disease resistance in plants. Front Plant Sci 10:550

Mysore KS, Ryu C-M (2004) Non host resistance: how much do we know? Trends Plant Sci 9 (2):97–104

Nekrasov V, Wang C, Win J, Lanz C, Weigel D, Kamoun S (2017) Rapid generation of a transgene-free powdery mildew resistant tomato by genome deletion. Sci Rep 7(1):482

Nelson R, Wiesner-Hanks T, Wisser R, Balint-Kurti P (2018) Navigating complexity to breed disease-resistant crops. Nat Rev Genet 19(1):21

Ortigosa A, Gimenez-Ibanez S, Leonhardt N, Solano R (2019) Design of a bacterial speck resistant tomato by CRISPR/Cas9-mediated editing of Sl JAZ 2. Plant Biotechnol J 17(3):665–673

Panstruga R, Schulze-Lefert P (2002) Live and let live: insights into powdery mildew disease and resistance. Mol Plant Pathol 3(6):495–502

Peng A, Chen S, Lei T, Xu L, He Y, Wu L, Yao L, Zou X (2017) Engineering canker-resistant plants through CRISPR/Cas9-targeted editing of the susceptibility gene Cs LOB 1 promoter in citrus. Plant Biotechnol J 15(12):1509–1519

Pennisi E (2015) Science and society. Gene drive turns mosquitoes into malaria fighters. Science 350(6264):1014

Pessina S, Lenzi L, Perazzolli M, Campa M, Dalla Costa L, Urso S, Valè G, Salamini F, Velasco R, Malnoy M (2016) Knockdown of MLO genes reduces susceptibility to powdery mildew in grapevine. Hortic Res 3:16016

Piffanelli P, Ramsay L, Waugh R, Benabdelmouna A, D'Hont A, Hollricher K, Jørgensen JH, Schulze-Lefert P, Panstruga R (2004) A barley cultivation-associated polymorphism conveys resistance to powdery mildew. Nature 430(7002):887

Prowse TAA (2017) Dodging silver bullets: good CRISPR gene-drive design is critical for eradicating exotic vertebrates. Proc R Soc B Biol Sci 284:799

Purvis A, Hector A (2000) Getting the measure of biodiversity. Nature 405(6783):212–219

Pyott DE, Sheehan E, Molnar A (2016) Engineering of CRISPR/Cas9-mediated potyvirus resistance in transgene-free Arabidopsis plants. Mol Plant Pathol 17(8):1276–1288

Qi Y, Zhang Y, Zhang F, Baller JA, Cleland SC, Ryu Y, Starker CG, Voytas DF (2013) Increasing frequencies of site-specific mutagenesis and gene targeting in Arabidopsis by manipulating DNA repair pathways. Genome Res 23(3):547–554

Ran FA, Hsu PD, Wright J, Agarwala V, Scott DA, Zhang F (2013) Genome engineering using the CRISPR-Cas9 system. Nat Protoc 8(11):2281

Rodriguez-Leal D, Lemmon ZH, Man J, Bartlett ME, Lippman ZB (2017) Engineering quantitative trait variation for crop improvement by genome editing. Cell 171:470–480

Rodriguez-Moreno L, Song Y, Thomma BP (2017) Transfer and engineering of immune receptors to improve recognition capacities in crops. Curr Opin Plant Biol 38:42–49

Romay G, Bragard C (2017) Antiviral defenses in plants through genome editing. Front Microbiol 8:47

Schloss PD, Handelsman J (2004) Status of the microbial census. Microbiol Mol Biol Rev 68 (4):686–691

Shanmugam K, Ramalingam S, Venkataraman G, Hariharan G (2019) The CRISPR/Cas9 system for targeted genome engineering in free-living fungi: advances and opportunities for lichenized fungi. Front Microbiol 10:62
Sovova T, Kerins G, Demnerová K, Ovesna J (2016) Genome editing with engineered nucleases in economically important animals and plants: state of the art in the research pipeline. Curr Issues Mol Biol 21(21):41–62
Strange RN, Scott PR (2005) Plant diseases: a threat to global food security. Annu Rev Phytopathol 43:83–116
Teng T, Tsai JH, Puyang X, Seiler M, Peng S, Prajapati S (2017) Splicing modulators act at the branch point adenosine binding pocket defined by the PHF5A–SF3b complex. Nat Commun 8:15522
Thomazella DPDT, Brail Q, Dahlbeck D, Staskawicz B (2016) CRISPR-Cas9 mediated mutagenesis of a DMR6 ortholog in tomato confers broad-spectrum disease resistance. BioRxiv:064824
Tripathi JN, Ntui VO, Ron M, Muiruri SK, Britt A, Tripathi L (2019) CRISPR/Cas9 editing of endogenous banana streak virus in the B genome of Musa spp. overcomes a major challenge in banana breeding. Commun Biol 2(1):46
Uniyal AP, Yadav SK, Kumar V (2019) The CRISPR–Cas9, genome editing approach: a promising tool for drafting defense strategy against begomoviruses including cotton leaf curl viruses. J Plant Biochem Biotechnol 2019:1–12
Verlaan MG, Hutton SF, Ibrahem RM, Kormelink R, Visser RG, Scott JW, Edwards JD, Bai Y (2013) The tomato yellow leaf curl virus resistance genes Ty-1 and Ty-3 are allelic and code for DFDGD-class RNA–dependent RNA polymerases. PLoS Genet 9(3):e1003399
Wang F, Wang C, Liu P, Lei C, Hao W, Gao Y, Liu Y-G, Zhao K (2016) Enhanced rice blast resistance by CRISPR/Cas9-targeted mutagenesis of the ERF transcription factor gene OsERF922. PLoS One 11(4):e0154027
Wang W, Pan Q, He F, Akhunova A, Chao S, Trick H, Akhunov E (2018) Transgenerational CRISPR-Cas9 activity facilitates multiplex gene editing in allopolyploid wheat. CRISPR J 1 (1):65–74
Woo JW, Kim J, Kwon SI, Corvalán C, Cho SW, Kim H, Kim S-G, Kim S-T, Choe S, Kim J-S (2015) DNA-free genome editing in plants with preassembled CRISPR-Cas9 ribonucleoproteins. Nat Biotechnol 33(11):1162
Yabuuchi E, Kosako Y, Yano I, Hotta H, Nishibuchi Y (1995) Microbiol Immunol 39:897–904
Yamaji Y, Maejima K, Komatsu K, Shiraishi T, Okano Y, Himeno M, Sugawara K, Neriya Y, Minato N, Miura C (2012) Lectin-mediated resistance impairs plant virus infection at the cellular level. Plant Cell 24(2):778–793
Zaidi SS, Tashkandi M, Mansoor S, Mahfouz MM (2016) Engineering plant immunity: using CRISPR/Cas9 to generate virus resistance. Front Plant Sci 7:1673
Zeng X, Luo Y, Vu NTQ et al (2020) CRISPR/Cas9-mediated mutation of OsSWEET14 in rice cv. Zhonghua 11 confers resistance to Xanthomonas oryzae pv. oryzae without yield penalty. BMC Plant Biol 20:313. https://doi.org/10.1186/s12870-020-02524-y
Zhang T, Zheng Q, Yi X, An H, Zhao Y, Ma S, Zhou G (2018) Establishing RNA virus resistance in plants by harnessing CRISPR immune system. Plant Biotechnol J 16(8):1415–1423
Zhou J, Peng Z, Long J, Sosso D, Liu B, Eom JS, Huang S, Liu S, Vera Cruz C, Frommer WB (2015) Gene targeting by the TAL effector PthXo2 reveals cryptic resistance gene for bacterial blight of rice. Plant J 82(4):632–643

CRISPR/Cas-Mediated Abiotic Stress Tolerance in Crops

6

Aftab Ahmad, Sidra Ashraf, Nayla Munawar, Amer Jamil, Abdul Ghaffar, and Muhammad Shahbaz

Abstract

To meet the food requirements of increasing human population in light of predicted climate change and extreme abiotic stress, crops with higher yield and enhanced abiotic stress tolerance are urgently needed. Previously, classical breeding and conventional genetic engineering led to limited success. However, these methods have several limitations, including time-consuming processes, ethical factors, environmental factors, and the especially complex nature of abiotic stresses. Over the last decade, CRISPR/Cas has emerged as powerful genome editing tool with immense potential to develop crops with enhanced stress tolerance. CRISPR has been successfully used for targeted genome modifications in model and crop plants. In this chapter, we briefly discuss design, delivery, and footprinting of CRISPR-based genome editing tools. We discuss how CRISPR/Cas-based mutagenesis makes it possible to decipher new genes to increase our understanding of stress-regulatory mechanisms in plants and develop stress-tolerant crops. In addition, we summarize the prospects of CRISPR-mediated deletions, insertions, and multiplex genome editing to develop

A. Ahmad (✉)
Center for Advanced Studies in Agriculture and Food Security (CASAFS), University of Agriculture, Faisalabad, Pakistan
e-mail: aftab.ahmad@uaf.edu.pk

S. Ashraf · A. Jamil
Department of Biochemistry, Faculty of Sciences, University of Agriculture, Faisalabad, Pakistan

N. Munawar
Department of Chemistry, United Arab Emirates University, Al-Ain, UAE

A. Ghaffar
Department of Industrial Biochemistry, Government College University, Faisalabad, Pakistan

M. Shahbaz
Department of Botany, Faculty of Sciences, University of Agriculture, Faisalabad, Pakistan

A. Ahmad et al. (eds.), *CRISPR Crops*,
https://doi.org/10.1007/978-981-15-7142-8_6

climate-resilient crops. We outline CRISPR applications to develop stress-tolerant crops using transcriptional regulation through CRISPRi and CRISPRa. Furthermore, we highlight the potential of CRISPR-based epigenome editing, base editing, and directed evolution to develop stress tolerance in crop plants. We also discuss how CRISPR-directed RNA editing would help in transcriptional control under abiotic stress conditions without creating DNA double-strand breaks. Finally, we discuss challenges and future prospects of CRISPR edited crops for sustainable agriculture.

6.1 Introduction

Crops substantially contribute to support human life on Earth. However, rapidly increasing human populations, decreasing arable land, and climate changes are putting more pressure on agricultural crops and resources to meet food requirements (Ray et al. 2013). By the end of 2050, the global food demand will increase by 100% compared to the beginning of the twenty-first century. Moreover, the impact of abiotic stresses combined with climate change is predicted to decrease crop yield by 20% by 2050 (Vaughan et al. 2018). Abiotic stresses, such as temperature, drought, salinity, and cold, compromise growth, yield, and crop production (Nguyen et al. 2018). Moreover, climate change is already intensifying the effect of abiotic stress in many agricultural countries in Asia and Africa. Hence, the adverse effect of these stresses will seriously comprise sustainable food supply and impact society with increased prices, unemployment, and poverty. Therefore, new, comprehensive technologies including climate-smart crops that can endure multiple abiotic stresses with better yield and nutrition must be developed (Scheben et al. 2017).

Classically, breeding strategies develop crops resilient to abiotic stresses (Chantre Nongpiur et al. 2016). Conventional breeding relies on genetic variability, screening, and germplasm crossing to introduce beneficial cultivar traits (Acquaah 2009). Although stress-tolerant crops have been achieved through conventional breeding, the process is time consuming and costly. Moreover, trait selection via breeding is not precise, because some unwanted traits could be transferred with desired ones. Previously, genetic engineering was the focus of several studies to improve crops against abiotic stresses. Using forward genetic approaches, such as T-DNA insertional mutagenesis and RNA interference (RNAi), numerous genes involved in abiotic stress tolerance were identified in model plants. However, successful gene transfer in crop plants is limited (Hussain et al. 2018). In addition, molecular approaches confirm that a complex network of genes, proteins, and metabolites modulate plant responses to abiotic stress (Nguyen et al. 2018). Thus, transgenic abiotic stress-tolerant crops are not available to farmers, though commercial insect- and herbicide-resistant transgenic crops are widely used. Moreover, transgenic crop development is hampered by issues such as antibiotic resistance, ethical factors, environmental factors, and complex crop genomes. Therefore, scientists continue searching for potential alternate approaches to overcome the limitations of conventional genetic engineering for future food security.

Revolutionary progresses in genome engineering have occurred recently. Newly developed genome editing technologies, including zinc-finger nucleases (ZFNs), transcription activator-like effector nucleases (TALENs), and clustered regularly interspaced short palindromic repeats associated with Cas9 (CRISPR/Cas9), have enabled scientists to efficiently and precisely target genome editing in different species (Wang et al. 2013). Over the last decade, advances in genome editing using CRISPR/Cas9 have dramatically reshaped our ability to precisely modify the cellular genome. CRISPR/Cas system is the new face of site-specific genome editing with applications such as gene editing, gene corrections, and replacement in diverse species. CRISPR/Cas9 has revolutionized genome editing and emerged as a breakthrough technology for desired modifications in diverse organisms, including plants (Sander and Joung 2014). CRISPR has been adapted universally in almost all life science disciplines. Genome editing through CRISPR/Cas system induces sequence-specific, double-stranded breaks that are subsequently repaired by DNA repair pathways, nonhomologous end joining (NHEJ), or homology-directed repair (HDR) (Adli 2018). Generally, NHEJ is error prone, results in small indels, while homologous end-joining inserts the desired sequence using specific donor sequences. Because of its simple design and efficiency, reengineering the CRISPR/Cas system has opened new horizons of genome editing in diverse biological contexts. CRISPR was successfully utilized for targeted genome modifications in model plants such as *Arabidopsis*, rice, tobacco, and crop species such as sorghum (Jiang et al. 2013), wheat (Upadhyay et al. 2013), maize (Liang et al. 2014), soybean (Cai et al. 2015; Jacobs et al. 2015), tomato, and woody plants such as apple and poplar (Osakabe et al. 2016). CRISPR efficiently engineers abiotic stress-tolerant crop plants, such as drought tolerance in maize (Shi et al. 2017), salt tolerance in rice (Hoang et al. 2016), cold responsiveness (Shen et al. 2017), and heavy metals (Basharat et al. 2018). The extraordinary multiplexing capability of this system demonstrates its potential for targeted crop redesign for abiotic stress tolerance (Arora and Narula 2017). Moreover, CRISPR-based forward genetic screens may be a powerful gene discovery platform using gRNAs to induce mutations in single and multiple gene families (Shalem et al. 2015). Tissue-specific and tunable expression through CRISPR/Cas may help extensively in developing abiotic stress-tolerant crops. In addition, by delivering CRISPR reagents such as ribonucleoproteins (RNPs) with NLS signals, CRISPR can address the major issue of transgene-free crops, thus bypassing the regulatory process (Sander and Joung 2014). During the past 5 years, several CRISPR-edited products such as mushrooms, maize, wheat, and rice with improved yield, drought resistance, and disease resistance are reported by researchers, and even some of them are commercially available in the market (Jain 2015). Here we discuss emerging CRISPR-based approaches for developing abiotic stress-tolerant crops.

6.2 CRISPR/Cas Systems: A Splendid Gift from Nature

6.2.1 CRISPR/Cas: An Adaptive Immune System in Bacteria and Archaea

CRISPR/Cas is an adaptive immune mechanism found in bacteria and archaea, which protects against viruses and plasmids and works via cooperation of diverse Cas proteins (Romay and Bragard 2017). It is an RNA-mediated defense mechanism based on acquiring DNA fragments from invading bacteriophages, which are transcribed into CRISPR RNAs (crRNAs) and used as guide sequences to cleave invading RNA or DNA (Garneau et al. 2010). The most widely used CRISPR/Cas system (CRISPR/Cas9) is derived from the *Streptococcus pyogenes* type II CRISPR system (*Sp*Cas9) and consists of two noncoding RNAs, a trans-activating RNA (tracr-RNA) and a precursor RNA (pre-crRNA), and one Cas9 nuclease. The pre-crRNA serves as guide sequences (spacers) for the Cas9 nuclease. The spacers are separated by identical direct repeats and processed into mature crRNA along with tracr-RNA (Jiang et al. 2015). In engineered CRISPR systems, these two RNA genes are redesigned as one gene, which expresses a single guide RNA (sgRNA) that mimics the crRNA–tracr-RNA complex. sgRNA is vital for Cas9 recruitment and binding specificity to the complementary DNA site (Jinek et al. 2012). The Cas9 binding and cleavage properties are determined by complementary pairing between sgRNA, target DNA, and a protospacer adjacent motif (PAM) at the 3′ end. PAM also allows self- and nonself-recognition in bacteria, thereby acting as self-defense against CRISPR/Cas system. However, PAM no longer limits the CRISPR system because recently identified Cas variants have variable PAM requirements. Using PAM and sgRNA, Cas9 can be directed to any sequence in the genome to create double-strand breaks (DSB). Cas9 contains two nuclease domains, HNH and RuvC, which cleave one DNA strand each. DSBs are rapidly repaired by endogenous repair mechanisms through either NHEJ or HDR. NHEJ causes deletion or insertion of a few bases at the DNA cleavage site, and the resultant allele is generally predicted to be inactivated. HDR depends upon the availability of donor template, which is used to generate a new or corrected allele. HDR efficiency is quite low in plants, but suppressing NHEJ may increase its efficiency (Doudna and Charpentier 2014).

CRISPR/Cas is a diverse endonuclease family found in bacteria and archaea, which protects against the invasion of foreign DNA elements (Horvath and Barrangou 2010). Currently, CRISPR is categorized into two major classes and six types, depending upon their components and mechanism (Leenay et al. 2016). Class I contains types I, III, and IV and requires many effector Cas proteins for target DNA/RNA cleavage. Class II comprises types II, V, and VI and depends upon single RNA-guided endonuclease to cleave invading DNA/RNA (Koonin and Makarova 2013). spCas9 remains the most widely adopted CRISPR/Cas system due to its simple PAM requirement (NGG) (where "N" can be any nucleotide base) (Adli 2018). Scientists have reengineered several CRISPR/Cas systems with different PAM requirements. Researchers have been continuously exploring new CRISPR systems, using new effector proteins and PAM requirements. Among the new additions, Cpf1 (Cas12) from *Acidaminococcus* (AsCpf1) and *Lachnospiraceae* (LbCpf1) are quite interesting because they require one sgRNA (no tracrRNA) and produce staggered ends (Fagerlund et al. 2015) instead of blunt ends (Adli 2018). Thus, novel CRISPR/Cas system applications and new proteins indicate that the best in CRISPR is yet to come.

6.2.2 Design and Delivery Tools for Efficient Genome Editing in Plants

CRISPR/Cas allows revolutionary genome editing because of its simplicity, broad scope, and easy design (Weber et al. 2015). The basic CRISPR/Cas mechanism is directed by a 20 bp sequence in the target genome, called the target site, followed by PAM (NGG in case of Cas9). A single sgRNA complementary to the target site recruits the Cas9 to the DNA target site and determines DSB site specificity. Annealing the 20-bp sgRNA to nonspecific or target-similar sites leads to off-target effects (Tsutsui and Higashiyama 2017). Therefore, sgRNA design is a major concern for many researchers, because binding may occur multiple times in the target genome. Some mismatch sites may become acceptable for Cas9 binding and cleavage. With simple design rules and extensive CRISPR/Cas applications in genome editing, researchers have developed several efficient and reliable in silico tools for designing and evaluating gRNAs (Jain 2015). The primary focus of these tools is to design gRNAs for specific CRISPR/Cas applications and evaluate them for on- and off-target specificities. While some tools may recommend potential sgRNAs for specific purposes, the user must select suitable sgRNAs for the intended purpose. Although several tools and databases exist, multiple strategies are used for design, predication, and validation. We suggest a comprehensive database that integrates species, sgRNAs used, target genes, tracing methods used, results, and off-target effects would help researchers compare CRISPR/Cas genome editing results between species. Details about designing sgRNAs are discussed in Chap. 3.

CRISPR/Cas has been used efficiently in multiple organisms including bacteria, yeast, animals, and plants with a wide range of applications. However, limited real-world delivery methods hamper genome editing. Efficient CRISPR reagents delivery, sgRNA specificity, and effective editing outcome recovery are the key steps of a CRISPR experiment. Most CRISPR applications are focused on creating DSBs and repair through NHEJ, thus resulting in gene disruption. However, there is limited progress using CRISPR for gene editing using HDR mechanism and donor templates (Chu et al. 2015). Effective HDR technology provides greater control of DNA repair mechanisms and editing outcomes, such as fusions, corrections, targeted replacement, and single-base substitutions. Based on already-established transformation methods, DNA, RNA, and proteins have been successfully used for CRISPR-mediated genome editing in plants (Fig. 6.1a) (Kouranova et al. 2016; Baltes et al. 2017). Plant delivery platforms include viral, bacterial, and physical methods that are grouped into direct and indirect methods (Fig. 6.1b). Direct methods involve delivery through plasmid-based DNA constructs into target cells using bacteria

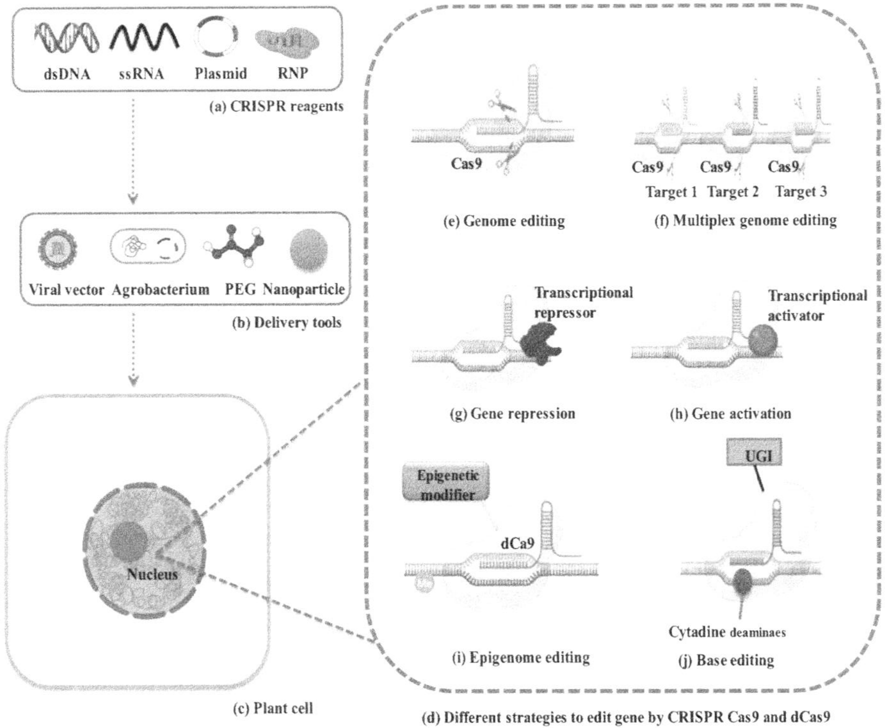

Fig. 6.1 CRISPR/Cas9 applications in plant cells. (**a**) DNA, RNA, plasmid, and protein CRISPR reagents can be used for plant genome editing. (**b**) Physical methods, viral and nonviral vectors are adopted for delivering CRISPR/Cas components. (**c**) CRISPR/Cas reagents are delivered to plant cell nuclei. (**d**) Representation of CRISPR/Cas9 and CRISPR/dCas9 activity to improve abiotic stress tolerance in plant cells: (**e**) Targeted modification or knockout of specific stress-responsive genes via genome editing. (**f**) Multiplex genome editing can simultaneously delete or knockout multiple genes. Multiple gRNAs are introduced

to target and cleave multiple DNA sequences. (**g** and **h**) Transcriptional repressors and activators are bound to dCas9 to regulate stress-responsive genes in plants. (**i**) Epigenetic modifiers are recruited by CRISPR/dCas9 to target site, which allows dynamic regulation of gene expression during abiotic stress in plants. (**j**) Base editing complexes developed by fusion of dCas9 with cytidine deaminase and uracil glycosylase inhibitors (UGI). This complex can produce single basepair mutations

(*Agrobacterium tumefaciens* or *A. rhizogenes*) (Baltes et al. 2017) or viruses (tobacco rattle virus (TRV) and geminiviruses) (Lowder et al. 2016). *Agrobacterium* is the most convenient and promising method for plant genetic engineering because of its ability to transfer its genome into the plant host (Wang et al. 2015c). Some viral vectors deliver CRISPR reagents into germline cells, thus bypassing tissue culture and regeneration requirements (Baltes et al. 2017). Currently for plant scientists, a major focus is developing a CRISPR germline delivery platform. Direct delivery methods can deliver multiple reagent types using physical (biolistic or electroporation), chemical (PEG-based protoplast transformation), or nanoparticles. Importantly, these methods do not involve biological delivery of CRISPR reagents. Detailed delivery tools have been presented in Chap. 3.

6.2.3 Cas9 Delivery as DNA, RNA, or Protein?

CRISPR reagents can be used in the form of DNA, RNA, and proteins for plant genome editing (Fig. 6.1a). Each have their own merits (Baltes et al. 2017). Reagent delivery methods can facilitate the commercial application of edited crops and bypass-regulatory processes. Moreover, the type of CRISPR reagents and the choice of delivery method will define the outcome of genome edited plants and affect the subsequent regulation and public perception of the edited plants. Plasmid vectors are the most common CRISPR reagents. However random genome integration is a major disadvantage and will place the edited plants under same regulations as conventional transgenic plants. Transgenic plants with random integration are not acceptable in the EU and other countries because of concerns related to horizontal gene transfer and potential disruption of endogenous genes. CRIPSR reagents can be transcribed in-vitro and delivered as RNA into the plant cell. To address the transgene issue in the final product, researchers use segregation to remove transgenes or use RNP as CRISPR reagents to produce transgene-free plants. Although RNPs delivery appears straightforward, delivering large RNP amounts to the target site is quite challenging. However, RNP-edited plants can bypass the regulatory process (Svitashev et al. 2016).

6.2.4 Tracking Footprints of CRISPR/Cas in Plant Genome Editing

Depending on sgRNA availability, Cas9 creates DSBs in the DNA that will be repaired by NEHJ or HDR, consequently leaving CRISPR footprints at the cleavage site. These footprints can be observed as base deletions or insertions at the target site (Epinat et al. 2003). Different methods have been used to analyze Cas protein footprints. The objectives of the CRISPR experiment and the Cas protein used will determine foot printing analysis method. Both molecular and phenotypic analyses are required for most experiments. Generally, sequencing is the best method to characterize the CRISPR/Cas results (Yadav et al. 2013). However, sequencing is expensive and time consuming and requires expert skills to validate the results. The most common tracking methods are T7E1A, HRM, PCR sequencing, and restriction site loss. The T7E1A assay is based on PCR amplification using proofreading polymerase. Because NHEJ is the dominant DSB repair method, the PCR products will be mixture of wild type and mutated strands. The PCR products are denatured and reannealed, which generates heteroduplexes that will be digested with endonuclease. The products can then be resolved by electrophoresis. PCR-restriction enzyme (PCR-RE) analysis is another simple and efficient method to detect cleavage site mutations. However, this method is based on the presence of the restriction site in the sgRNA. CRISPR-induced mutations will result in restriction site loss at the cleavage site (Shan et al. 2013). The target region is amplified by PCR and digested with restriction endonucleases. Restriction enzymes will cleave the wild-type amplicon, while the mutated amplicon will be not digested due the loss of the restriction site. Although PCR-RE can detect monoallelic and biallelic mutations, it cannot detect homozygous or heterozygous biallelic mutations. A similar method, RGEN-RFLP, is based on amplicon digestion with purified Cas9, which will digest the wild-type fragment, but not the mutated one (Kim et al. 2014). PCR followed by polyacrylamide gel electrophoresis (PAGE) is another efficient and cost-effective method for tracking CRISPR/Cas footprints. Specific primers that span cleavage site are used in PCR and amplicons are denatured, annealed, and finally resolved on PAGE. Using PAGE, heteroduplexes will run more slowly than homoduplexes (Zhu et al. 2014).

High-resolution melt (HRM) analysis is an economical and efficient method that uses size and base-dependent fluorescence changes during amplicon melting (D'Agostino et al. 2016). For HRM, amplification and melting can be performed in the same tube without purification, quantification, or sequencing (Samarut et al. 2016). However, HRM is not an ideal method to quantify mutations at cleavage sites nor is it as sensitive as T7E1A, RE, and RNA-guided engineered nuclease (RGEN) assays. Drop-off assays are another PCR-based method to quantify indels through droplet digital PCR. It uses two fluorophore-labeled duplex primer probes to quantify mutations. PCR-based methods are quick and efficient, but deep sequencing is a reliable method to further quantify specificity and off-target effects. TA-NGS is very sensitive and allows indel quantification via target amplification with PCR and next-generation amplicon sequencing (Sentmanat et al. 2018). Although TA-NGS is PCR-based, it is more expensive than other methods. For in-vitro genome-wide CRISPR footprint detection, digenome and circle sequencing have been used. For in-vivo mutation detection, high-throughput genome-wide translocation sequencing (HTGTS), integrase-deficient lentiviral vectors (IDLV) sequencing, and breaks labeling and enrichment on streptavidin and sequencing (BLESS) are also recommended. Finally, whole-genome sequencing is recommended for CRISPR/Cas global footprints and off-target analysis. For details about footprinting tools, please see Chap. 3.

6.2.5 The Key Role of Promoters in Abiotic Stress-Tolerant Crop Development

Promoters and cis-regulatory elements are critical for controlling gene expression. Several abiotic stress-regulatory genes have been characterized (Chen and Zhu 2004) and transformed to develop stress tolerance in plants. However, most of these genes are driven by constitutive promoters such as CamV35S, ubiquitin, or actin. Expression under constitutive promoters sometimes compromises yield under stress conditions. For example, when the DRED/CBF3 gene was expressed under a constitutive promoter, the plant gained salt tolerance, but growth under normal conditions was severely affected. However, when expressed under the stress-inducible *RD29A* promoter, the transgenic plants had a normal phenotype. Similarly, ERA expression, which regulates ABA signaling, results in stress tolerance when expressed under a stress-inducible promoter (Cominelli et al. 2013). Therefore, spatiotemporal gene expression is important for stress tolerance and to avoid pleiotropic effects.

Abiotic stress such as drought, salt, and heat mainly target plant leaves and roots. Plants respond to stress conditions with these two organs (Gewin 2010). Roots are mainly involved in water absorption, so manipulation of root-specific expression is vital for stress-tolerant crop development (Ghanem et al. 2011). Root-specific expression of the Na^+ transporter HKT1 in mature *Arabidopsis* root stele confers salt tolerance by increasing Na^+ influx into root cells and reducing Na^+ accumulation in shoots. However, HKT1 expression under the CamV35S promoter results in higher Na^+ accumulation in shoots, consequently reducing growth (Yang et al. 2008). Similarly, guard cell-specific expression is also required for engineering stress tolerance in crops. However, no guard cell-specific promoters are used to develop stress-tolerant crops. Shoot-specific promoters are also required for developing stress-tolerant crops. For example, tissue-specific FTA downregulation using the AtHPA promoter results in drought stress tolerance (Wang et al. 2009). These findings highlight the importance of tissue-specific promoters for engineering genome edited abiotic stress-tolerant plants.

CRISPR was first used in plants in 2013 (Liu et al. 2017). Most initial studies were performed using constitutive promoters. However, CRISPR holds potential for spatiotemporal editing or gene modification in plants, because tissue-specific promoters have been used to drive Cas9 in *Arabidopsis* (Tsutsui and Higashiyama 2017). Tissue-specific promoters in CRISPR/Cas can also avoid pleotropic effects. Spatiotemporal control may also reduce off-target effects in plants (Jain 2015). Therefore, spatiotemporal control will be critical for future CRISPR crops. Spatiotemporal expression in CRISPR can be achieved at the transcriptional and posttranscriptional levels using chemical or light-inducible promoters. Several optogenetic tools based on CRISPR/Cas have been developed in human cells. Distinct promoters, such as the doxycycline promoter, can be used to drive Cas9 in mice; however, promoter leak may occur. In posttranslational induction, a chemically responsive intein was inserted to inhibit nuclease activity. Upon chemical induction, the intein was cleaved and Cas9 becomes active. However, this Cas9 activation is irreversible (de Solis et al. 2016). In CRISPR inactivation (CRISPRi) and CRISPR activation (CRISPRa), the dCas9 is fused with light- or chemical-inducible systems for spatiotemporal gene control. Light-inducible CRISPRa systems are used to activate endogenous genes upon exposure to blue light. This provides reversible CRISPR/Cas control. Furthermore, split-Cas9 systems were developed to control spatiotemporal Cas expression (Polstein and Gersbach 2015). However, in split systems, CRISPR/Cas activity is sometimes compromised. Split-Cas9 systems can overcome Cas9 size limitations. Light-inducible and split-Cas9 system are also helpful to develop abiotic stress-tolerant crops with spatiotemporal gene expression. Moreover, dCas9 coupled with epigenome modifiers can redefine methylation patterns and chromatin status under abiotic stress. However, little attention is given to CRISPR-driven epigenetic modifications in the development of abiotic stress-tolerant crops. Therefore, spatiotemporal systems are a vital part of future CRISPR-edited crops.

6.3 Fundamental CRISPR/Cas9 Tools for Developing Abiotic Stress Tolerance in Crops

When developing agricultural products for abiotic stress through genome editing, three basic prerequisites must be considered: (1) the new trait to be engineered; (2) the modifications (indels, regulations, or rewriting of epigenetic signature) are required to achieve that trait; and (3) how the trait will be introduced into the intended crop. Most abiotic stresses are complex multigenic traits with substantial interaction between signaling, regulatory, and metabolic pathways. The precise mechanisms regulated by abiotic stresses are not yet fully understood (Nakashima et al. 2009; Hirayama and Shinozaki 2010). Moreover, inefficient transformation systems and complex genomes in certain crops such as cotton and wheat are additional constraints for the development of abiotic stress tolerance in crops. Several genes were identified and transformed in crops to improve abiotic stress tolerance (Roy et al. 2011). However, most single-gene knockout (KO) or overexpression (OX) mutants do not produce a phenotype, which makes it difficult to functionally characterize that gene in abiotic stress-regulatory systems. Additionally, functional gene redundancy in complex genomes also limits functional characterization in abiotic stress.

Since the emergence of CRISPR/Cas9 genome editing platforms, agricultural crops with enhanced yield and grain size, improved biotic and abiotic stress tolerance (Abdelrahman et al. 2018), and improved oil contents have been developed. CRISPR/Cas is an efficient and robust system for functional genomics studies, gene interaction studies, and characterization of noncoding DNA and RNA during abiotic stress in plants. More importantly, domestication of new crops, germline mutations, and speed breeding can be achieved with CRISPR/Cas. Multiplex genome editing through CRISPR/Cas can also be successfully performed in model and crop plants (Li et al. 2015; Ma et al. 2015; Zhu et al. 2014), which could be helpful to address the issue of gene redundancy. CRISPR-based approaches such as CRISPR/Cas13,

CRISPRi, and CRISPRa could also help to characterize genes in abiotic stress tolerance, where KO is not required at the DNA level (Osakabe et al. 2016). In the following section, we will discuss different fundamental tools based on CRISPR/Cas, such as CRISPR-based targeted mutagenesis, multiplex genome editing, epigenetic modifications, base editing, gene regulation, directed evolution, and RNA editing for engineering abiotic stress tolerance in crops (Fig. 6.1d).

6.4 CRISPR-Based Targeted Mutagenesis as Functional Genomics Tools for Abiotic Stress Tolerance

Despite the advancement in genome sequencing and genetic engineering, only limited success in developing abiotic stress-tolerant crops has been achieved so far. Trait complexity and lack of functionally characterized genes are the major limitations in crop genetic improvement for abiotic stress (Scheben et al. 2017). For example, less than 1% of maize genes are functionally characterized (Andorf et al. 2015). Several genes related to signaling, transcription factors, secondary metabolites, and membrane-protective roles have been identified. However, several gene aspects including posttranscriptional and posttranslational modifications, protein–DNA interactions, and protein–protein interactions also contribute to trait complexity. Many of these aspects are not fully understood. Therefore, further gene characterization, expression data related to protein expression, tissue-specific regulation patterns, and posttranscriptional modifications are important for the development of successful and economically feasible crops. Several genes associated with abiotic stress tolerance can be modified by random T-DNA mutations (Alonso et al. 2003) or targeting induced local lesions in genomes (TILLING) methods (Till et al. 2007), but these techniques are labor intensive and time consuming. Using genome editing through sequence-specific nucleases, researchers have been focusing on CRISPR-based targeted mutagenesis to decipher the complex network of regulatory genes in abiotic stress. CRISPR is an efficient, simple, and flexible tool to manipulate any genomic sequence, depending upon the availability of PAM and sgRNA (Sander and Joung 2014). Moreover, CRISPR/Cas can be programmed to offer several sequence-specific modifications and thus reveal their functions. CRISPR can also be used to create mutations in single or multiple genes simultaneously and at the genome-wide scale. Indeed, genome-wide screening sgRNA libraries can be constructed to target the entire genome. sgRNA libraries can be mobilized in binary vectors and used to screen plant phenotypes against abiotic stress (Sedeek et al. 2019). The phenotype-specific genes can then be traced by sgRNA cloning and sequencing. CRISPR/Cas offers multiple advantages over chemical and T-DNA mutations, including targeted mutations, gene multiplexing, biallelic mutations, and quick gene discovery. In addition, both loss-of-function and gain-of-function mutations are possible with CRISPR/Cas. Particularly, Cas9 and Cas12a offer an efficient forward genetics platform to generate and screen mutants for abiotic stress tolerance (Zetsche et al. 2015). Another important CRISPR/Cas9 feature is the ability to produce transgene-free edited plants that are resistant to abiotic stress.

Once the mutations are created, CRISPR/Cas9 can be removed through segregation. However, abiotic stress selection can be performed while the Cas9 is still expressed. If permanent mutations result in pleiotropic effects, dCas9 or Cas3 can be used. Additionally, RNA editing using Cas13a and Cas13b can also be used to reduce pleiotropic effects. CRISPR-based genome-wide screening is used in rice (Lu and Zhu 2017; Meng et al. 2017). Thus, CRISPR/Cas is a powerful platform to identify abiotic stress-regulatory network genes and to develop climate-resilient crops.

6.4.1 CRISPR-Based Targeted Gene Knockout for Abiotic Stress Tolerance

Gene knockout abolishes gene expression, thus causing profound phenotypic and cellular effects in plants (Weckwerth et al. 2004). Previously, RNAi was among the best approaches to silence endogenous genes in model systems and crop species (Matzke et al. 2001). Besides the identification of new genes through genome-wide screening, CRISPR can knock out or knock down negative abiotic stress-regulatory genes to engineer stress-tolerant crops (Abdelrahman et al. 2018). CRISPR has a tremendous potential to generate single- or multiple-gene knockout organisms (Fig. 6.1e, f). Targeted gene knockout via CRISPR is well-established in model plants and crops, where it is used to develop abiotic stress-tolerant plants. For example, the *Oryza sativa* E3 ubiquitin ligase SRFP1 (*OsSRF1*) was knocked out using CRISPR to confer salt tolerance (Fang et al. 2015). In addition, the *Arabidopsis* spindly gene (*SPY-3*) could also be a potential CRISPR knockout target to generate drought- and salt-tolerant phenotypes. Several other genes, such as MODD and TaDREB2, also negatively regulate abiotic stress tolerance (Tang et al. 2016a; Kim et al. 2018). These results show that CRISPR-mediated gene knockout could be the ideal approach to generate abiotic stress tolerance in crops.

In addition to regular genes, CRISPR can also be employed to study microRNAs (miRNAs), which posttranscriptionally regulate genes and modulate abiotic stress adaptation in crops. miRNAs are 20–22-nucleotide noncoding RNAs, which regulate their targets through cleavage (Farhat et al. 2019). miRNA expression affects various abiotic stress-responsive genes. Because miRNAs target transcription factors, fine-tuning miRNA expression via CRISPR could be a valid strategy to generate abiotic stress tolerance in plants (Tang and Chu 2017). MIR169a and MIR827a (Zhao et al. 2012) were deleted in *Arabidopsis* by targeting Cas9 to both sides of the miRNA. CRISPR was used to target two other miRNAs, MIR1514 and MIR1509 in soybean (Jacobs et al. 2015). Additionally, the miR156 recognition site in IPA (ideal plant architecture 1) was targeted in rice, leading to an IPA-like phenotype (Li et al. 2016b). Several other miRNAs, including MIR159, MIR164, MIR166, MIR169, MIR393, and MIR397, are differentially expressed under stress in crops. These applications of CRISPR/Cas highlight the role of miRNA in the development of abiotic stress tolerance in crops (Joshi et al. 2017).

6.4.2 Site-Specific Insertions Leading to Abiotic Stress Tolerance

Although gene disruption can improve abiotic stress tolerance, climate resilience can be conferred by precise modifications to several genes. Precise gene modifications involve sequence-specific insertions, nucleotide changes, or nucleotide replacement at predetermined sites using the HDR repair pathway. Targeted insertions can address most limitations of random integration using *Agrobacterium* or biolistic transformation. Random integration through *Agrobacterium* or biolistic approaches poses several challenges, such as (1) integration may induce negative effects; (2) the need to screen transformants with stable transformations; and (3) breeding of two or more linked traits. Site-specific integration allows user-defined integration, which does not disturb agronomic traits and can introduce new traits while maintaining already-existing functional genes (Bortesi et al. 2016). Precise gene alterations can be achieved with donor DNA as a repair template at the DSB site. However, precise donor DNA integration into the plant genome is challenging because NHEJ is the dominant repair system (Bortesi et al. 2016). HDR activation frequency is quite low compared to NHEJ. HDR efficiency can be improved by the simultaneous delivery of Cas9 and donor template. In addition, HDR efficiency can be increased by repressing NHEJ. However, NHEJ can also be used for precise genome editing, as demonstrated by Li et al., who used NHEJ to target two adjacent introns of the *ESPS* gene and provided repair template of the intermediate exon (Li et al. 2016a). The amount of donor template at the target site can also be improved with geminivirus based delivery (Baltes et al. 2014). Cermak et al. achieved efficient gene targeting in tomato using a similar approach (Čermák et al. 2015). In addition, biolistic donor template delivery can be used in maize (Svitashev et al. 2016) in soybean (Li et al. 2015). Similarly, HDR was used for *Argos* promoter swapping to confer drought tolerance in maize, while HDR was used to manipulate the *ALS* gene in rice (Bortesi et al. 2016).

Lack of efficient transformation and regeneration systems is an additional challenge to improve crops with the low-frequency HDR system (Altpeter et al. 2016). Moreover, in-plant transformation has been used to simultaneously transform with both repair template and CRISPR constructs. In this system, gene targeting occurs at the cell cycle stage. Further, this method was used for site-specific insertion in *Arabidopsis* (Schiml et al. 2014).

6.4.3 Multiplex Genome Editing for Abiotic Stress Tolerance

Plant abiotic stress responses are determined by multigene networks, such as signaling, transcription regulation, biochemical, physiological, and metabolic pathways (Vinocur and Altman 2005; Nakashima et al. 2009). Pyramiding multiple genes to develop abiotic stress tolerance is a time consuming, expensive, and laborious process. CRISPR-mediated multiplex genome editing offers an efficient and inexpensive method to target multiple genes simultaneously. CRISPR-based editing, regulation, repair, or replacement of multiple genes is comparable to gene pyramiding. Stress-responsive genes can be manipulated by these methods to develop future abiotic stress-tolerant crops (Scheben et al. 2017).

Transcription factors bind cis-regulatory elements (CREs) in gene promoter regions to regulate transcription. For example, ABRE for ABA (Hossain et al. 2010), G-box for bHLH (Wang et al. 2018), DRE (Li et al. 2005), ERE (Mizoi et al. 2012), and MYC (Ji et al. 2012) are well-characterized CRE, which regulates abiotic stress-responsive genes. CRISPR can play an important role in the characterization of CRE, because multiplex CRISPR can be programmed to bind multiple CRE-regulating abiotic stress-responsive genes. Single, double, and triple mutants can be generated using multiplex CRISPR to identify the most important CRE elements for stress regulation (Ha 2015). Once identified, KO- or HDR-based integration can be used to redesign crops with enhanced tolerance. Furthermore, new CREs may help redesign promoters to improve abiotic stress tolerance in crops.

In complex plant genomes, gene characterization can be difficult due to functional gene redundancy. Sometimes single-gene knockouts in complex genomes of crops do not produce the desired phenotype, so functional characterization of new genes becomes difficult (Jain 2015). Therefore, functional characterization studies need to target multiple genes simultaneously. Multiplex CRISPR/Cas offers a robust platform to overcome functional gene redundancy and characterize new genes for abiotic stress tolerance.

Advances in functional genomics using *Arabidopsis* and rice as model systems helped scientists to characterize a large gene set. However, several genes and their interaction with other genes remains uncharacterized (Radivojac et al. 2013). Similarly, with advancements in next-generation sequencing platforms, whole genome sequences for most crops are available. Thus, the next job is to characterize their genes. Moreover, noncoding genome region functions are not yet clear, though multiple evidences indicate a role in gene regulation (Haudry et al. 2013). Multiplex genome editing represents an important platform to reveal functional gene characterization and interactome analysis in abiotic stress responses. CRISPR-based multiplex genome editing is used to edit multiple genes simultaneously in model and crop species (Fig. 6.1f) (Scheben et al. 2017). In multiplex genome editing, editing efficiency sometimes decreases with increased construct size. For example, Xie et al. used an endogenous t-RNA-based multiplex approach to simultaneously express eight sgRNAs (Xie et al. 2015). A t-RNA base system was also used in maize (Qi et al. 2016). Tang et al. used PolIII to express multiple sgRNAs and Cas9 (Tang et al. 2016b). A multiplex CRISPR system called pYLCRISPR was used to mutate five C-aminobutyric acid (GABA) genes in tomato, which increased GABA 19-fold in leaves. This system highlights the importance of CRISPR/Cas for metabolic engineering to develop abiotic stress-tolerant crops (Li et al. 2018b).

Multiplex CRISPR/Cas systems could also be used to edit gene families and networks to develop abiotic stress tolerance. Moreover, multiplex activation and repression is a powerful platform to regulate multiple genes. For example, four MPK genes were edited simultaneously to develop abiotic stress-tolerant rice (Moustafa et al. 2014). Similarly, in *Physco-patens*, multiplex CRISPR was used to edit nine

KAI2L gene family members and four *AP2/ERF* gene family members, which control ABA signaling to confer abiotic stress tolerance.

6.4.4 Gene Repression and Activation by CRISPRi/a

Transcription factors are often encoded by early stress-responsive genes, functioning as molecular switches and playing crucial roles in stress-responsive gene regulation in plants (Yadav et al. 2013). Molecular studies have characterized several transcription factor types such as bHLH, WRKY, CBF, and DREB, which are vital for controlling plant adaptation to abiotic stress (Gujjar et al. 2014). Functional characterization of key transcription factors and their interacting genes are crucial for dissecting abiotic stress-regulatory mechanisms and consequently developing stress-tolerant crops. Therefore, manipulating transcription factor expression has emerged as a popular approach to develop abiotic stress-tolerant crops (Khan et al. 2018). Apart from creating DSBs and NHEJ- or HDR-mediated repair, Cas9 can be deactivated (dCas9) and fused with activators (VP64, P65, or VPR) or repressors (KRAB). dCas9 retains DNA-binding ability, thus allowing modulation of specific promoters. dCas9 fused with activators or repressors are called CRISPRa and CRISPRi, respectively. CRISPRi/a can also be scaled up to study transcriptional changes at the genome-wide level. CRISPRi allows targeted, reversible gene knockdown (La Russa and Qi 2015). CRISPRi and CRISPRa, respectively, downregulates or upregulates gene expression by enhancing or suppressing RNA polymerase (Yadav et al. 2013). Therefore, CRISPRi and CRISPRa hold great potential to explore and regulate stress-regulatory genes and develop abiotic stress-tolerant crops (Fig. 6.1g, h).

CRISPRi/a has been successfully used in plants to modulate expression by over 1000-fold (La Russa and Qi 2015). Tang et al. used CRISPR to reduce *OsNRAMP5* expression to reduce cadmium accumulation in rice (Tang et al. 2017). In addition, CRISPRa could be used to enhance positive regulators of abiotic stress tolerance, such as major latex-like protein (MLP43), a positive ABA regulator, leading to drought tolerance in *Arabidopsis* (Wang et al. 2015b). Similarly, *MODD* deactivates *OsbZIP46*, negatively regulating ABA signaling and drought stress in rice. Thus, *MODD* suppression by CRISPRi may improve drought tolerance (Tang et al. 2016a). The ABA synthesis gene, *NCED* (9-cis-epoxy carotenoid dioxygenase), positively regulates drought tolerance, so its activation by CRISPRa could also improve drought tolerance (Iuchi et al. 2001).

Although CRISPR offers immense potential to characterize new genes by creating knockouts, sometimes gene knockout may compromise plant growth and development. In addition, sometimes permanent changes are not required at the genomic level. Rather, downregulating genes at the transcriptional level in a tissue-specific manner is often advantageous. For example, gossypol is a defense molecule in cotton leaves, making genome-wide gossypol knockdown in cotton is detrimental for plant. So, CRISPR can reduce gossypol expression specifically in cottonseeds while maintaining expression in leaves. In such cases, CRISPRi offers tissue-specific gene regulation at the transcriptional level. In addition, sometimes gene KO or downregulation does not produce the predicted phenotype or can be lethal. Under such conditions, gain-of-function mutations using CRISPRa offer an alternative approach. In CRISPRa, CRISPR/dCas9 is fused with transcriptional activators that target the promoter and enhancer region of the candidate gene (Dominguez et al. 2016). To activate a gene using a single sgRNA, dCas9 must recruit multiple enhancers. Several systems have been developed using direct fusion of dCas9 with VPR, P65, and Rta, alone or in combination. dCas9 is also fused with protein scaffolds (Suntag) or RNA scaffold SAM20 (systematic activation mediator complex) (La Russa and Qi 2015). For CRISPRi, dCas9 fusion with KRAB (Kruppel associated) domains suppresses gene expression up to 99% (Gilbert et al. 2013). Synthetic activators or repressors can also modulate endogenous gene expression in plants (Piatek et al. 2015). CRISPRa has been used to target *AREB1* to improve drought stress in *Arabidopsis*. To activate expression, the endogenous *AREB1* promoter was targeted with a CRISPRa/dCas9HAT system. The modified plants displayed a confirmed drought-tolerant phenotype. The results showed that *AREB1* and *RD29A* were positively regulated, with increased gene expression compared with wild-type plants (Paixão et al. 2019). These findings highlight that CRISPRi and CRISPRa are powerful tools that allow functional gene characterization and reveal the complex nature of abiotic stress in plants.

6.4.5 CRISPR-Mediated Epigenome Editing for Abiotic Stress in Plants

Epigenetic gene regulation via DNA methylation, histone modification, small RNA production, or transposon element processing play a role in abiotic stress responses in plants (Cominelli et al. 2013). Epigenetic phenomena induce heritable changes in chromatin architecture and consequently alter gene expression without causing permanent DNA alterations. Epigenetic mechanisms are highly dynamic processes, which fine tune gene expression during abiotic stress in plants (Sudan et al. 2018). DNA methylation is catalyzed by cytosine transferase, which adds methyl groups at cytosine residues primarily within CpG islands, CHG, or CHH (H is A, T, or G) dinucleotides in plants. In mammals, methylation occurs at the fifth cytosine residue (5-methyl cytosine) of CpG islands. Two DNA methyl transferases, DNMT3A and DNMT3B, catalyze de novo methylation, while ten-eleven translocation (TET) proteins catalyze endogenous demethylation. 5′-Cytosinemethylation at promoters or distal regulatory elements are associated with transcriptional repression. Methylation provides a platform for protein complex binding. Thus, modifying histone scaffolds consequently regulates gene expression (Adli 2018). Various posttranscriptional modifications such as acetylation, methylation, and phosphorylation induce gene expression, while sumoylation and biotinylation repress gene expression. In addition, small noncoding RNAs also regulate transcription through formation of RNA scaffolds, which modifies chromatin structure and recruitment of histone and methyl transferases (Chinnusamy and Zhu 2009). Activation of these pathways regulates gene expression by altering chromatin conformation under abiotic stress. Promoter and coding region methylation modulate gene expression in organ- and tissue-specific patterns under salt stress (Sudan et al. 2018).

Chromatin marks can be dynamically controlled by specifically recruiting epigenetic writers, erasers, or readers to regulate gene expression (Strahl and Allis 2000). Although CRISPR has been used in plant science for multiple applications, CRISPR/dCas-mediated epigenetic modification applications are relatively few (Fig. 6.1i). Epigenetic modifiers such as eukaryotic DNMT3A (Liu et al. 2016) or prokaryotic MQ3 (Lei et al. 2017) can be used in the CRISPR/dCas system for targeted epigenetic modifications to modulate gene expression under abiotic stress. Previously, antisense-mediated histone acetylation showed increased photosynthesis under drought stress. In addition, transient H3 and H4 acetylation occurs in *Arabidopsis* and tobacco under high salinity (Sokol et al. 2007). These results indicate the importance of epigenetic marks in abiotic stress and highlight the potential of CRISPR/Cas for sequence-specific epigenetic modification to develop abiotic stress tolerance in plants.

6.4.6 Abiotic Stress-Tolerant Crops Using CRISPR-Mediated Gene Stacking

Most commercial cotton and maize in the USA contain multiple insect and herbicide resistance traits. Even crops with five different engineered traits are available, indicating the importance of multiple traits for future crops. Abiotic stresses are complex traits, which require multiple gene stacking to develop broad spectrum resistance against multiple abiotic stresses (Bhat et al. 2016). Pyramiding multiple traits is an expensive, lengthy, and laborious process. The solution to this problem is co-localizing multiple transgenes, so that they behave as a single locus. This can be achieved by constructing single vectors containing multiple transgenes, called stacks. However, vector size limits this process (Biswal et al. 2019). CRISPR offers an efficient solution to stack multiple genes at a single locus using the HDR system (Ainley et al. 2013). For example, the HPT gene can be targeted to chromosome 4 in soybean by providing HPT as donor templates flanked by sequences homologous to the target site (Li et al. 2015). Moreover, CRISPR/Cas may also be helpful in removing poor agronomic traits or replacing them with beneficial traits like abiotic stress tolerance to address the future food security. An alternative to gene stacking is the development of complex trait loci (CTLs) through CRISPR/Cas. CTLs are like gene stacks, where several genes are co-localized and separated by 50 Kb- to 1 Mb-long regions, rather than several hundred bp regions in stacks (Leong et al. 2018). However, CTL development using CRISPR is an inefficient process.

6.4.7 Base Editing Through CRISPR/Cas: A New Tool for Engineering Abiotic Stress Tolerance in Crops

CRISPR/Cas has been reshaping biological research, including plant sciences, with the development of designer crops. Some important agronomic traits are controlled by point mutations or changes in a few nucleotides (Doebley et al. 2006; Li et al. 2017a). CRISPR-based HDR is a potential approach to replace or correct point mutations and improve crops even against abiotic stress. However, low HDR efficiency is a major challenge in plants (Wang et al. 2015a). Genome editing-based, programmable site-specific changes in one or few bases in genome, may introduce elite characters in crop varieties. Previously, the CRISPR system was only suitable for gene knockout or knock-in, but researchers were unable to convert bases. These challenges prompted researchers to seek alternative approaches, allowing precise and stable base editing through CRISPR/Cas.

Base editing via CRISPR/Cas is a novel approach for programmed nucleotide substitution without creating DSBs or requiring donor templates (Fig. 6.1j) (Komor et al. 2016). In base editing, dCas9 or Cas9 nickase is fused with a cytosine or adenine deaminase domain, which allows nucleotide substitution. All four substitutions (C to T, G to A, A to G, and T to C) can be programmed with available base editors. Cytosine base editors (CBEs) can convert C-G to T-A, while adenine base editors (ABEs) change A-T to G-C. RNA base editing is also possible with RBEs, which convert A to I. The first base editor developed was the cytidine deaminase, APOBEC ("apolipoprotein B mRNA editing enzyme, catalytic polypeptide-like") (Komor et al. 2016), which converts C:G base pairs to T:A. APOBEC fused with dCas9 first converts cytosine to uracil. The resulting U:G will be repaired as U:A and consequently repaired as T:A. ABEs deaminate A to I, which binds with C and can be repaired as G. In addition to these DNA base editors, an RNA editing system, ADAR2, deaminates adenosine to inosine, which is read as G by translational machinery. dCas13 protein fused with ADAR2 is used for base editing and is called REPAIR (RNA editing for programmable A to I (G) *replacement*) (Cox et al. 2017). In RNA base editing, no permanent change is encoded in the DNA, so it is safer than DNA base editing for reversible regulation. Base editing can revert a single base without cleaving DNA or creating indels, thus providing a powerful platform to improve abiotic stress tolerance in crops.

Recently, base editing has emerged as an efficient tool for precise genome editing. Successful applications of base editing in plants and animals have highlighted the versatility of base editing in basic and applied research (Hua et al. 2018; Zong et al. 2017). Base editing has been successfully applied in rice, wheat, and tomato (Lu and Zhu 2017; Li et al. 2018a). In addition, base editing has also been used in rice to edit Sub1A-2, which is sensitive to submergence. This indicates that single base edits may change a sensitive allele to tolerant one (Mickelbart et al. 2015).

6.4.8 RNA Editing with Cas13 and Abiotic Stress Tolerance

RNA plays several important roles in cells, including gene regulation, translation, protein translocation, and splicing (Matsui et al. 2019). During abiotic stress in plants, RNA processing, stability, and decay are involved in the regulation of gene expression via posttranscriptional regulation. Modifications of stress-related transcripts are crucial for the adaptation to abiotic stress. In addition, noncoding RNAs are vital for gene regulation, RNA methylation, histone and chromatin modification, and RNA–RNA interaction. RNA splicing is one of the first responses to abiotic stress. Indeed, most splicing factor-related mutations show altered ABA sensitivity and abiotic stress tolerance (Okamoto et al. 2016; Laloum et al. 2018).

Different RNA types are predicted to play important roles in abiotic stresses. For example, miRNA and siRNA regulate abiotic stress in plants. In addition, ncRNA also regulates splicing, epigenetic modifications, and gene repression in cells (Mirlohi and He 2016).

Although ncRNA plays diverse roles in plants, fundamental studies are constrained by few ncRNA mutant lines. Therefore, CRISPR/Cas13 could be an efficient tool to study ncRNA because it directly cleaves ncRNA in-vivo. Cas13 can also target RNA in a tissue-specific manner when complete RNA knockout may be lethal. Therefore, Cas13 offers an efficient platform to reveal ncRNA functions (Wolter and Puchta 2018). In addition, tRNA cleavage is enhanced under stress conditions, thus inhibiting translation (Thompson and Parker 2009). CRISPR/dCas13 could also regulate translation by coupling with translational repressors or enhancers. Moreover, stress granules are produced in plants exposed to abiotic and biotic stress due to inhibited mRNA translation. Stress granule formation protects stress-related mRNAs, because repeat translation could result in energy loss. While CRISPR/Cas type II is mostly used for DNA editing, an RNA editing system involving CRISPR/Cas13 (Type VI) was identified. In addition, another RNA editing system has been identified, called the type III CRISPR/Cas system. Domain organization between Cas13, Cas9, and Cas12 is different, but like CRISPR/Cas9 and Cas12a, Cas13a is also activated by single crRNA. Cas13 cleaves the target at U-rich regions and also non-specific RNA results in collateral cleavage, thus controlling infection (Abudayyeh et al. 2016). CRISPR/Cas13 has been used to confer resistance against RNA viruses (Aman et al. 2018) in *benthamiana*. Moreover, diagnostic tools have also been developed based on CRISPR/Cas13 (Gootenberg et al. 2018). Because tRNA modifications regulate translation, dCas13 is helpful in posttranscriptional tRNA modifications. Alternate splicing plays a critical role in posttranscriptional gene regulation. Errors in the splicing process may cause serious consequences in plants and animals (Scotti and Swanson 2016). mRNA splicing patterns can be controlled by fusing splicing factors with CRISPR/dCas13 to promote different splice isoforms that can help plants tolerate abiotic stress. Moreover, the precise targeting and multiplexing properties of CRISPR/dCas13 could also promote transcript plasticity and remove pathogenic isoforms. Finally, dCas13 fusion with an hnRNP-A1 glycine-rich domain affects alternate splicing and exon exclusion.

6.4.9 Directed Evolution to Develop Abiotic Stress Tolerance

Evolution is the basis of all diverse protein functions and properties in living organisms. However, these functions are just a small fraction of the all biological possibilities. By modifying the fundamental protein sequence, we can go beyond natural evolution and produce entirely new functions and phenotypes useful for agriculture and medicine (Brustad and Arnold 2011). Random or precise mutagenesis can alter the gene sequence, consequently producing modified proteins with the desired functions. Combining directed evolution with CRISPR offers a powerful approach to discover and integrate entirely new and diverse traits in plants. CRISPR-enabled directed evolution may accelerate the identification and development of new traits vital for maximizing crop potential and resilience to climate change. Recently, Butt et al. used CRISPR-based directed evolution in rice to target the splicing factor 3b subunit 1 gene (*OsSF3B1*) (Butt et al. 2019). *SF3B1* encodes a protein that is involved in processing gene transcripts. Targeted *SF3B1* inhibition by herboxidiene (GEX1A) results in transcriptome-wide splicing inhibition. Using CRISPR/Cas-directed evolution, Butt et al. demonstrated new OsSF3B1 variants that are tolerant to GEX1A using a targeted mutagenesis approach. Platforms based on sequence diversification and selective pressure, such as EvlovR, use error-prone PCR. Nevertheless, they represent powerful methods for future directed evolution in plants. Multiplexing with EvlovR may help to engineer multiple combined traits simultaneously in crops. Therefore, CRISPR-based directed evolution will be a powerful tool for evolving and engineering new traits in crops, such as herbicide tolerance, insect resistance, improved photosynthesis, enhanced yield, and enhanced tolerance to biotic and abiotic stresses (Butt et al. 2019). Directed evolution was also used to isolate enhanced salt-tolerant variants of the DS-26-16 gene from *Dunaliella salina*. Random mutagenesis through deoxynosine triphosphate-mediated error-prone PCR was used to produce mutant libraries (Guo et al. 2016). However, this was not a CRISPR-based method. These few examples show the tremendous potential of CRISPR-based directed evolution in crop improvement for future food security.

6.4.10 Highlights of Practical CRISPR Applications for Abiotic Stress Tolerance

In recent years, CRISPR/Cas has emerged as a breakthrough technology with a continuously expanding toolkit, which provides new resources for genetic crop improvement. Several research papers and reviews highlight the importance of CRISPR/Cas to engineer abiotic stress tolerance in crops to address the challenges of food security. Detailed CRISPR protocols for model and crop plants have been published (Belhaj et al. 2013; Shan et al. 2014), and CRISPR applications are expanding to characterize de novo genes and engineering already-identified genes for crop improvement under climate change. Due to its simple design, cost effectiveness, and modular nature, CRISPR is routinely used for genome editing in plants, which offers new solutions for transgene-free crop development. Although CRISPR has become a common method for genetic improvement in plants, its potential to improve abiotic stress tolerance in crops is not yet fully explored.

Abiotic stress such as salt, drought, heat, and heavy metals negatively impacts crop yield, growth, and productivity. It has been predicted that more than 70% of agricultural production is reduced by the direct impact of abiotic stress. Within plants, abiotic stress is a quantitative trait controlled by hundreds of genes. During abiotic stress, several physiological, biochemical, morphological, and molecular changes occur (Collins et al. 2008). Generally, plants respond to abiotic stress via signaling, regulatory, and metabolic pathways. Stress-induced gene expression can be classified into three general categories: (1) enzymes or structural functions, (2) transcription factors to regulate different genes, and (3) unknown functions (Bhatnagar-Mathur et al. 2008). Using transgenic approaches, genes regulating diverse pathways such as osmo-protectants (Majumder et al. 2009), late embryogenesis abundant (LEA) (Banerjee and Roychoudhury 2016), chaperones (Hu et al. 2009), ROS (reactive oxygen species) scavenging (Gururani et al. 2013), detoxifying enzymes (Yousuf et al. 2012), ion transporters (Brini and Masmoudi 2012), protein kinases (Sinha et al. 2011), Ca^{2+} regulators (Franz et al. 2011), aquaporins (Ayadi et al. 2011), transcription factors (Agarwal et al. 2006), and ncRNAs (Contreras-Cubas et al. 2012) have been engineered to develop climate resilience in model and crop plants.

Several reports describe the successful use of CRISPR-mediated genome editing in model plants and crops for abiotic stress tolerance (Osakabe et al. 2016). Among abiotic stresses, salinity is a major stress type that seriously threatens crop yield (Liang et al. 2018). Removing excess salt is difficult, and areas with high salt levels in soil are increasing due to climate change and tsunamis. CRISPR has been used to mutate the SUMO protease gene *OsOTS1* in rice, which plays an important role in salt and drought stress. *OsOTS1* mutants show hypersensitive phenotypes and reduced biomass under salt stress, indicating its importance in salt-regulatory mechanism in rice. In addition, *OST2* mutation through CRISPR/Cas also showed altered stomata phenotypes (Sadanandom et al. 2019). Like salt stress, drought stress is also complex and challenging. CRISPR offers an efficient platform for developing drought-tolerant crops. Mitogen-activated kinase 3 (*SlMAPK3*) was mutated with CRISPR in tomato, resulting in severe drought-like symptoms in edited plants (Wang et al. 2017b). Moreover, a recent study used CRISPR to target ABA-activated protein kinase 2 in rice and showed enhanced sensitivity to drought stress (Lou et al. 2017). Another study edited maize with CRISPR/Cas to replace the auxin-regulated gene involved in organ size (*ARGOS8*). CRISPR variants showed increased *ARGOS8* expression and enhanced yield under drought stress (Shi et al. 2017). Furthermore, cold and heat stress-regulatory genes were edited with CRISPR/Cas to engineer cold and heat stress tolerance in plants. Tomato *slagamous-like 6* (*SlAGL6*) gene was disrupted with CRISPR, and mutated plants produced parthenocarpic fruit under heat stress (Klap et al. 2017). In addition, mutated Ca^{2+}-dependent phospholipid-binding protein (*OsANN3*) in Japonica rice showed enhanced electrical conductivity and hypersensitivity to cold stress, indicating the importance of *OsANN3* in cold stress (Shen et al. 2017). Moreover, *CBF1*, *CBF2*, and *CBF3* were also targeted with CRISPR, and single, double, and triple mutants were confirmed (Cho et al. 2017).

Herbicide-tolerant transgenic plants have revolutionized agriculture, allowing herbicides to kill weeds without damaging transgenic plants (Lombardo et al. 2016). These plants have greatly reduced the deleterious effects of herbicides and decreased weed control costs. 5-Enolpyruvylshikimate-3-phosphate (*EPSPS*) is a target of many herbicides. Thus, mutations in this gene allow edited plants to grow under herbicide pressure. CRISPR-mediated precise mutations in *EPSPS* and *ALS* genes result in the growth of edited plants in the presence of herbicides (Han and Kim 2019). Finally, CRISPR has also been used to develop heavy metal tolerance in plants. CRISPRi and CRISPRa have also been used to regulate heavy metal transporters in plants (Basharat et al. 2018) (Fig. 6.1). A complete list of CRISPR-targeted genes and potential genes that could be engineered to develop abiotic stress tolerance in plants are found in Table 6.1.

6.5 Prospects and Challenges

In recent years, advancements in next-generation sequencing have provided insight into the location and structure of functional genes in plant genomes. The next challenge is to functionally characterize these genes in vivo. At the same time, we have witnessed a rapid advancement in genome engineering tools, especially CRISPR/Cas, that have provided scientists tremendous control over genetic elements. CRISPR has increased our understanding of functional genomics and entirely changed the way we create engineered crops. One of the most anticipated CRISPR applications is to repair or replace inferior traits in crops using donor templates and HDR repair mechanisms. Moreover, creating mutations to produce knockout or knock-in mutants through CRISPR/Cas9 is vital for functional genomics studies and abiotic stress tolerance. CRISPR can also be used to insert a tag or restriction site at the region of interest. Further, CRISPR/Cas can be repurposed to modulate gene expression by deactivating the nuclease domain and creating activator or repressor domain fusion proteins. One interesting future CRISPR/Cas application is directed evolution to rewrite the genetic code and explore genetic properties beyond natural evolution. Epigenetic control and studying *cis*-regulatory elements through CRISPR will further reshape crop design in the future for biotic and abiotic stress tolerance. As the first wave of CRISPR edited crops are approaching markets, we will get a glimpse of CRISPR-based genetic crop improvement to improve world food security. It is clear that CRISPR will be involved in every aspect of next-generation crops to achieve food security. However, to use the full potential of CRISPR/Cas, certain challenges must be overcome. CRISPR is a revolutionary tool, which has transformed every aspect of biological sciences, including plant sciences. It has immense potential to develop new crop varieties to address the future demands of an increasing global population. However, with this rapid pace, genome editing has limitations to achieve the full spectrum of crop improvement. NHEJ repair is the preferred repair pathway, and HDR efficiency is quite low in plants. Thus, HDR

Table 6.1 Genes targeted by CRISPR/Cas to improve the stress tolerance or potential target genes for CRISPR/Cas based genome editing to improve strsss tolerance in plants. Target gene, model plant or crop, CRISPR/Cas tool used and intended modifications have been listed

Crop	Target	Target sequence	CRISPR/Cas tool	Editing type	Trait modification	Reference
Arabidopsis thaliana	Gene	*UGT79-B2 (UDP glycosyltransferase 79B2 gene)* and *UGT79-B3 (UDP glycosyltransferase 79B3)*	CRISPR/Cas9	Mutation	Oxidative stress, drought, cold, and salt tolerance	Li et al. (2017b)
	Protein pump	*AVP1 (Arabidopsis thaliana vacuolar H+-pyrophosphatase)*	CRISPR/dCas9	AVP1 overexpression	Drought tolerance	Park et al. (2017)
	Transcriptional factor	*AREB1 (Abscisic acid-responsive element binding protein1)*	CRISPRa	Regulation	Drought tolerance	Paixão et al. (2019)
	Gene	*ALS (Acetolactate synthase gene)*	CRISPR/Cas9	Base editing	Herbicide tolerance	Chen et al. (2017)
	Gene	*mir169a (microRNA 169a)*	CRISPR/Cas9	Knockout	Drought tolerance	Zhao et al. (2016)
Oryza sativa	Gene	*OsMODD (Mediator of OsbZIP46 deactivation and degradation)*	CRISPR/Cas9 and CRISPRi	Knockout	Drought tolerance	Tang et al. (2016a)
	Gene	*OsNAC14 (rice NAC transcription factor)*	CRISPR/dCas	*OsNAC14* overexpression	Drought tolerance	Shim et al. (2018)
	Gene	*OsSRFP1 (Oryza sativa Stress-related RING Finger Protein 1)*	CRISPR/Cas9	Knockdown	Salt tolerance	Fang et al. (2015)
	Gene	*SAPK2 (Serine/threonine-protein kinase 2)*	CRISPR/Cas9	Mutation	Drought and salinity sensitivity	Lou et al. (2017)
	Gene	*OsNCED (Rice 9-cis-epoxycarotenoid dioxygenase)*	CRISPR/Cas9	Mutation	Absence of *OsNCED* makes plant sensitive to drought and salt stress	Huang et al. (2018)
	Promoter	*OsRAV2 (Rice RAV 2)*	CRISPR/Cas9	Mutation in promoter	Functional inactivation of *OsRAV2* promoter reduces salt tolerance	Duan et al. (2016)

	Gene	Rice Phosphate transporter 4 (*OsPT4*)	CRISPR/Cas9	Mutation	Reduces arsenic uptake from soil	Ye et al. (2017)
	Gene	*OsNRAMP5 (Natural resistance-associated macrophage protein 5)*	CRISPR/Cas9	Knockout	Absence of *OsNRAMP5* reduces cadmium uptake from soil	Tang et al. (2017)
	Gene	*OsARM1 (ARSENITE-RESPONSIVE MYB1)*	CRISPR/Cas9	Knockout	Absence of *OsARM1* reduces arsenic uptake from soil	Wang et al. (2017a)
	Gene	*OsHAK1 (Potassium transporter 1)*	CRISPR/Cas9	Mutation	Reduces cesium uptake from soil	Nieves-Cordones et al. (2017)
	Gene	*OsLCT1 (Low affinity cation transporter 1)*	CRISPR/Cas9	Mutation	Heavy metal tolerance	Songmei et al. (2019)
	Gene	*OsRAMP5 (Natural resistance-associated macrophage protein 5)*	CRISPR/Cas9	Mutation	Heavy metal tolerance	Songmei et al. (2019)
	Gene	*OsANN3 (rice annexin gene)*	CRISPR/Cas9	Knockout	Absence of *OsANN3* activity makes plant sensitive to cold	Shen et al. (2017)
	Gene	*OsCOLD1 (CHILLING-TOLERANCE DIVERGENCE 1)*	CRISPR/Cas9-HDR	A>T SNP alteration in exon 4	Cold tolerance	Ma et al. (2015)
	Gene	*OsRR22 (Rice transcription factor gene)*	CRISPR/Cas9	Mutation	Salt tolerance	Zhang et al. (2019)
	Gene	*ALS (Acetolactate synthase gene)*	CRISPR/Cas9	Mutation	Herbicide tolerance	Sun et al. (2016)
	Gene	*EPSPS (5-enolpyruvylshikimate-3-phosphate synthase)*	CRISPR/Cas9	Mutation	Herbicide tolerance	Li et al. (2016a)

(continued)

Table 6.1 (continued)

Crop	Target	Target sequence	CRISPR/Cas tool	Editing type	Trait modification	Reference
	Gene	*OsNRAMP5 (Natural resistance-associated macrophage protein 5)*	CRISPR/Cas9	Knockout	Reduced cadmium accumulation	Yang et al. (2019)
Solanum lycopersicum	Gene	*SlMAPK3 (Mitogen-activated protein kinase 3)*	CRISPR/Cas	Mutation	Absence of *SlMAPK3* activity makes plant susceptible to drought stress	Wang et al. (2017b)
	Gene	*SlNPR1 (Nonexpressor of pathogenesis-related1)*	CRISPR/Cas9	Mutation	Drought tolerance	Li et al. (2019)
Zea mays	Promoter	*ARGOS8 (Zea mays auxin regulated gene involved in organ size)*	CRISPR/Cas	Promoter Replacement	High *ARGOS8* expression increases drought tolerance	Shi et al. (2017)
Glycine max	Gene	*ALS (Acetolactate synthase gene)*	CRISPR/Cas9	Mutation	Improved herbicide tolerance	Chilcoat et al. (2017)
Linum usitatissimum	Gene	*EPSPS (5-enolpyruvylshikimate-3-phosphate synthase)*	CRISPR/Cas9	Mutation	Glyphosate tolerance	Sauer et al. (2016)

efficiency must be improved for targeted gene insertion. In addition, HDR involves targeted mutagenesis for crop improvement without leaving any traces of inserted transgenes. Off-target effects and specificity are also limitations that must be improved. Efficient transformation and targeted delivery of genome editing reagents are additional challenges. Finally, tissue specificity and spatiotemporal gene expression are very important for crop improvement.

References

Abdelrahman M, Al-Sadi AM, Pour-Aboughadareh A, Burritt DJ, Tran L-SP (2018) Genome editing using CRISPR/Cas9–targeted mutagenesis: an opportunity for yield improvements of crop plants grown under environmental stresses. Plant Physiol Biochem 131:31–36

Abudayyeh OO, Gootenberg JS, Konermann S, Joung J, Slaymaker IM, Cox DB, Shmakov S, Makarova KS, Semenova E, Minakhin L (2016) C2c2 is a single-component programmable RNA-guided RNA-targeting CRISPR effector. Science 353(6299):aaf5573

Acquaah G (2009) Principles of plant genetics and breeding. Wiley, Hoboken

Adli M (2018) The CRISPR tool kit for genome editing and beyond. Nat Commun 9(1):1–13

Agarwal PK, Agarwal P, Reddy M, Sopory SK (2006) Role of DREB transcription factors in abiotic and biotic stress tolerance in plants. Plant Cell Rep 25(12):1263–1274

Ainley WM, Sastry-Dent L, Welter ME, Murray MG, Zeitler B, Amora R, Corbin DR, Miles RR, Arnold NL, Strange TL (2013) Trait stacking via targeted genome editing. Plant Biotechnol J 11 (9):1126–1134

Alonso JM, Stepanova AN, Leisse TJ, Kim CJ, Chen H, Shinn P, Stevenson DK, Zimmerman J, Barajas P, Cheuk R (2003) Genome-wide insertional mutagenesis of Arabidopsis thaliana. Science 301(5633):653–657

Altpeter F, Springer NM, Bartley LE, Blechl AE, Brutnell TP, Citovsky V, Conrad LJ, Gelvin SB, Jackson DP, Kausch AP (2016) Advancing crop transformation in the era of genome editing. Plant Cell 28(7):1510–1520

Aman R, Ali Z, Butt H, Mahas A, Aljedaani F, Khan MZ, Ding S, Mahfouz M (2018) RNA virus interference via CRISPR/Cas13a system in plants. Genome Biol 19(1):1

Andorf CM, Cannon EK, Portwood JL, Gardiner JM, Harper LC, Schaeffer ML, Braun BL, Campbell DA, Vinnakota AG, Sribalusu VV (2015) MaizeGDB update: new tools, data and interface for the maize model organism database. Nucleic Acids Res 44(D1):D1195–D1201

Arora L, Narula A (2017) Gene editing and crop improvement using CRISPR-Cas9 system. Front Plant Sci 8:1932

Ayadi M, Cavez D, Miled N, Chaumont F, Masmoudi K (2011) Identification and characterization of two plasma membrane aquaporins in durum wheat (Triticum turgidum L. subsp. durum) and their role in abiotic stress tolerance. Plant Physiol Biochem 49(9):1029–1039

Baltes NJ, Gil-Humanes J, Cermak T, Atkins PA, Voytas DF (2014) DNA replicons for plant genome engineering. Plant Cell 26(1):151–163

Baltes NJ, Gil-Humanes J, Voytas DF (2017) Genome engineering and agriculture: opportunities and challenges. In: Progress in molecular biology and translational science, vol 149. Elsevier, London, pp 1–26

Banerjee A, Roychoudhury A (2016) Group II late embryogenesis abundant (LEA) proteins: structural and functional aspects in plant abiotic stress. Plant Growth Regul 79(1):1–17

Basharat Z, Novo L, Yasmin A (2018) Genome editing weds CRISPR: what is in it for phytoremediation? Plants 7(3):51

Belhaj K, Chaparro-Garcia A, Kamoun S, Nekrasov V (2013) Plant genome editing made easy: targeted mutagenesis in model and crop plants using the CRISPR/Cas system. Plant Methods 9 (1):39

Bhat JA, Ali S, Salgotra RK, Mir ZA, Dutta S, Jadon V, Tyagi A, Mushtaq M, Jain N, Singh PK (2016) Genomic selection in the era of next generation sequencing for complex traits in plant breeding. Front Genet 7:221

Bhatnagar-Mathur P, Vadez V, Sharma KK (2008) Transgenic approaches for abiotic stress tolerance in plants: retrospect and prospects. Plant Cell Rep 27(3):411–424

Biswal AK, Mangrauthia SK, Reddy MR, Yugandhar P (2019) CRISPR mediated genome engineering to develop climate smart rice: challenges and opportunities. In: Semin cell dev biol. Elsevier, London

Bortesi L, Zhu C, Zischewski J, Perez L, Bassié L, Nadi R, Forni G, Lade SB, Soto E, Jin X (2016) Patterns of CRISPR/Cas9 activity in plants, animals and microbes. Plant Biotechnol J 14 (12):2203–2216

Brini F, Masmoudi K (2012) Ion transporters and abiotic stress tolerance in plants. ISRN Mol Biol 2012:927436

Brustad EM, Arnold FH (2011) Optimizing non-natural protein function with directed evolution. Curr Opin Chem Biol 15(2):201–210

Butt H, Eid A, Momin AA, Bazin J, Crespi M, Arold ST, Mahfouz MM (2019) CRISPR directed evolution of the spliceosome for resistance to splicing inhibitors. Genome Biol 20(1):73

Cai Y, Chen L, Liu X, Sun S, Wu C, Jiang B, Han T, Hou W (2015) CRISPR/Cas9-mediated genome editing in soybean hairy roots. PLoS One 10(8):e0136064

Čermák T, Baltes NJ, Čegan R, Zhang Y, Voytas DF (2015) High-frequency, precise modification of the tomato genome. Genome Biol 16(1):232

Chantre Nongpiur R, Lata Singla-Pareek S, Pareek A (2016) Genomics approaches for improving salinity stress tolerance in crop plants. Curr Genomics 17(4):343–357

Chen WJ, Zhu T (2004) Networks of transcription factors with roles in environmental stress response. Trends Plant Sci 9(12):591–596

Chen Y, Wang Z, Ni H, Xu Y, Chen Q, Jiang L (2017) CRISPR/Cas9-mediated base-editing system efficiently generates gain-of-function mutations in Arabidopsis. Sci China Life Sci 60 (5):520–523

Chilcoat D, Liu Z-B, Sander J (2017) Use of CRISPR/Cas9 for crop improvement in maize and soybean. In: Progress in molecular biology and translational science, vol 149. Elsevier, London, pp 27–46

Chinnusamy V, Zhu J-K (2009) Epigenetic regulation of stress responses in plants. Curr Opin Plant Biol 12(2):133–139

Cho S, Yu S-i, Park J, Mao Y, Zhu J-K, Yun D-J, Lee B-h (2017) Accession-dependent CBF gene deletion by CRISPR/Cas system in Arabidopsis. Front Plant Sci 8:1910

Chu VT, Weber T, Wefers B, Wurst W, Sander S, Rajewsky K, Kühn R (2015) Increasing the efficiency of homology-directed repair for CRISPR-Cas9-induced precise gene editing in mammalian cells. Nat Biotechnol 33(5):543

Collins NC, Tardieu F, Tuberosa R (2008) Quantitative trait loci and crop performance under abiotic stress: where do we stand? Plant Physiol 147(2):469–486

Cominelli E, Conti L, Tonelli C, Galbiati M (2013) Challenges and perspectives to improve crop drought and salinity tolerance. New Biotechnol 30(4):355–361

Contreras-Cubas C, Palomar M, Arteaga-Vázquez M, Reyes JL, Covarrubias AA (2012) Non-coding RNAs in the plant response to abiotic stress. Planta 236(4):943–958

Cox DB, Gootenberg JS, Abudayyeh OO, Franklin B, Kellner MJ, Joung J, Zhang F (2017) RNA editing with CRISPR-Cas13. Science 358(6366):1019–1027

D'Agostino Y, Locascio A, Ristoratore F, Sordino P, Spagnuolo A, Borra M, D'Aniello S (2016) A rapid and cheap methodology for CRISPR/Cas9 zebrafish mutant screening. Mol Biotechnol 58 (1):73–78

de Solis CA, Ho A, Holehonnur R, Ploski JE (2016) The development of a viral mediated CRISPR/ Cas9 system with doxycycline dependent gRNA expression for inducible in vitro and in vivo genome editing. Front Mol Neurosci 9:70

Doebley JF, Gaut BS, Smith BD (2006) The molecular genetics of crop domestication. Cell 127 (7):1309–1321

Dominguez AA, Lim WA, Qi LS (2016) Beyond editing: repurposing CRISPR–Cas9 for precision genome regulation and interrogation. Nat Rev Mol Cell Biol 17(1):5

Doudna JA, Charpentier E (2014) The new frontier of genome engineering with CRISPR-Cas9. Science 346(6213):1258096

Duan Y-B, Li J, Qin R-Y, Xu R-F, Li H, Yang Y-C, Ma H, Li L, Wei P-C, Yang J-B (2016) Identification of a regulatory element responsible for salt induction of rice OsRAV2 through ex situ and in situ promoter analysis. Plant Mol Biol 90(1-2):49–62

Epinat JC, Arnould S, Chames P, Rochaix P, Desfontaines D, Puzin C, Patin A, Zanghellini A, Pâques F, Lacroix E (2003) A novel engineered meganuclease induces homologous recombination in yeast and mammalian cells. Nucleic Acids Res 31(11):2952–2962

Fagerlund RD, Staals RH, Fineran PC (2015) The Cpf1 CRISPR-Cas protein expands genome-editing tools. Genome Biol 16(1):251

Fang H, Meng Q, Xu J, Tang H, Tang S, Zhang H, Huang J (2015) Knock-down of stress inducible OsSRFP1 encoding an E3 ubiquitin ligase with transcriptional activation activity confers abiotic stress tolerance through enhancing antioxidant protection in rice. Plant Mol Biol 87 (4-5):441–458

Farhat S, Jain N, Singh N, Sreevathsa R, Das PK, Rai R, Yadav S, Kumar P, Sarkar A, Jain A (2019) CRISPR-Cas9 directed genome engineering for enhancing salt stress tolerance in rice. In: Semin cell dev bio. Elsevier, London

Franz S, Ehlert B, Liese A, Kurth J, Cazalé A-C, Romeis T (2011) Calcium-dependent protein kinase CPK21 functions in abiotic stress response in Arabidopsis thaliana. Mol Plant 4(1):83–96

Garneau JE, Dupuis M-È, Villion M, Romero DA, Barrangou R, Boyaval P, Fremaux C, Horvath P, Magadán AH, Moineau S (2010) The CRISPR/Cas bacterial immune system cleaves bacteriophage and plasmid DNA. Nature 468(7320):67

Gewin V (2010) Food: an underground revolution. Nat News 466(7306):552–553

Ghanem ME, Hichri I, Smigocki AC, Albacete A, Fauconnier M-L, Diatloff E, Martinez-Andujar C, Lutts S, Dodd IC, Pérez-Alfocea F (2011) Root-targeted biotechnology to mediate hormonal signalling and improve crop stress tolerance. Plant Cell Rep 30(5):807–823

Gilbert LA, Larson MH, Morsut L, Liu Z, Brar GA, Torres SE, Stern-Ginossar N, Brandman O, Whitehead EH, Doudna JA (2013) CRISPR-mediated modular RNA-guided regulation of transcription in eukaryotes. Cell 154(2):442–451

Gootenberg JS, Abudayyeh OO, Kellner MJ, Joung J, Collins JJ, Zhang F (2018) Multiplexed and portable nucleic acid detection platform with Cas13, Cas12a, and Csm6. Science 360 (6387):439–444

Gujjar RS, Akhtar M, Singh M (2014) Transcription factors in abiotic stress tolerance. Indian J Plant Physiol 19(4):306–316

Guo Y, Dong Y, Hong X, Pang X, Chen D, Chen X (2016) Directed evolution of Dunaliella salina Ds-26-16 and salt-tolerant response in Escherichia coli. Int J Mol Sci 17(11):1813

Gururani MA, Upadhyaya CP, Baskar V, Venkatesh J, Nookaraju A, Park SW (2013) Plant growth-promoting rhizobacteria enhance abiotic stress tolerance in Solanum tuberosum through inducing changes in the expression of ROS-scavenging enzymes and improved photosynthetic performance. J Plant Growth Regul 32(2):245–258

Ha CD (2015) The role of cis-regulatory elements in abiotic-stress responses in plants. Tap Chi Sinh Hoc 37(3):370–383

Han Y-J, Kim J-I (2019) Application of CRISPR/Cas9-mediated gene editing for the development of herbicide-resistant plants. Plant Biotechnol Rep 2019:1–11

Haudry A, Platts AE, Vello E, Hoen DR, Leclercq M, Williamson RJ, Forczek E, Joly-Lopez Z, Steffen JG, Hazzouri KM (2013) An atlas of over 90,000 conserved noncoding sequences provides insight into crucifer regulatory regions. Nat Genet 45(8):891

Hirayama T, Shinozaki K (2010) Research on plant abiotic stress responses in the post-genome era: past, present and future. Plant J 61(6):1041–1052

Hoang TML, Tran TN, Nguyen TKT, Williams B, Wurm P, Bellairs S, Mundree S (2016) Improvement of salinity stress tolerance in rice: challenges and opportunities. Agronomy 6 (4):54

Horvath P, Barrangou R (2010) CRISPR/Cas, the immune system of bacteria and archaea. Science 327(5962):167–170

Hossain MA, Cho J-I, Han M, Ahn C-H, Jeon J-S, An G, Park PB (2010) The ABRE-binding bZIP transcription factor OsABF2 is a positive regulator of abiotic stress and ABA signaling in rice. J Plant Physiol 167(17):1512–1520

Hu W, Hu G, Han B (2009) Genome-wide survey and expression profiling of heat shock proteins and heat shock factors revealed overlapped and stress specific response under abiotic stresses in rice. Plant Sci 176(4):583–590

Hua K, Tao X, Yuan F, Wang D, Zhu J-K (2018) Precise A· T to G· C base editing in the rice genome. Mol Plant 11(4):627–630

Huang Y, Guo Y, Liu Y, Zhang F, Wang Z, Wang H, Wang F, Li D, Mao D, Luan S (2018) 9-cis-Epoxycarotenoid dioxygenase 3 regulates plant growth and enhances multi-abiotic stress tolerance in rice. Front Plant Sci 9:162

Hussain B, Lucas SJ, Budak H (2018) CRISPR/Cas9 in plants: at play in the genome and at work for crop improvement. Brief Funct Genomics 17(5):319–328

Iuchi S, Kobayashi M, Taji T, Naramoto M, Seki M, Kato T, Tabata S, Kakubari Y, Yamaguchi-Shinozaki K, Shinozaki K (2001) Regulation of drought tolerance by gene manipulation of 9-cis-epoxycarotenoid dioxygenase, a key enzyme in abscisic acid biosynthesis in Arabidopsis. Plant J 27(4):325–333

Jacobs TB, LaFayette PR, Schmitz RJ, Parrott WA (2015) Targeted genome modifications in soybean with CRISPR/Cas9. BMC Biotechnol 15(1):16

Jain M (2015) Function genomics of abiotic stress tolerance in plants: a CRISPR approach. Front Plant Sci 6:375

Ji X, Wang Y, Liu G (2012) Expression analysis of MYC genes from Tamarix hispida in response to different abiotic stresses. Int J Mol Sci 13(2):1300–1313

Jiang W, Zhou H, Bi H, Fromm M, Yang B, Weeks DP (2013) Demonstration of CRISPR/Cas9/sgRNA-mediated targeted gene modification in Arabidopsis, tobacco, sorghum and rice. Nucleic Acids Res 41(20):e188–e188

Jiang W, Zhao X, Gabrieli T, Lou C, Ebenstein Y, Zhu TF (2015) Cas9-assisted targeting of chromosome segments CATCH enables one-step targeted cloning of large gene clusters. Nat Commun 6:8101

Jinek M, Chylinski K, Fonfara I, Hauer M, Doudna JA, Charpentier E (2012) A programmable dual-RNA–guided DNA endonuclease in adaptive bacterial immunity. Science 337 (6096):816–821

Joshi R, Gupta P, Singla-Pareek SL, Pareek A (2017) Biomass production and salinity response in plants: role of MicroRNAs. Nucleic Acids Res 22(4):448–457

Khan S-A, Li M-Z, Wang S-M, Yin H-J (2018) Revisiting the role of plant transcription factors in the battle against abiotic stress. Int J Mol Sci 19(6):1634

Kim JM, Kim D, Kim S, Kim J-S (2014) Genotyping with CRISPR-Cas-derived RNA-guided endonucleases. Nat Commun 5:3157

Kim D, Alptekin B, Budak H (2018) CRISPR/Cas9 genome editing in wheat. Funct Integr Genomics 18(1):31–41

Klap C, Yeshayahou E, Bolger AM, Arazi T, Gupta SK, Shabtai S, Usadel B, Salts Y, Barg R (2017) Tomato facultative parthenocarpy results from Sl AGAMOUS-LIKE 6 loss of function. Plant Biotechnol J 15(5):634–647

Komor AC, Kim YB, Packer MS, Zuris JA, Liu DR (2016) Programmable editing of a target base in genomic DNA without double-stranded DNA cleavage. Nature 533(7603):420

Koonin EV, Makarova KS (2013) CRISPR-Cas: evolution of an RNA-based adaptive immunity system in prokaryotes. RNA Biol 10(5):679–686

Kouranova E, Forbes K, Zhao G, Warren J, Bartels A, Wu Y, Cui X (2016) CRISPRs for optimal targeting: delivery of CRISPR components as DNA, RNA, and protein into cultured cells and single-cell embryos. Hum Gene Ther 27(6):464–475

La Russa MF, Qi LS (2015) The new state of the art: Cas9 for gene activation and repression. Mol Cell Biol 35(22):3800–3809

Laloum T, Martín G, Duque P (2018) Alternative splicing control of abiotic stress responses. Trends Plant Sci 23(2):140–150

Leenay RT, Maksimchuk KR, Slotkowski RA, Agrawal RN, Gomaa AA, Briner AE, Barrangou R, Beisel CL (2016) Identifying and visualizing functional PAM diversity across CRISPR-Cas systems. Mol Cell 62(1):137–147

Lei Y, Zhang X, Su J, Jeong M, Gundry MC, Huang Y-H, Zhou Y, Li W, Goodell MA (2017) Targeted DNA methylation in vivo using an engineered dCas9-MQ1 fusion protein. Nat Commun 8:16026

Leong K-YB, Chan Y-H, Abdullah WM, Lim S-HE, Lai K-S (2018) The CRISPR/Cas9 System for Crop Improvement: Progress and Prospects. Next Generation Plant Breed 2019:129

Li X-P, Tian A-G, Luo G-Z, Gong Z-Z, Zhang J-S, Chen S-Y (2005) Soybean DRE-binding transcription factors that are responsive to abiotic stresses. Theor Appl Genet 110(8):1355–1362

Li Z, Liu Z-B, Xing A, Moon BP, Koellhoffer JP, Huang L, Ward RT, Clifton E, Falco SC, Cigan AM (2015) Cas9-guide RNA directed genome editing in soybean. Plant Physiol 169 (2):960–970

Li J, Meng X, Zong Y, Chen K, Zhang H, Liu J, Li J, Gao C (2016a) Gene replacements and insertions in rice by intron targeting using CRISPR–Cas9. Nat Plants 2(10):16139

Li M, Li X, Zhou Z, Wu P, Fang M, Pan X, Lin Q, Luo W, Wu G, Li H (2016b) Reassessment of the four yield-related genes Gn1a, DEP1, GS3, and IPA1 in rice using a CRISPR/Cas9 system. Front Plant Sci 7:377

Li J, Sun Y, Du J, Zhao Y, Xia L (2017a) Generation of targeted point mutations in rice by a modified CRISPR/Cas9 system. Mol Plant 10(3):526–529

Li P, Li YJ, Zhang FJ, Zhang GZ, Jiang XY, Yu HM, Hou BK (2017b) The Arabidopsis UDP-glycosyltransferases UGT79B2 and UGT79B3, contribute to cold, salt and drought stress tolerance via modulating anthocyanin accumulation. Plant J 89(1):85–103

Li C, Zong Y, Wang Y, Jin S, Zhang D, Song Q, Zhang R, Gao C (2018a) Expanded base editing in rice and wheat using a Cas9-adenosine deaminase fusion. Genome Biol 19(1):59

Li R, Li R, Li X, Fu D, Zhu B, Tian H, Luo Y, Zhu H (2018b) Multiplexed CRISPR/Cas9-mediated metabolic engineering of γ-aminobutyric acid levels in Solanum lycopersicum. Plant Biotechnol J 16(2):415–427

Li R, Liu C, Zhao R, Wang L, Chen L, Yu W, Zhang S, Sheng J, Shen L (2019) CRISPR/Cas9-Mediated SlNPR1 mutagenesis reduces tomato plant drought tolerance. BMC Plant Biol 19 (1):38

Liang Z, Zhang K, Chen K, Gao C (2014) Targeted mutagenesis in Zea mays using TALENs and the CRISPR/Cas system. J Genet Genomics 41(2):63–68

Liang W, Ma X, Wan P, Liu L (2018) Plant salt-tolerance mechanism: a review. Biochem Biophy Res 495(1):286–291

Liu XS, Wu H, Ji X, Stelzer Y, Wu X, Czauderna S, Shu J, Dadon D, Young RA, Jaenisch R (2016) Editing DNA methylation in the mammalian genome. Cell 167(1):233–247

Liu X, Wu S, Xu J, Sui C, Wei J (2017) Application of CRISPR/Cas9 in plant biology. APSB 3:292–302

Lombardo L, Coppola G, Zelasco S (2016) New technologies for insect-resistant and herbicide-tolerant plants. Trends Biotechnol 34(1):49–57

Lou D, Wang H, Liang G, Yu D (2017) OsSAPK2 confers abscisic acid sensitivity and tolerance to drought stress in rice. Front Plant Sci 8:993

Lowder L, Malzahn A, Qi Y (2016) Rapid evolution of manifold CRISPR systems for plant genome editing. Front Plant Sci 7:1683

Lu Y, Zhu J-K (2017) Precise editing of a target base in the rice genome using a modified CRISPR/Cas9 system. Mol Plant 10(3):523–525

Ma Y, Dai X, Xu Y, Luo W, Zheng X, Zeng D, Pan Y, Lin X, Liu H, Zhang D (2015) COLD1 confers chilling tolerance in rice. Cell 160(6):1209–1221

Majumder AL, Sengupta S, Goswami L (2009) Osmolyte regulation in abiotic stress. In: Abiotic stress adaptation in plants. Springer, New York, pp 349–370

Matsui A, Nakaminami K, Seki M (2019) Biological function of changes in RNA metabolism in plant adaptation to abiotic stress. Plant Cell Physiol 60(9):1897–1905

Matzke M, Matzke AJ, Kooter JM (2001) RNA: guiding gene silencing. Science 293 (5532):1080–1083

Meng X, Yu H, Zhang Y, Zhuang F, Song X, Gao S, Gao C, Li J (2017) Construction of a genome-wide mutant library in rice using CRISPR/Cas9. Mol Plant 10(9):1238–1241

Mickelbart MV, Hasegawa PM, Bailey-Serres J (2015) Genetic mechanisms of abiotic stress tolerance that translate to crop yield stability. Nat Rev Genet 16(4):237–251

Mirlohi S, He Y (2016) Small RNAs in plant response to abiotic stress. In: Abiotic and biotic stress in plants—recent advances and future perspectives. InTech, London, pp 63–80

Mizoi J, Shinozaki K, Yamaguchi-Shinozaki K (2012) AP2/ERF family transcription factors in plant abiotic stress responses. BBA Gene Regul Mech 1819(2):86–96

Moustafa K, AbuQamar S, Jarrar M, Al-Rajab AJ, Trémouillaux-Guiller J (2014) MAPK cascades and major abiotic stresses. Plant Cell Rep 33(8):1217–1225

Nakashima K, Ito Y, Yamaguchi-Shinozaki K (2009) Transcriptional regulatory networks in response to abiotic stresses in Arabidopsis and grasses. Plant Physiol 149(1):88–95

Nguyen HC, Lin KH, Ho SL, Chiang CM, Yang CM (2018) Enhancing the abiotic stress tolerance of plants: from chemical treatment to biotechnological approaches. Physiol Plant 164 (4):452–466

Nieves-Cordones M, Mohamed S, Tanoi K, Kobayashi NI, Takagi K, Vernet A, Guiderdoni E, Périn C, Sentenac H, Véry AA (2017) Production of low-Cs+ rice plants by inactivation of the K + transporter Os HAK 1 with the CRISPR-Cas system. Plant J 92(1):43–56

Okamoto M, Matsui A, Tanaka M, Morosawa T, Ishida J, Iida K, Mochizuki Y, Toyoda T, Seki M (2016) Sm-like protein-mediated RNA metabolism is required for heat stress tolerance in Arabidopsis. Front Plant Sci 7:1079

Osakabe Y, Watanabe T, Sugano SS, Ueta R, Ishihara R, Shinozaki K, Osakabe K (2016) Optimization of CRISPR/Cas9 genome editing to modify abiotic stress responses in plants. Sci Rep 6:26685

Paixão JFR, Gillet F-X, Ribeiro TP, Bournaud C, Lourenço-Tessutti IT, Noriega DD, de Melo BP, de Almeida-Engler J, Grossi-de-Sa MF (2019) Improved drought stress tolerance in Arabidopsis by CRISPR/dCas9 fusion with a Histone AcetylTransferase. Sci Rep 9(1):8080

Park J-J, Dempewolf E, Zhang W, Wang Z-Y (2017) RNA-guided transcriptional activation via CRISPR/dCas9 mimics overexpression phenotypes in Arabidopsis. PLoS One 12(6):e0179410

Piatek A, Ali Z, Baazim H, Li L, Abulfaraj A, Al-Shareef S, Aouida M, Mahfouz MM (2015) RNA-guided transcriptional regulation in planta via synthetic dC as9-based transcription factors. Plant Biotechnol J 13(4):578–589

Polstein LR, Gersbach CA (2015) A light-inducible CRISPR-Cas9 system for control of endogenous gene activation. Nat Chem Biol 11(3):198

Qi W, Zhu T, Tian Z, Li C, Zhang W, Song R (2016) High-efficiency CRISPR/Cas9 multiplex gene editing using the glycine tRNA-processing system-based strategy in maize. BMC Biotechnol 16 (1):58

Radivojac P, Clark WT, Oron TR, Schnoes AM, Wittkop T, Sokolov A, Graim K, Funk C, Verspoor K, Ben-Hur A (2013) A large-scale evaluation of computational protein function prediction. Nat Methods 10(3):221

Ray DK, Mueller ND, West PC, Foley JA (2013) Yield trends are insufficient to double global crop production by 2050. PLoS One 8(6):e66428

Romay G, Bragard C (2017) Antiviral defenses in plants through genome editing. Front Microbiol 8:47
Roy B, Noren S, Mandal AB, Basu AK (2011) Genetic engineering for abiotic stress tolerance in agricultural crops. Biotechnology 10(1):1–22
Sadanandom A, Srivastava AK, Zhang C (2019) Targeted mutagenesis of the SUMO protease, overly tolerant to Salt1 in rice through CRISPR/Cas9-mediated genome editing reveals a major role of this SUMO protease in salt tolerance. bioRxiv:555706
Samarut É, Lissouba A, Drapeau P (2016) A simplified method for identifying early CRISPR-induced indels in zebrafish embryos using high resolution melting analysis. BMC Genomics 17 (1):547
Sander JD, Joung JK (2014) CRISPR-Cas systems for editing, regulating and targeting genomes. Nat Biotechnol 32(4):347
Sauer NJ, Narváez-Vásquez J, Mozoruk J, Miller RB, Warburg ZJ, Woodward MJ, Mihiret YA, Lincoln TA, Segami RE, Sanders SL (2016) Oligonucleotide-mediated genome editing provides precision and function to engineered nucleases and antibiotics in plants. Plant Physiol 170 (4):1917–1928
Scheben A, Wolter F, Batley J, Puchta H, Edwards D (2017) Towards CRISPR/Cas crops–bringing together genomics and genome editing. New Phytol 216(3):682–698
Schiml S, Fauser F, Puchta H (2014) The CRISPR/C as system can be used as nuclease for in planta gene targeting and as paired nickases for directed mutagenesis in A rabidopsis resulting in heritable progeny. Plant J 80(6):1139–1150
Scotti MM, Swanson MS (2016) RNA mis-splicing in disease. Nat Rev Genet 17(1):19
Sedeek KE, Mahas A, Mahfouz M (2019) Plant genome engineering for targeted improvement of crop traits. Front Plant Sci 10:114
Sentmanat MF, Peters ST, Florian CP, Connelly JP, Pruett-Miller SM (2018) A survey of validation strategies for CRISPR-Cas9 editing. Sci Rep 8(1):888
Shalem O, Sanjana NE, Zhang F (2015) High-throughput functional genomics using CRISPR–Cas9. Nat Rev Genet 16(5):299–311
Shan Q, Wang Y, Li J, Zhang Y, Chen K, Liang Z, Zhang K, Liu J, Xi JJ, Qiu J-L (2013) Targeted genome modification of crop plants using a CRISPR-Cas system. Nat Biotechnol 31(8):686
Shan Q, Wang Y, Li J, Gao C (2014) Genome editing in rice and wheat using the CRISPR/Cas system. Nat Protoc 9(10):2395
Shen C, Que Z, Xia Y, Tang N, Li D, He R, Cao M (2017) Knock out of the annexin gene OsAnn3 via CRISPR/Cas9-mediated genome editing decreased cold tolerance in rice. J Plant Biol 60 (6):539–547
Shi J, Gao H, Wang H, Lafitte HR, Archibald RL, Yang M, Hakimi SM, Mo H, Habben JE (2017) ARGOS 8 variants generated by CRISPR-Cas9 improve maize grain yield under field drought stress conditions. Plant Biotechnol J 15(2):207–216
Shim JS, Oh N, Chung PJ, Kim YS, Choi YD, Kim J-K (2018) Overexpression of OsNAC14 improves drought tolerance in rice. Front Plant Sci 9:310
Sinha AK, Jaggi M, Raghuram B, Tuteja N (2011) Mitogen-activated protein kinase signaling in plants under abiotic stress. Plant Signal Behav 6(2):196–203
Sokol A, Kwiatkowska A, Jerzmanowski A, Prymakowska-Bosak M (2007) Up-regulation of stress-inducible genes in tobacco and Arabidopsis cells in response to abiotic stresses and ABA treatment correlates with dynamic changes in histone H3 and H4 modifications. Planta 227(1):245–254
Songmei L, Jie J, Yang L, Jun M, Shouling X, Yuanyuan T, Youfa L, Qingyao S, Jianzhong H (2019) Characterization and evaluation of OsLCT1 and OsNramp5 mutants generated through CRISPR/Cas9-mediated mutagenesis for breeding low Cd rice. Rice Sci 26(2):88–97
Strahl BD, Allis CD (2000) The language of covalent histone modifications. Nature 403(6765):41
Sudan J, Raina M, Singh R (2018) Plant epigenetic mechanisms: role in abiotic stress and their generational heritability. Biotech 8(3):172

Sun Y, Zhang X, Wu C, He Y, Ma Y, Hou H, Guo X, Du W, Zhao Y, Xia L (2016) Engineering herbicide-resistant rice plants through CRISPR/Cas9-mediated homologous recombination of acetolactate synthase. Mol Plant 9(4):628–631
Svitashev S, Schwartz C, Lenderts B, Young JK, Cigan AM (2016) Genome editing in maize directed by CRISPR–Cas9 ribonucleoprotein complexes. Nat Commun 7:13274
Tang J, Chu C (2017) MicroRNAs in crop improvement: fine-tuners for complex traits. Nat Plants 3 (7):17077
Tang N, Ma S, Zong W, Yang N, Lv Y, Yan C, Guo Z, Li J, Li X, Xiang Y (2016a) MODD mediates deactivation and degradation of OsbZIP46 to negatively regulate ABA signaling and drought resistance in rice. Plant Cell 28(9):2161–2177
Tang X, Zheng X, Qi Y, Zhang D, Cheng Y, Tang A, Voytas DF, Zhang Y (2016b) A single transcript CRISPR-Cas9 system for efficient genome editing in plants. Mol Plant 9 (7):1088–1091
Tang L, Mao B, Li Y, Lv Q, Zhang L, Chen C, He H, Wang W, Zeng X, Shao Y (2017) Knockout of OsNramp5 using the CRISPR/Cas9 system produces low Cd-accumulating indica rice without compromising yield. Sci Rep 7(1):14438
Thompson DM, Parker R (2009) Stressing out over tRNA cleavage. Cell 138(2):215–219
Till BJ, Cooper J, Tai TH, Colowit P, Greene EA, Henikoff S, Comai L (2007) Discovery of chemically induced mutations in rice by TILLING. BMC Plant Biol 7(1):19
Tsutsui H, Higashiyama T (2017) pKAMA-ITACHI vectors for highly efficient CRISPR/Cas9-mediated gene knockout in Arabidopsis thaliana. Plant Cell Physiol 58(1):46–56
Upadhyay SK, Kumar J, Alok A, Tuli R (2013) RNA-guided genome editing for target gene mutations in wheat. G3: genes, genomes. Genetics 3(12):2233–2238
Vaughan MM, Block A, Christensen SA, Allen LH, Schmelz EA (2018) The effects of climate change associated abiotic stresses on maize phytochemical defenses. Phytochem Rev 17 (1):37–49
Vinocur B, Altman A (2005) Recent advances in engineering plant tolerance to abiotic stress: achievements and limitations. Curr Opin Biotechnol 16(2):123–132
Wang Y, Beaith M, Chalifoux M, Ying J, Uchacz T, Sarvas C, Griffiths R, Kuzma M, Wan J, Huang Y (2009) Shoot-specific down-regulation of protein farnesyltransferase (α-subunit) for yield protection against drought in canola. Mol Plant 2(1):191–200
Wang Y, Li Z, Xu J, Zeng B, Ling L, You L, Chen Y, Huang Y, Tan A (2013) The CRISPR/Cas system mediates efficient genome engineering in Bombyx mori. Cell Res 23(12):1414
Wang B, Li K, Wang A, Reiser M, Saunders T, Lockey RF, Wang J-W (2015a) Highly efficient CRISPR/HDR-mediated knock-in for mouse embryonic stem cells and zygotes. Biotechniques 59(4):201–208
Wang Y, Yang L, Chen X, Ye T, Zhong B, Liu R, Wu Y, Chan Z (2015b) Major latex protein-like protein 43 (MLP43) functions as a positive regulator during abscisic acid responses and confers drought tolerance in Arabidopsis thaliana. J Exp Bot 67(1):421–434
Wang Z-P, Xing H-L, Dong L, Zhang H-Y, Han C-Y, Wang X-C, Chen Q-J (2015c) Egg cell-specific promoter-controlled CRISPR/Cas9 efficiently generates homozygous mutants for multiple target genes in Arabidopsis in a single generation. Genome Biol 16(1):144
Wang F-Z, Chen M-X, Yu L-J, Xie L-J, Yuan L-B, Qi H, Xiao M, Guo W, Chen Z, Yi K (2017a) OsARM1, an R2R3 MYB transcription factor, is involved in regulation of the response to arsenic stress in rice. Front Plant Sci 8:1868
Wang L, Chen L, Li R, Zhao R, Yang M, Sheng J, Shen L (2017b) Reduced drought tolerance by CRISPR/Cas9-mediated SlMAPK3 mutagenesis in tomato plants. J Agric Food Chem 65 (39):8674–8682
Wang P, Su L, Gao H, Jiang X, Wu X, Li Y, Zhang Q, Wang Y, Ren F (2018) Genome-wide characterization of bHLH genes in grape and analysis of their potential relevance to abiotic stress tolerance and secondary metabolite biosynthesis. Front Plant Sci 9:64

Weber J, Öllinger R, Friedrich M, Ehmer U, Barenboim M, Steiger K, Heid I, Mueller S, Maresch R, Engleitner T (2015) CRISPR/Cas9 somatic multiplex-mutagenesis for high-throughput functional cancer genomics in mice. Proc Natl Acad Sci 112(45):13982–13987

Weckwerth W, Loureiro ME, Wenzel K, Fiehn O (2004) Differential metabolic networks unravel the effects of silent plant phenotypes. Proc Natl Acad Sci 101(20):7809–7814

Wolter F, Puchta H (2018) The CRISPR/Cas revolution reaches the RNA world: Cas13, a new Swiss Army knife for plant biologists. Plant J 94(5):767–775

Xie K, Minkenberg B, Yang Y (2015) Boosting CRISPR/Cas9 multiplex editing capability with the endogenous tRNA-processing system. Proc Natl Acad Sci 112(11):3570–3575

Yadav NR, Taunk J, Rani A, Aneja B, Yadav RC (2013) Role of transcription factors in abiotic stress tolerance in crop plants. Climate change and plant abiotic stress tolerance. Wiley, Hoboken, pp 605–640

Yang Y, Costa A, Leonhardt N, Siegel RS, Schroeder JI (2008) Isolation of a strong Arabidopsis guard cell promoter and its potential as a research tool. Plant Methods 4(1):6

Yang C-H, Zhang Y, Huang C-F (2019) Reduction in cadmium accumulation in japonica rice grains by CRISPR/Cas9-mediated editing of OsNRAMP5. J Integr Agric 18(3):688–697

Ye Y, Li P, Xu T, Zeng L, Cheng D, Yang M, Luo J, Lian X (2017) OsPT4 contributes to arsenate uptake and transport in rice. Front Plant Sci 8:2197

Yousuf PY, Hakeem KUR, Chandna R, Ahmad P (2012) Role of glutathione reductase in plant abiotic stress. In: Abiotic stress responses in plants. Springer, New York, pp 149–158

Zetsche B, Gootenberg JS, Abudayyeh OO, Slaymaker IM, Makarova KS, Essletzbichler P, Volz SE, Joung J, Van Der Oost J, Regev A (2015) Cpf1 is a single RNA-guided endonuclease of a class 2 CRISPR-Cas system. Cell 163(3):759–771

Zhang A, Liu Y, Wang F, Li T, Chen Z, Kong D, Bi J, Zhang F, Luo X, Wang J (2019) Enhanced rice salinity tolerance via CRISPR/Cas9-targeted mutagenesis of the OsRR22 gene. Mol Breed 39(3):47

Zhao M, Tai H, Sun S, Zhang F, Xu Y, Li W-X (2012) Cloning and characterization of maize miRNAs involved in responses to nitrogen deficiency. PLoS One 7(1):e29669

Zhao Y, Zhang C, Liu W, Gao W, Liu C, Song G, Li W-X, Mao L, Chen B, Xu Y (2016) An alternative strategy for targeted gene replacement in plants using a dual-sgRNA/Cas9 design. Sci Rep 6:23890

Zhu X, Xu Y, Yu S, Lu L, Ding M, Cheng J, Song G, Gao X, Yao L, Fan D (2014) An efficient genotyping method for genome-modified animals and human cells generated with CRISPR/Cas9 system. Sci Rep 4:6420

Zong Y, Wang Y, Li C, Zhang R, Chen K, Ran Y, Qiu J-L, Wang D, Gao C (2017) Precise base editing in rice, wheat and maize with a Cas9-cytidine deaminase fusion. Nat Biotechnol 35 (5):438

Key Applications of CRISPR/Cas for Yield and Nutritional Improvement

7

Muhammad Salman Mubarik, Sultan Habibullah Khan, and Muhammad Sajjad

Abstract

The ever-growing human population and climate shifts are creating a dire need to the development of crop plants with high yield and improved nutritional quality. Conventional breeding tools cannot satisfy the future needs. Increased agricultural production through advanced technologies is urgently required to increase yield and easy access to nutritious food worldwide. Providentially, next-generation genome editing technologies like CRISPR/Cas system pave the way toward a new horizon for crop plant improvement and subsequently transform classical plant breeding. In this chapter, applications of CRISPR/Cas-based genome editing for yield and quality improvement are highlighted.

7.1 Introduction

Food security is the most imperative challenge currently facing worldwide. Increased production and nutritional quality will be a major issue for feeding an ever-growing world population. The ultimate challenge facing mankind is ensuring access to adequate and nutritious food. It is estimated that about 80.5 million people

M. S. Mubarik (✉)
Centre of Agricultural Biochemistry and Biotechnology (CABB), University of Agriculture, Faisalabad, Pakistan

S. H. Khan
Centre of Agricultural Biochemistry and Biotechnology (CABB), University of Agriculture, Faisalabad, Pakistan

Center for Advanced Studies in Agriculture and Food Security (CASAFS), University of Agriculture, Faisalabad, Pakistan
e-mail: sultan@uaf.edu.pk

M. Sajjad
Department of Biosciences, COMSATS University Islamabad (CUI), Islamabad, Pakistan

A. Ahmad et al. (eds.), *CRISPR Crops*,
https://doi.org/10.1007/978-981-15-7142-8_7

are persistently undernourished (Behera et al. 2019). Logical and tangible measures should be taken to meet future demands to address global food security challenges, as the challenges are enormous and multifaceted.

The green revolution of the twentieth century caused enormous increase in cereal yields around the globe. Since then, the global production of cereal crops has been stagnated and current trends show that yields will not be enough to meet growing food demand. With current low rate of genetic gain for yield potential, disease resistance, and adaptation to climate changes in the major food crops, many developing countries are facing serious challenges of food security. The adoption and application of newly emerging crop improvement technologies is indispensable to speed up genetic gains for yield potential, biotic and abiotic stress tolerance in major food crops.

Fast track crop improvement is inevitable to meet the growing demand for food under a changing climate for agriculture. By 2050, more than 9 billion people are expected to add up in the global population and crop demand may increase by more than 100% (Tilman et al. 2011). However, with the current pace of crop improvement, yields of the major crops will increase by just 38–67% (Ray et al. 2013). Simultaneously, increased occurrence of biotic and abiotic stresses due to changing climate may lead to increased crop yield losses (Ahmad et al. 2019). In post green revolution era a major increase in crop produce was achieved by improving agronomic practices such as the application of fertilizers and the use of insecticide/pesticides (Matson et al. 1997). However, these agronomic practices mainly rely on limited resources, environmental condiotions and are unlikely to enhance further crop yields (Scheben et al. 2017). As the cultivable land is continuously decreasing due to growing population pressure, urbanization and land degradation, therefore, engineering crops to tolerate stresses and better utilize available resources is way forward approach accompanying agronomic management. Besides, resource-efficient and eco-friendly crop varieties will significantly mitigate environmental impacts by reducing the use of fertilizer and pesticide required for achieving optimal yields.

To speed up genetic improvement in crop plants, successful application of the current and newly emerging state-of-the-art biotechnology approaches in crop breeding is essential. Crop engineering is moving into a new level of exponential breakthrough that can help us to achieve the required crop yield and quality improvement. Editing plant genomes with site-specific endonucleases is an important tool to supplement traditional recombinant DNA technology.

The evolution of genome editing systems offers an amazing opportunity for crop plant improvement because they make possible to manipulate DNA at any desired positions without upsetting the background genotype (Zlobin et al. 2020). Genome editing technologies such as ZFNs and TALENs have made it possible for molecular biologists to more precisely target any DNA sequences. However, these technologies have certain limitations; expensive, time-consuming, and involve intricate steps of protein engineering for each target DNA. Unlike these first-generation technologies, the next-generation CRISPR/Cas-based genome editing tools require straightforward designing and simple cloning, with the same Cas protein (s) being possibly available for use with several guide RNAs targeting multiple genomic sites (Mubarik et al. 2016).

CRISPR/Cas genome editing technology and its classification as a molecular breeding tool has the potential to significantly improve the crop yield and nutritional quality by engineering new traits in a commercial crops. Scientists believe this tide of engineering crop plants will continue to rise, and future advances in genome engineering research will ultimately achieve the milestones to cope with the global food security issue.

7.2 Plant Yield in the Context of Sustainable Agriculture

Since the end of World War II, agriculture scenario has been changed globally. Crop productivity has risen due to advances in agriculture mechanization and increased use of fertilizers, insecticides, and pesticides. Agricultural sustainability stands on the basis of meeting the present needs without losing the capability of upcoming generations to meet their requirements. Therefore, long-term stewardship of agricultural resources has prime importance to future economic gain. Stewardship of agricultural resources involves improving or retaining the quality of these resources and exploiting them in a way that let them to be restored for the future.

The concepts of agriculture sustainability date back to the hoariest remaining texts from Rome, Greece, and China (Pretty 2008). Current agricultural sustainability is based on developing agricultural technologies and practices that do not have harmful effects on the ecosystem and lead to increases in crop productivity with positive side effects on the environment. Sustainability incorporates models of both resilience (the ability of crop plants to safeguard biotic and abiotic stresses), yield persistence, and addresses many wider social, economic, and environmental outcomes.

The key principles for agriculture sustainability include:

1. Consistent use of biological and ecological practices into food production processes such as; soil regeneration and nutrient cycling, allelopathy, nitrogen fixation, predation, and parasitism.
2. Non-renewable inputs that affect the farmer's health and ecosystem should be curtailed.
3. Use of the scientific knowledge and farmer skills in a more productive way, therefore, developing self-belief and replacing human capital for expensive external inputs.

The agricultural sustainability, even though, does not mean to preclude any new technologies on conceptual grounds. If a new technology works out to enhance agricultural productivity and does not affect the environment, then it is apt to have sustainability gains.

7.3 Important Yield Components/Traits

In agricultural research, yield potential is the topmost important objective throughout the history of crop domestication (Pearsall 2008). In cereals, grain yield is the product of three major components: the number of grains per plant, grain weight, and spikes/panicles/cobs per unit area (Sajjad et al. 2017; Rustgi et al. 2013). The three major yield components are influenced directly or indirectly by other agronomic, physiological, and phenological components such as days to flowering/blooming, plant biomass, root/shoot biomass, leaf area, chlorophyll contents, and plant height. Each yield component is underpinned by complex genetic interactions at multiple loci (Wang et al. 2012). Plant height and grain weight are more stably inherited compared to number of grains and root/shoot biomass which are more variable in nature and affected by interaction between genotype and environment (Dhungana et al. 2007).

Previously, researchers identified various traits positively associated with grain yield and introduced those traits into crop plants to develop improved cultivars. Moreover, the introduction of desirable traits in breeding programmes to increase photosynthesis is also useful to augment biomass production. The association of solar radiation and crop duration are the two key components that have subsidized most to boosted total photosynthesis and biomass production and grain yield of crop plants (Richards 2000).

Classical plant breeding, genetic engineering, and agronomic practices have collectively contributed to the current yield potential of food crops. Both plant genetics and agronomy have been important to manipulate yield-related traits, with crop rotation, integration of resistant genes, and the use of chemical sprays. The interaction between crop breeding and crop husbandry has been very important and will stay on. It has been mostly evident in rice and wheat by extensive adoption of dwarfing genes which have decreased plant lodging and permitted higher rates of fertilizers to be used.

Elevated photosynthesis rate through increased stomatal conductance is a major physiological contributor to cereal grain yield. Genetic loci controlling stomatal conductance could be utilized as a secondary selection principle for yield enhancement (Ahmed et al. 2019). It is reported that crop plants have the extra capacity for additional photosynthesis gains mainly during reproductive growth.

During reproductive growth, crops are restricted by sink size rather than source availability. However, experiments with CO_2 fertilization also suggested that crops are source limited at the beginning of the stem elongation phase (Miralles and Slafe 2007). Substantial increase in crop growth rate was observed at this stage and when stems, roots leaves, and ears are all vigorously growing. Perhaps, during reproductive growth stage inadequate assimilates are accessible to crop plants. Selection of plants having high photosynthesis rate during reproductive growth may allow to select genotypes with high and a sustained stomatal conductance into the late afternoon. A substantial association has been found between stomatal conductance and crop yield in a high-yielding group of spring wheat (Fischer et al. 1998).

Data collected from wheat and maize—the most extensively cultivated crops—suggested that except for pre-anthesis stage, under favourable environments crops have spare resources for photosynthesis, and photosynthate production is not reducing. Therefore, during the last century, yield potential of the major crops turns out to be doubled without any change in the per unit area of photosynthesis rate. Most of these remarkable genetic gains have been accomplished through direct selection of the promising crop plants. Still, a quest for fundamental processes that may be responsible for variation in crop limiting factors and overall yield must go on as the knowledge grown will lead to attaining genetic gains more rapidly.

7.4 Molecular Basis of Plant Yield

Plant yield is the outcome of both growth and development during the entire growing season of the plant. Yield is formed constantly from sowing day to harvesting of the crop, with directly or indirectly almost all genes impacting the yield. The two fundamental genetic elements to be considered include genes controlling yield itself and the genes corresponding to adaptation to biotic or abiotic constraints for yield. The molecular basis of yield in most important crops has been engineered since the start of agriculture farming (Evans et al. 1998) either by increasing adaptation during the early ages of agricultural development or by enhancing yield with clear success throughout the last century (Araus et al. 2008).

Even if the increase of yield has taken place constantly after the rediscovery of Mendel's laws illustrating the molecular basis of inheritance. Consequently, continuous attempts have been made to identify the genes controlling important yield traits like number of seed/grain and their weight, plant height, disease resistance, and photoperiod sensitivity. Considering the yield potential, most of the developments in plant genetics have been focused on discovering essential genes responsible for variation in important traits influencing yield adaptation, rather than yield potential. In the last couple of decades, remarkable innovations in molecular biology techniques have been occurred which has tremendous potential to discover and locate genes of most complex quantitative characters such as yield in a plant genome.

Recently, the use of advanced functional genomics tools rapidly increased our knowledge of the genetic basis and developmental mechanisms of important yield components such as grain weight and size. A large number of QTLs/genes underpinning grain weight, size and grains per spike have been isolated, mapped, and functionally characterized in rice, maize (Liu et al. 2017), barley (Sun et al. 2019), and wheat (Zhang et al. 2018; Zhai et al. 2018; Sajjad et al. 2017). Most of the genes in common wheat were isolated using their orthologous genes in rice as reference genes. For a single mapping population reported QTLs associated with yield in a certain environmental condition are not essentially the same as those reported in a different environment. This does not invalidate the suitability and effectiveness of marker-assisted selection (MAS) in plant breeding for increasing the efficiency and reliability in selection for desirable traits.

Plant breeders and agronomists have made realistic efforts to comprehend that yield is an outcome of the expression of multiple traits. In wheat the dissection of yield into numerical components including; yield is measured as the product of plants per unit land area, number of spikes per plant, number of spikelet's per spike, number of grains per spikelet, and average grain weight. Though this method seems rational, it is unrealistic from a physiological point of view to expecting the impact of single trait in overall yield (Fischer et al. 1998). An ultimate challenge for crop yield improvement is that yield traits are negatively correlated with each other and nearly invariable. Therefore, if one trait can be improved through breeding, others may be affected negatively. There is a dire need to manipulate these components individually using current state-of-the-art technologies to get maximum yield benefits.

7.5 CRISPR/Cas-Based Strategies for Yield Improvement

Adverse climatic changes, deterioration of soil health and air quality significantly affect the overall crop yield. To improve crop yield in the changing climate, scientists are exploring the new avenues to engineer crop plants for higher yield potential and tolerate severe and unpredictable environmental conditions. The advancement in sequencing technologies and the availability of sequence information for major cultivated crops and recent additions in plant genome editing tools have opened new horizons to breed for almost any given desirable trait (Table 7.1).

In the last couple of years, the use of genome editing tools has magnificently shown precise gene editing in both plant and animal systems. In contrast to the earlier transgenic approaches, which lead to arbitrary DNA insertions and frequently random phenotypes, recently emerged genome editing tools produce well-defined mutants, thus becoming a powerful tool in plant breeding and functional genomics research. Those improved cultivars can be utilized to develop new crop varieties through new breeding programmes with more farmer's acceptability and comparatively less stringent regulatory processes compared to traditional transgenics (Waltz 2018).

As compared to first-generation genome editing technologies, CRISPR/Cas9 system requires straightforward designing and simple cloning protocols, for different gRNAs targeting multiple genomic sites being potentially used with the same Cas9 protein. CRISPR/Cas9 gene editing tools have been embraced in different crop species for several traits including yield enhancement and resistance against biotic and abiotic stresses (Ricroch et al. 2017). Eliminating negative regulators of yield controlling genes is an encouraging approach for crop genetic improvement. Therefore, most common and simplest application of CRISPR/Cas9 is the knocking out of genes that confer undesirable traits.

For constant crop improvement, gene editing of yield-related traits is the foremost target. Yield is considered as complex plant trait due to the involvement of multiple factors. Knocking-out negative regulators of yield-related traits such as seed/grain number, seed/grain size (Li et al. 2016), seed/grain weight and stem/tiller number (Shen et al. 2018), produced plants of expected phenotypes having loss-of-function

Table 7.1 Application of CRISPR/Cas9 system in crop plants for the improvement of yield-related traits

Crop species	Target gene (s) for editing	Target traits	References
Barley	HvPM19	A positive regulator of grain dormancy	Lawrenson et al. (2015)
Brassica napus	BnALC	Increased shatter resistance to avoid seed loss during mechanical harvest	Braatz et al. (2017)
	BnaRGA, BnaDA1, BnaFUL, BnaDA2	Crop improvement	Yang et al. (2017)
Brassica oleracea	BolC.GA4.a	Dwarf stature and reduced fruit dehiscence	Lawrenson et al. (2015)
Cotton	GhCLA1	An albino phenotype	Wang et al. (2019)
Maize	ZmLG1	Upright architecture leaves	Li et al. (2016)
	ARGOS	Improved maize grain yield under drought stress conditions	Shi et al. (2017)
Rice	BADH2	Fragrance	Shao et al. (2017)
	OsYSA and OsROC5	Albino-free rice seedlings	Lowder et al. (2015)
	OsGSTU, OsAnP, and OsMRP15	Decreased anthocyanin accumulation	Ma et al. (2015)
	OsWaxy	Reduced synthesis of amylase	Zhang et al. (2018)
	Rc	White pericarp	Civan and Brown (2017)
	OsNramp5	Low Cd content in grains	Tang et al. (2017)
	SBEI and SBEIIb	High amylose rice	Sun et al. (2016)
Soybean	GmFT2a	Delayed flowering time	Cai et al. (2018)
Tomato	SP5G	Rapid flowering and enhancing the compact determinate growth habit of tomatoes in the field	Soyk et al. (2017)
	CLV-WUS	Improved yield	Rodriguez-Leal et al. (2017)
Wheat	TaGW2-A1, -B1, and -D1	Controlled grain weight	Zhai et al. (2018)

mutations, proving that CRISPR/Cas9 is a valuable tool for improving yield-associated traits. In rice, three grain weight-related genes (*OsGW2*, *OsGW5*, and *OsTGW6*) were simultaneously knockout for trait pyramiding, which significantly increased grain weight. In triple mutant rice, significantly up to 30% increase in seed size was reported (Viana et al. 2019). These results showed that CRISPR/Cas9 can effectively be utilized to engineer crop plants for elevated yields in other cereal crops.

As most of the yield-associated traits are quantitative in nature and governed by many loci, only knocking out single factors may not be sufficiently raising yield in the field. Recently, a method developed by Huang et al. (2018) for the extensive gene's identification, that contribute to overall yield, a complex quantitative trait, by combining CRISPR/Cas9 technology, whole-genome sequencing, and pedigree analysis. They selected and sequenced 30 cultivars of the Green Revolution miracle rice variety IR8 and found more than 50 genes that are conserved in all high-yielding cultivars for gene editing via knockout (Cas9) or knockdown (dCas9), respectively.

Another powerful method for increasing crop yield is hybrid breeding. The maternal male-sterile line is a prerequisite for developing a high-yielding hybrid variety. In hybrid breeding programmes, sterility is the main barrier to fully exploit the heterosis mechanism. CRISPR/Cas9 mediated gene knockouts made enormous progress including a thermo-sensitive male-sterile line (tm5) in rice and maize (Xie et al. 2017), csa and ms45 photosensitive genic male-sterile lines in rice and wheat, respectively (Li et al. 2019). In hybrid rice, male fertility was rescued by knocking out two copies of their fertility related SC genes (Shen et al. 2017). Likewise, Shen et al. (2017) demonstrated that knockout of the *Orf2* (toxin gene) improved the fertility of rice hybrids and responsible for the newly found selfish-gene suicide mechanism in rice. Genome editing considered as a successful tool for improving other traits, such as shortening growth times and improving haploid breeding which combinatorially enhance the overall crop yield (Hickey et al. 2019; Li et al. 2016).

Several yield-related traits are governed by change in gene expression, single-nucleotide substitutions, or the addition of new gene functions. CRISPR variants offer site-specific gene editing such as gene knock-ins, gene knockouts, and base-editing facilitates crop breeding by introducing new alleles devoid of linkage drag or producing allelic variants that do not occur naturally (Shimatani et al. 2017). Besides gene knockout and knock-in, modulating gene expression through CRISPRa and CRISPRi are useful approaches for studying gene functions and can considerably accelerate crop breeding. Gene expression can be affected at transcriptional, post-transcriptional, and post-translational levels. These processes are under the control of a series of *cis* and *trans* acting regulatory elements, which can be customized by CRISPR based strategies. It is anticipated that the application of CRISPR/Cas system in cultivated crop plants will substantially improve overall yield in a coming year and will be the important tool for researchers to engineer crop plants for safeguarding seriously demanding food security.

7.6 Importance of Plant Quality

The improvement of plant quality has been a fundamental human quest since crop cultivation began. The key components of food that impact human nutrition are carbohydrates, fats, proteins, minerals, water, vitamins, and fibre. The carbohydrates and fats are required for calorific energy, while proteins, minerals, and water play an important role in the development of body tissues and structure maintenance.

Proteins are the most vital nutritional part of a human diet after fulfilling calorific energy requirements.

With the world's increasing population, there is an urgent need not only to increase the yield of existing crops but simultaneously improve their qualitative aspects as well. There are compelling reasons for attempting to increase the quality and quantity of nutrients available in crop plants through molecular biology tools. What used to be thought of as a "nutrition gap" has now come to be considered in terms of malnutrition. This is only right since calorie nutrition is inseparable. Now there are still unanswered questions as to the precise nutrient requirements of humans as a function of age, health, and ambient conditions. There are, also, some indications that the incidence of malnutrition-related diseases is increasing in different parts of the world.

Malnutrition is a challenging condition affected by production, supply, marketing, and policy in the food system which has a foremost impact on human health and the country's economy (Philipson et al. 2014). Biofortification has been considered as one of the potential solutions for the improvement of crops phytonutrient contents (Hefferon and Makhzoum 2015). Another approach that may complement nutrient biofortification is the integration of underutilized and abandoned crop plants into cultivation to address the growing food security and local challenges imposed by changing climatic conditions. About 30 crops provide approximately 95% of the food requirements out of the 7000 or more plant species that are being cultivated by humans for food, with four of them (wheat, maize, rice, and potato) accountable for about 60% of the total energy requirements (Jain and Gupta 2013).

Using natural plant variations, breeders successfully developed crop plants with improved quality and nutritional contents. The diversity of fruits and vegetables with pigments, nutrients, and flavour qualities at any major grocery store has been achieved by the exploitation of natural variation through plant breeding. But in most of the cases available natural variation through plant breeding is not sufficient, and it is necessary to get the genes from other sources. A popular example of such engineering is Golden rice, in which β-carotene gene was introduced in rice endosperm (Zhu et al. 2020). Traditional transgenics development is a single piece solution and needs new methods to be incorporated and adopted. Application of biotechnological approaches in plant biology research has allowed people to grow and shift from a state of caloric deficiency to a significantly improved, albeit erratically distributed, condition of nourishment.

7.7 Genetics and Complexity of Quality Traits

For many plant traits, little information we know about exactly how many genes are involved, where the genes are present, effects of those genes, and how their interaction. Fundamental plant breeding and molecular biology revolutionized new techniques for dissecting multi-locus systems. Significant advancement has been made at the molecular level by the development and introduction of molecular markers in breeding programmes (Michelmore and Shaw 1988). New approaches based on multi-locus models will offer the method of choice for the future research.

Quantitative traits are governed by multiple genes, but the distribution of quality traits seems to be continuous due to presence of many genotypes in the population. However, if multiple genes affect the one trait, it can still exhibit continuous variation even though the environmental factors influence that particular trait. In most of the circumstances, genetic dissection is nearly impossible from the trait data alone because both genetic and environmental factors are active. For genetic analysis of complex traits, molecular markers provide new insights for plant scientists with new and advanced technology (Tanksley 1993).

The identification of DNA markers and associated genes with specific traits is achieved through QTL mapping. Various characteristics affect the precision of QTL mapping include; genotyping errors, replication levels of phenotypic information, environmental effects, population sizes and types. In genetic mapping studies, the presence of QTL controlling quality traits in crop plants has been thoroughly investigated in various crop plants through linkage and linkage disequilibrium methods. For the comparison between different plant species a reference based on accumulated information of the QTL accounting for quality traits has also been identified. Therefore, to obtain unbiased estimates of QTL effects, independent or cross validations have been suggested.

In a number of major crops, such as wheat, rice, maize, soybean, barley, sugar beet, sunflower, and forage plants, the potential and constraints of genome-wide estimates have been assessed (Desta and Ortiz 2014; Korte and Farlow 2013; Heffner et al. 2009; Bernardo 2008). Genome-wide prediction is considered as a powerful tool to accelerate the selection gain in plant breeding. Moreover, to predict the plant quality traits, the potential to manipulate the epistasis needs to be explored comprehensively.

7.8 CRISPR/Cas-Based Approaches for Improving the Nutritional Quality of Crops

Although yield is the precedence for both conventional plant breeding and modern plant genetic engineering, the objective of enhancing the nutritional quality of plants is taking growing attention (Mohan and Suprasanna 2011). For human health and growth, minerals, vitamins, essential amino acids, and fatty acids are key elements, and plant foods rich in vital nutrients can considerably improve human health and life expectancy (Walker et al. 2003). Various human diseases like cancer and heart disease can also be prevented by dietary supplements of specific nutrients (Knauf and Facciotti 1995).

Traditional breeding approaches have been immensely effective in improving the nutritional value of food and feed; however, this process is time-exhausting, and the insufficient genetic potential of most crop plants has left little room for further improvement by conventional approaches. During the last two decades, the possibilities for crop improvement have been extended by extensive whole-genome sequencing, gene mapping and identification, innovations in plant transformation, and advances in genetic engineering tools. Captivated efforts are now ongoing to engineer metabolic pathways to manipulate plant nutrients. These efforts depend on a complete knowledge of plant metabolic pathways and genes encoding their essential enzymes.

Current approaches in crop breeding aiming to increase the nutrition of crop plants by offering balanced and diversify diets with suitable levels of minerals and vitamins that enhance human health. Current state-of-the-art biotechnology methodologies make it possible to manipulate the important enzymes in particular metabolic pathways, thereby enhancing the key nutrients contents such as iron and vitamins and reducing the unfavourable compounds such as acrylamide-forming amino acids and phytic acids. The nutrition deficiency problem has been addressed by producing several biofortified crops such as wheat, rice, and maize (Bouis and Welch 2019).

Accompanied by the improvement of plant resistance, nutritional value of crop plants has been improved by CRISPR technology (Table 7.2). Stable and heritable mutations have been generated with CRISPR/Cas9 system, which can easily segregate out to prevent further modifications in the genome. As a result, only in a single generation homozygous modified transgene-free plants have been developed. To date, in a number of cereal, vegetable, and fruit plants, a characteristic genome editing works has been done for the improvement of quality in terms of improving starch contents, storage quality, fragrance, nutritional value, etc., (Ma et al. 2015; Sun et al. 2016; Waltz 2016; Zhang et al. 2018).

In terms of nutritional value, potato is one of the world's most essential food crops. Some wild types of potato have more nutritional value; but, due to sexual incompatibility, traditional genetic techniques would take several years for their breeding. Three different regions of the potato GBSS gene were targeted transiently and with DNA-free genome editing via Cas9-gRNA ribonucleic protein (RNP) delivery, resulting in different levels of mutations and have more commercial significance since potato starch (amylopectin) was improved in quality (Andersson et al. 2018). The aroma is an important quality trait, and rice varieties with desired aroma when cooked have risen commercial importance. With the application of CRISPR/Cas9 system, in a major planting area of China, a research group added the fragrance trait to more than 30 elite rice cultivars. In maize, phytate contents were reduced by knockdown of the ZmIPK gene expression involved in phytate biosynthesis (Sedeek et al. 2019).

Tomato despite being a highly nutritive plant also has a key neurotransmitter (γ-aminobutyric acid) of the human central nervous system (CNS) (Sequeira et al. 2019). Five key genes including; GABA-TP1, TP2, TP3, SlySSADH, and SlyCAT9 were knocked out using the CRISPR-Cas9 system to alter very specific metabolic pathway to produce a high amount of GABA in tomatoes (Li et al. 2018a, 2018b). Comparing to wild-type, mutant tomato contains 19-fold higher GABA content in the leaves, making these mutants commercially more attractive. CRISPR/Cas9 has been used to generate functional knockout mutants of a-gliadin genes and carotenoid biosynthesis pathway genes to develop wheat with hypoimmunogenic gluten and

Table 7.2 Application of CRISPR/Cas9 system in crop plants for quality improvement

Crop species	Target gene (s) for editing	Target traits	References
Apple	PDS	Carotenoid biosynthesis	Nishitani et al. (2016)
Arabidopsis	FAD2	Increase oleic acid content	Jiang et al. (2017)
Brassica napus	FAD2	Increased the content of oleic acid	Okuzaki et al. (2018)
Cassava	MePDS	Carotenoid biosynthesis	Odipio et al. (2017)
C. sativa	CsFAD2	Increased oleic acid content	Jiang et al. (2017)
Maize	ZmAgo18a, ZmAgo18b, a1 and a4	Anthocyanin biosynthesis	Char et al. (2017)
	tga1	Conferring naked kernels	Wang et al. (2015)
Pennycress	FAE1 and FAE2	Increased oil quality	McGinn et al. (2019)
	EaDAcT	Produced low-viscosity acetyl triacylglycerol-containing seed oil	McGinn et al. (2019)
	DGAT	Improved oil content	Sedbrook et al. (2014)
Potato	GBSS and StGBSS	Starch quality	Andersson et al. (2018), Wang et al. (2015)
	St16DOX	Reduce accumulation of steroidal glycoalkaloids: α-solanine and α-chaconine	Nakayasu et al. (2018)
Rice	BADH2	Fragrance	Shao et al. (2017)
	OsYSA and OsROC5	Albino-free rice seedlings	Lowder et al. (2015)
	OsGSTU, OsAnP, and OsMRP15	Decreased anthocyanin accumulation	Ma et al. (2015)
	OsWaxy	Reduced synthesis of amylase	Zhang et al. (2018, 2018b)
	Rc	White pericarp	Civan and Brown (2017)
	OsNramp5	Low Cd content in grains	Tang et al. (2017)
	SBEI and SBEIIb	High amylose rice	Sun et al. (2016)
Soybean	01gDDM1, 11gDDM1, Glyma04g36150, miR1509, miR1514, Glyma06g18790	Improved quality	Jacobs et al. (2015)
	GmPDS11, GmPDS18	Carotenoid biosynthesis	Du et al. (2016)
Sweet orange	CsPDS	Carotenoid biosynthesis	Wang et al. (2014)
Tomato	SlClV1, SlCLV2, SlCLV3, SlRRA3a	Increased fruit size	Xu et al. (2015)
	SP	Increased shelf life	Zsogon et al. (2018)

(continued)

Table 7.2 (continued)

Crop species	Target gene (s) for editing	Target traits	References
	SlPDS, SlPIF4	Carotenoid biosynthesis	Pan et al. (2016)
	O, FW2.2, and CycB	Larger fruit size, increased fruit weight and increased lycopene accumulation, respectively	Zsogon et al. (2018)
	RIN	Long shelf life	Ito et al. (2015)
	SGR1	Improved lycopene content in tomato fruit	Li et al. (2018a, 2018b)
	ALC	Long shelf life tomato lines	Yu et al. (2017)
	PSY1	Yellow-coloured tomato	Ito et al. (2017)
	MYB12	Pink-coloured tomato	Wang et al. (2019)
	ANT2	Purple-coloured tomato	Cermak et al. (2015)
	PL	Long shelf life tomato	Uluisik et al. (2016)
	ALC	Long shelf life tomato	Yu et al. (2017)
	MPK20	Repression of genes controlling sugar metabolism	Chen et al. (2018)
	GAD2, GAD3	Increase in GABA content	Nonaka et al. (2017)
	GABA-TP1, GABA-TP2, GABA-TP3, CAT9, SSADH	Increase in GABA content	Li et al. (2018a, 2018b)
	SGR1, LCY-E, Blc, LCY-B1, LCY-B2	Increase in lycopene content	Li et al. (2018a, 2018b)
	ALMT9	Decrease in malate content	Ye et al. (2017)
Wheat	α-Gluten-encoding	Low gluten wheat (reduces celiac disease in human)	Sanchez-Leon et al. (2018)

tomato with enhanced lycopene content, respectively (Jouanin et al. 2019; Li et al. 2018a, 2018b).

The application of genome editing has also been extended to modify seed oil content to produce improved oils for human consumption, as well as biofuels. Improved knowledge of the metabolic pathways and the characterization of the genes encoding enzymes related to fatty acid biosynthesis possibly make possible to modify seed oil contents (Thelen and Ohlrogge 2002). In two allohexaploid oilseed crops *Camelina sativa* and *Brassica napus*, three FAD2 homeolog genes were knocked out to achieve substantial increase of the oleic acid level and decreased level of undesirable polyunsaturated fatty acids (Okuzaki et al. 2018). These studies demonstrated that CRISPR/Cas9 system has a great potential to produce and commercialize elite crop plants with enhanced nutritional quality.

7.9 Concluding Remarks

In the last few years, numerous applications of CRISPR/Cas-based genome editing have been found in a number of plant species. Employing CRISPR in agriculture is simple, swift, more accurate and perhaps less costly than conventional genetic engineering approaches. In the recent past, several studies have proven the efficiency and fidelity of the CRISPR/Cas system-based gene editing in agriculture. The applications of next-generation CRISPR/Cas mediated gene editing in plants are growing very fast compared to first-generation ZFNs and TALENs gene editing tools, which is mainly due to simple cloning and high efficacy of CRISPR/Cas-based gene edits. Primarily, several reports annotating proof-of-concept or technical improvement studies in the model plants such as tobacco and Arabidopsis. More than 50 studies were published in 2017, described the use of CRISPR/Cas system in plant genome editing, and in the successive year the number increased to 106, demonstrating that in only one year the number of reports has been roughly doubled. The number of crop species being used in these studies was also increased. These results advocate that the CRISPR/Cas system plays a crucial role in agriculture research and that its usage will continue to expand.

Concerning the future of genome editing in agriculture for improved yield and quality traits for food, feed, and other uses, it is clear that CRISPR/Cas technology has become a major tool in applied research and is starting to play an important role in crop research and development. These trends seem irreversible. To go a step further and to have a major impact on agriculture, some major facts need to be addressed: appropriate knowledge on gene (s) responsible for traits to be targeted, improve efficiency of HR-mediated gene editing, elite genotypes should be handy to genome editing, onerous GMO regulations should be exempted for CRISPR/Cas edited crop plants. We are forecasting that with this pace of development CRISPR edited crops in labs with increased yield and improved quality will be seen in the field very soon.

References

Ahmad B, Raina A, Khan S (2019) Impact of biotic and abiotic stresses on plants and their responses. In: Disease resistance in crop plants. Springer, Cham, pp 1–19

Ahmed HG, Sajjad M, Li M et al (2019) Selection criteria for drought-tolerant bread wheat genotypes at seedling stage. Sustainability 11(9):2584

Andersson M, Turesson H, Olsson N et al (2018) Genome editing in potato via CRISPR-Cas9 ribonucleoprotein delivery. Physiol Plant 164(4):378–384

Araus JL, Slafer GA, Royo C et al (2008) Breeding for yield potential and stress adaptation in cereals. Crit Rev Plant Sci 27(6):377–412

Behera BK, Rout PK, Behera S (2019) Health for all. In: Move towards zero. Springer, Singapore, pp 91–112

Bernardo R (2008) Molecular markers and selection for complex traits in plants: learning from the last 20 years. Crop Sci 48(5):1649–1664

Bouis H, Welch R (2019) Reducing mineral and vitamin deficiencies through biofortification: progress under harvest plus. In: Sustaining global food security: the nexus of science and policy. CSIRO Publishing, Canberra, p 64

Braatz J, Harloff HJ, Mascher M et al (2017) CRISPR-Cas9 targeted mutagenesis leads to simultaneous modification of different homoeologous gene copies in polyploid oilseed rape (Brassica napus). Plant Physiol 174(2):935–942

Cai Y, Chen L, Liu X et al (2018) CRISPR/Cas9-mediated targeted mutagenesis of GmFT2a delays flowering time in soya bean. Plant Biotechnol J 16(1):176–185

Cermak T, Baltes NJ, Cegan R, Zhang Y, Voytas DF (2015) High-frequency, precise modification of the tomato genome. Genome Biol 16(1):232

Char SN, Neelakandan AK, Nahampun H et al (2017) An Agrobacterium-delivered CRISPR/Cas9 system for high-frequency targeted mutagenesis in maize. Plant Biotechnol J 15(2):257–268

Chen L, Yang D, Zhang Y et al (2018) Evidence for a specific and critical role of mitogen-activated protein kinase 20 in uni-to-binucleate transition of microgametogenesis in tomato. New Phytol 219(1):176–194

Civan P, Brown TA (2017) Origin of rice (Oryza sativa L.) domestication genes. Genet Resour Crop Evid 64(6):1125–1132

Desta ZA, Ortiz R (2014) Genomic selection: genome-wide prediction in plant improvement. Trends Plant Sci 19(9):592–601

Dhungana P, Eskridge KM, Baenziger PS et al (2007) Analysis of genotype-by-environment interaction in wheat using a structural equation model and chromosome substitution lines. Crop Sci 47(2):477–484

Du H, Zeng X, Zhao M et al (2016) Efficient targeted mutagenesis in soybean by TALENs and CRISPR/Cas9. J Biotechnol 217:90–97

Evans LT, Evans LT, Evans LT (1998) Feeding the ten billion: plants and population growth. Cambridge University Press, Cambridge

Fischer RA, Rees D, Sayre KD et al (1998) Wheat yield progress associated with higher stomatal conductance and photosynthetic rate, and cooler canopies. Crop Sci 38(6):1467–1475

Hefferon KL, Makhzoum A (2015) Biotechnological approaches for nutritionally enhanced food crop production. Adv Food Biotechnol 10(6):944–952

Heffner EL, Sorrells ME, Jannink JL (2009) Genomic selection for crop improvement. Crop Sci 49 (1):1–12

Hickey LT, Hafeez AN, Robinson H et al (2019) Breeding crops to feed 10 billion. Nat Biotechnol 37(7):744–754

Huang J, Li J, Zhou J et al (2018) Identifying a large number of high-yield genes in rice by pedigree analysis, whole-genome sequencing, and CRISPR-Cas9 gene knockout. Proc Natl Acad Sci 115 (32):E7559–E7567

Ito Y, Nishizawa-Yokoi A, Endo M et al (2015) CRISPR/Cas9-mediated mutagenesis of the RIN locus that regulates tomato fruit ripening. Biochem Biophy Res 467(1):76–82

Ito Y, Nishizawa-Yokoi A, Endo M et al (2017) Re-evaluation of the rin mutation and the role of RIN in the induction of tomato ripening. Nat Plants 3(11):866–874

Jacobs TB, La-Fayette PR, Schmitz RJ et al (2015) Targeted genome modifications in soybean with CRISPR/Cas9. BMC Biotechnol 15(1):16

Jain SM, Gupta SD (2013) Biotechnology of neglected and underutilized crops. Springer, Berlin

Jiang WZ, Henry IM, Lynagh PG et al (2017) Significant enhancement of fatty acid composition in seeds of the allohexaploid, Camelina sativa, using CRISPR/Cas9 gene editing. Biotechnol J 15 (5):648–657

Jouanin A, Schaart JG, Boyd LA et al (2019) Outlook for coeliac disease patients: towards bread wheat with hypoimmunogenic gluten by gene editing of α-and γ-gliadin gene families. BMC Plant Biol 19(1):333

Knauf VC, Facciotti D (1995) Genetic engineering of foods to reduce the risk of heart disease and cancer. In: Nutrition and biotechnology in heart disease and cancer. Springer, Boston, pp 221–228

Korte A, Farlow A (2013) The advantages and limitations of trait analysis with GWAS: a review. Plant Methods 9(1):29
Lawrenson T, Shorinola O, Stacey N et al (2015) Induction of targeted, heritable mutations in barley and Brassica oleracea using RNA-guided Cas9 nuclease. Genome Biol 16(1):258
Li M, Li X, Zhou Z et al (2016) Reassessment of the four yield-related genes Gn1a, DEP1, GS3, and IPA1 in rice using a CRISPR/Cas9 system. Front Plant Sci 7:377
Li R, Li R, Li X et al (2018a) Multiplexed CRISPR/Cas9-mediated metabolic engineering of γ-aminobutyric acid levels in Solanum lycopersicum. Biotechnol J 16(2):415–427
Li X, Wang Y, Chen S et al (2018b) Lycopene is enriched in tomato fruit by CRISPR/Cas9-mediated multiplex genome editing. Front Plant Sci 9:559
Li J, Li Y, Ma L (2019) CRISPR/Cas9-based genome editing and its applications for functional genomic analyses in plants. Small Methods 3(3):1800473
Liu J, Cheng X, Liu P et al (2017) miR156-targeted SBP-box transcription factors interact with DWARF53 to regulate TEOSINTE BRANCHED1 and BARREN STALK1 expression in bread wheat. Plant Physiol 174(3):1931–1948
Lowder LG, Zhang D, Baltes NJ et al (2015) A CRISPR/Cas9 toolbox for multiplexed plant genome editing and transcriptional regulation. Plant Physiol 169(2):971–985
Ma X, Zhang Q, Zhu Q et al (2015) A robust CRISPR/Cas9 system for convenient, high-efficiency multiplex genome editing in monocot and dicot plants. Mol Plant 8(8):1274–1284
Matson PA, Parton WJ, Power AG et al (1997) Agricultural intensification and ecosystem properties. Science 277(5325):504–509
McGinn M, Phippen WB, Chopra R et al (2019) Molecular tools enabling pennycress (Thlaspi arvense) as a model plant and oilseed cash cover crop. Biotechnol J 17(4):776–788
Michelmore RW, Shaw DV (1988) Character dissection. Nature 335(6192):672–673
Miralles DJ, Slafe GA (2007) Sink limitations to yield in wheat: how could it be reduced? J Agric Sci 145(2):139–149
Mohan JS, Suprasanna P (2011) Induced mutations for enhancing nutrition and food production. Gene Conserve 10:41
Mubarik MS, Khan SH, Ahmad A et al (2016) Disruption of phytoene desaturase gene using transient expression of Cas9: gRNA complex. Int J Agric Biol 18(5):990–996
Nakayasu M, Akiyama R, Lee HJ, Osakabe K, Osakabe Y, Watanabe B, Sugimoto Y, Umemoto N, Saito K, Muranaka T, Mizutani M (2018) Generation of α-solanine-free hairy roots of potato by CRISPR/Cas9 mediated genome editing of the St16DOX gene. Plant Physiol Biochem 131:70–77
Nishitani C, Hirai N, Komori S et al (2016) Efficient genome editing in apple using a CRISPR/Cas9 system. Sci Rep 6(1):1–8
Nonaka S, Someya T, Zhou S et al (2017) An Agrobacterium tumefaciens strain with gamma-aminobutyric acid transaminase activity shows an enhanced genetic transformation ability in plants. Sci Rep 7:42649
Odipio J, Alicai T, Ingelbrecht I et al (2017) Efficient CRISPR/Cas9 genome editing of phytoene desaturase in cassava. Front Plant Sci 8:1780
Okuzaki A, Ogawa T, Koizuka C et al (2018) CRISPR/Cas9-mediated genome editing of the fatty acid desaturase 2 gene in Brassica napus. Plant Physiol Biochem 131:63–69
Pan C, Ye L, Qin L et al (2016) CRISPR/Cas9-mediated efficient and heritable targeted mutagenesis in tomato plants in the first and later generations. Sci Rep 6:24765
Pearsall DM (2008) Plant domestication and the shift to agriculture in the Andes. In: The handbook of South American archaeology. Springer, New York, pp 105–120
Philipson T, Linthicum MT, Snider JT (2014) Tutorial on health economics and outcomes research in nutrition. JPEN Parenter Enter 38:5–16
Pretty J (2008) Agricultural sustainability: concepts, principles and evidence. Philos Trans R Soc B 363(1491):447–465
Ray DK, Mueller ND, West PC et al (2013) Yield trends are insufficient to double global crop production by 2050. PLoS One 8(6):e66428

Richards RA (2000) Selectable traits to increase crop photosynthesis and yield of grain crops. J Exp Bot 51:447–458
Ricroch A, Clairand P, Harwood W (2017) Use of CRISPR systems in plant genome editing: toward new opportunities in agriculture. Top Life Sci 1(2):169–182
Rodriguez-Leal D, Lemmon ZH, Man J et al (2017) Engineering quantitative trait variation for crop improvement by genome editing. Cell 171(2):470–480
Rustgi S, Shafqat MN, Kumar N et al (2013) Genetic dissection of yield and its component traits using high-density composite map of wheat chromosome 3A: bridging gaps between QTLs and underlying genes. PLoS One 8(7):e0070526
Sajjad M, Ma X, Khan SH et al (2017) TaFlo2-A1, an ortholog of rice Flo2, is associated with thousand grain weight in bread wheat (Triticum aestivum L.). BMC Plant Biol 17(1):164
Sanchez-Leon S, Gil-Humanes J, Ozuna CV et al (2018) Low gluten, non-transgenic wheat engineered with CRISPR/Cas9. Plant Biotechnol J16(4):902–910
Scheben A, Wolter F, Batley J et al (2017) Towards CRISPR/Cas crops–bringing together genomics and genome editing. New Phytol 216(3):682–698
Sedbrook JC, Phippen WB, Marks MD (2014) New approaches to facilitate rapid domestication of a wild plant to an oilseed crop: example pennycress (Thlaspi arvense L.). Plant Sci 227:122–132
Sedeek KE, Mahas A, Mahfouz M (2019) Plant genome engineering for targeted improvement of crop traits. Front Plant Sci 10:114
Sequeira A, Shen K, Gottlieb A et al (2019) Human brain transcriptome analysis finds region-and subject-specific expression signatures of GABA AR subunits. Commun Biol 2(1):1–14
Shao GN, Xie LH, Jiao GA et al (2017) CRISPR/CAS9-mediated editing of the fragrant gene Badh2 in rice. Chin J Rice Sci 31(2):216–222
Shen R, Wang L, Liu X et al (2017) Genomic structural variation-mediated allelic suppression causes hybrid male sterility in rice. Nat Commun 8(1):1–10
Shen L, Wang C, Fu Y et al (2018) QTL editing confers opposing yield performance in different rice varieties. J Integr Plant Biol 60(2):89–93
Shi J, Gao H, Wang H et al (2017) ARGOS 8 variants generated by CRISPR-Cas9 improve maize grain yield under field drought stress conditions. Plant Biotechnol J 15(2):207–216
Shimatani Z, Kashojiya S, Takayama M et al (2017) Targeted base editing in rice and tomato using a CRISPR-Cas9 cytidine deaminase fusion. Nat Biotechnol 35(5):441–443
Soyk S, Muller NA, Park SJ et al (2017) Variation in the flowering gene SELF PRUNING 5G promotes day-neutrality and early yield in tomato. Nat Genet 49(1):162
Sun Y, Zhang X, Wu C et al (2016) Engineering herbicide-resistant rice plants through CRISPR/Cas9-mediated homologous recombination of acetolactate synthase. Mol Plant 9(4):628–631
Sun D, Wang Q, Sun G et al (2019) Dissecting the genetic basis of grain size and weight in barley (Hordeum vulgare L.) by QTL and comparative genetic analyses. Front Plant Sci 10:469
Tang L, Mao B, Li Y et al (2017) Knockout of OsNramp5 using the CRISPR/Cas9 system produces low Cd-accumulating indica rice without compromising yield. Sci Rep 7(1):1–12
Tanksley SD (1993) Mapping polygenes. Annu Rev Genet 27(1):205–233
Thelen JJ, Ohlrogge JB (2002) Metabolic engineering of fatty acid biosynthesis in plants. Metab Eng 4(1):12–21
Tilman D, Balzer C, Hill J et al (2011) Global food demand and the sustainable intensification of agriculture. Proc Natl Acad Sci 108(50):20260–20264
Uluisik S, Chapman NH, Smith R et al (2016) Genetic improvement of tomato by targeted control of fruit softening. Nat Biotechnol 34(9):950–952
Viana VE, Pegoraro C, Busanello C et al (2019) Mutagenesis in rice: the basis for breeding a new super plant. Front Plant Sci 2019:10
Walker AR, Walker BF, Adam F (2003) Nutrition, diet, physical activity, smoking, and longevity: from primitive hunter-gatherer to present passive consumer—how far can we go? Nutrition 19 (2):169–173
Waltz E (2016) CRISPR-edited crops free to enter market, skip regulation. Nat Biotechnol 34 (6):582

Waltz E (2018) With a free pass, CRISPR-edited plants reach market in record time. Nat Biotechnol 36:6–7

Wang P, Xing Y, Li Z et al (2012) Improving rice yield and quality by QTL pyramiding. Mol Breed 29(4):903–913

Wang YJ, Zheng HP, Zhang B et al (2014) Cloning and respond of a cold shock domain protein (CnCSDP) gene to cold stress in noble scallop Chlamys nobilis (Bivalve: Pectinidae). Mol Biol Rep 41(12):7985–7994

Wang H, Studer AJ, Zhao Q et al (2015) Evidence that the origin of naked kernels during maize domestication was caused by a single amino acid substitution in tga1. Genetics 200(3):965–974

Wang T, Zhang H, Zhu H (2019) CRISPR technology is revolutionizing the improvement of tomato and other fruit crops. Hortic Res 6(1):1–3

Xie H, Zhu M, Tang G et al (2017) Progress in gene editing technique and its prospect in rice quality research. Agric Sci Technol 18(12):2246–2275

Xu C, Liberator KL, MacAlister CA et al (2015) A cascade of arabinosyltransferases controls shoot meristem size in tomato. Nat Genet 47(7):784–792

Yang H, Wu JJ, Tang T et al (2017) CRISPR/Cas9-mediated genome editing efficiently creates specific mutations at multiple loci using one sgRNA in Brassica napus. Sci Rep 7(1):1–13

Ye J, Wang X, Hu T et al (2017) An InDel in the promoter of Al-activated malate transporter9 selected during tomato domestication determines fruit malate contents and aluminum tolerance. Plant Cell 29(9):2249–2268

Yu QH, Wang B, Li N et al (2017) CRISPR/Cas9-induced targeted mutagenesis and gene replacement to generate long-shelf life tomato lines. Sci Rep 7(1):1–9

Zhai H, Feng Z, Du X et al (2018) A novel allele of TaGW2-A1 is located in a finely mapped QTL that increases grain weight but decreases grain number in wheat (Triticum aestivum L.). Theor Appl Genet 131(3):539–553

Zhang J, Zhang H, Botella JR et al (2018) Generation of new glutinous rice by CRISPR/Cas9-targeted mutagenesis of the Waxy gene in elite rice varieties. J Integr Plant Biol 60(5):369–375

Zhang Y, Li D, Zhang D et al (2018b) Analysis of the functions of Ta GW 2 homoeologs in wheat grain weight and protein content traits. Plant J 94(5):857–866

Zhu Q, Wang B, Tan J et al (2020) Plant synthetic metabolic engineering for enhancing crop nutritional quality. Plant Commun 1(1):100017

Zlobin NE, Lebedeva MV, Taranov VV (2020) CRISPR/Cas9 genome editing through in planta transformation. Crit Rev Biotechnol 2020:1–16

Zsogon A, Cermak T, Naves ER et al (2018) De novo domestication of wild tomato using genome editing. Nat Biotechnol 36(12):1211–1216

Applications of CRISPR/Cas Beyond Simple Traits in Crops

8

Sultan Habibullah Khan, Muhammad Zubair Ghouri, Sabin Aslam, Muhammad Salman Mubarik, Zulqurnain Khan, Muhammad Qadir Ahmad, and Muhammad Sajjad

Abstract

Advancements in genome manipulation have transformed the whole face of synthetic biology. Technologies like CRISPR/Cas based gene editing offer myriad ways to tweak genomes and help plant scientists to fulfil what they have been wishing for since the advent of molecular biology during the mid-nineties. The past decade witnessed an enormous adoption of this technology in almost all branches of biology from agriculture to medicine and environment. After successfully deploying gene editing to modify simple traits, scientists are now embarked on more ambitious adventures in genomics to combat challenges of food security in the wake of increasing population and climate change

S. H. Khan (✉)
Centre of Agricultural Biochemistry and Biotechnology (CABB), University of Agriculture, Faisalabad, Pakistan

Center for Advanced Studies in Agriculture and Food Security (CASAFS), University of Agriculture, Faisalabad, Pakistan
e-mail: sultan@uaf.edu.pk

M. Z. Ghouri
Center for Advanced Studies in Agriculture and Food Security (CASAFS), University of Agriculture, Faisalabad, Pakistan

S. Aslam · M. S. Mubarik
Centre of Agricultural Biochemistry and Biotechnology (CABB), University of Agriculture, Faisalabad, Pakistan

Z. Khan
Department of Biotechnology, Institute of Plant Breeding and Biotechnology (IPBB), MNS University of Agriculture, Multan, Pakistan

M. Q. Ahmad
Department of Plant Breeding and Genetics, Bahauddin Zakariya University, Multan, Pakistan

M. Sajjad
Department of Biosciences, COMSATS University Islamabad (CUI), Islamabad, Pakistan

A. Ahmad et al. (eds.), *CRISPR Crops*,
https://doi.org/10.1007/978-981-15-7142-8_8

adversaries. The chapter outlines use of gene editing in tailoring crops beyond simple traits aiming to harvest the desired diversity lost during domestication and manipulating complex traits which evolved over evolutionary timescale like photosynthesis, nitrogen fixation, and epigenetics.

8.1 Versatility of CRISPR/Cas System to Manipulate Complex Traits

After the first observation of clustered regularly interspaced short palindromic repeats (CRISPR) in *Escherichia coli* in 1987 (Ishino et al. 1987), detailed sequence and bioinformatics studies soon uncovered CRISPR-associated (Cas) genes (Jansen et al. 2002a, b) and by the year 2011, CRISPR-Cas9 system was successfully engineered to work as a genome editing tool in eukaryotes (Deltcheva et al. 2011). Cas9 was the first CRISPR-associated protein characterized and used to induce site-specific double strands breaks in DNA based on its interaction with crRNA (CRISPR RNA) (Garneau et al. 2010). Due to remarkable applications in biological research, genome editing has been in the limelight in terms of its widespread use in major fields of biology and a rapid boom in research publications. The preceding decade witnessed not only a steep rise in its applications, but also identification and characterization of more Cas proteins with different functions and substrate affinities enabling biologists to manipulate genetic instructions in multiple ways (Fig. 8.1).

Classification of Cas proteins is continuously changing as more and more new variants are being added rapidly. There are two main classes which are further divided into five types and sixteen subtypes (Burstein et al. 2017). A multi-subunit crRNA effector complex is present in Class I (Types I, III, and IV), while Class II (types II and V) contains a single subunit crRNA effector complex (Shmakov et al. 2018).

Cas1 protein along with alpha helical fold (an integrase) cleaves the specific sites and this cleavage allows the new spacers to be inserted into CRISPR arrays (Nuñez et al. 2014). In archaea, Cas1 enzyme may also catalyze the movement of "Casposons" (genetic elements like transposons). Integrase activity of Cas1 in casposon prefers strong target site for duplication (Krupovic et al. 2016). In several toxin–antitoxin systems, Cas2 protein performs a function which is homologous to mRNA interferase (Makarova et al. 2011a, b). Cas1 protein forms a complex with Cas2 protein which is responsible for adaptation module in CRISPR-Cas mechanism. Cas2 protein possesses DNase and RNase catalytic activities and involves in spacer acquisition (Nam et al. 2012a, b). Cas3 is a signature protein in Type I CRISPR-Cas system which is a single stranded DNA nuclease and ATP dependent helicase. It contains seven motifs and codes for superfamily 2 helicase which is stimulated by single stranded DNA (ssDNA) and causes separation of RNA–DNA duplexes and double stranded DNA (Sinkunas et al. 2011). Amino terminal end of Cas3 protein has endonuclease domain and fusion of helicase domain with endonuclease domain is responsible for cutting of target DNA at specific sites. Cas3 is encoded by another gene known as cas3 (Mulepati and Bailey 2011). Cas4 has exonuclease activity and binds with RecBCD to form a complex which stimulates

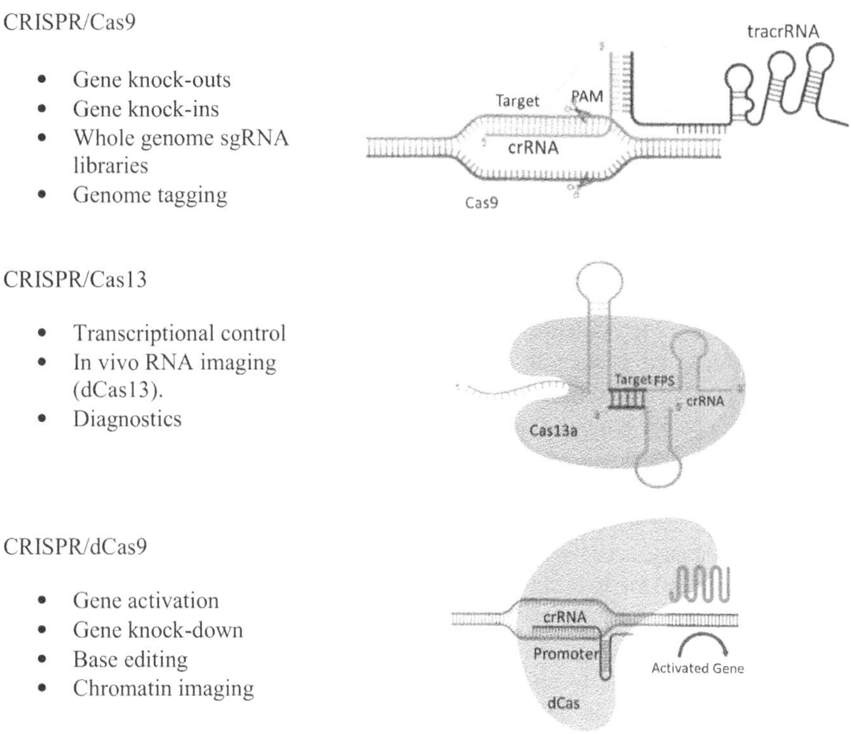

Fig. 8.1 Functional diversity of Cas proteins

homologous recombination of double strand breaks. DNA metabolism processes, e.g. chromosome segregation, regulation of transcription, and homologous recombination may be stimulated by Cas3 and Cas4 proteins. Activity of different CRISPR-Cas proteins may involve in mechanism of CRISPR loci genesis. Compared to other CRISPR-Cas proteins, Cas4 proteins are less common (Jansen et al. 2002a, b).

Cas5 is a small subunit protein which catalyzes the crRNA processing. Its subtype Cas5d is involved in processing of pre CRISPR-RNAs. Cas6 has nuclease activity which ensures the processing of crRNA. It is not effector complex subunit (Nam et al. 2012a, b). Cas7 is mostly functional in crRNA-effector complexes and plays a significant role in co-transcriptional RNA cleavage during interference (Samai et al. 2015). Cas8 is the only protein which has no significant sequence similarity with other subtype proteins. It possesses diverged sequence and considered as a functional homologue of Cas10 protein (Makarova et al. 2011a, b). During interference stage, single stranded DNA molecule can be degraded by nuclease activity of some Cas8 proteins (Cass et al. 2015).

Cas9 is a Type II signature protein which is crRNA dependent nuclease. It contains various domains to form crRNA effector complex and mediate DNA cleavage. Cas9 protein core contains both nuclease-based domains (HNH and

RuvC domain) and G-rich PAM preferred sequences (Jinek et al. 2012a, b). In Cas9 crRNA effector complex, the target site for DNA cleavage is 3 bp upstream of the PAM region (Garneau et al. 2010). Cas9 has its PAM sequence at the end of protospacer. DNA cleavage is only processed in the presence of both tracrRNA and crRNA. TracrRNA is required for processing of pre CRISPR-RNA. HNH and RuvC domains catalyze the degradation of the target DNA strand and non-target DNA strand, respectively. Inactivation of either RuvC or HNH domains converts Cas9 into DNA nickase. Fusion of crRNA and tracrRNA results in the formation of single guided RNA and this single guided RNA along with Cas9 mediates RNA-guided DNA cleavage (Carroll 2012). Cas10 is a signature gene in type III CRISPR systems. It contains nucleic acid polymerase and nucleotide cyclase activities which are present in the form of an RNA recognition motif (RRM) (Shao et al. 2013). In interference stage, double stranded DNA is cleaved by catalytic site of active cyclase domain (Samai et al. 2015).

Cas12a (Cpf1) is classified into type V of Class II. It is a large protein with 1300 amino acids which contains a RuvC like nuclease domain but lacks the HNH domain (Schunder et al. 2013). T-rich PAM sequence is preferred by Cas12a for target site. Cas12a has nuclease activity which generates overhangs by cutting DNA molecules (Zetsche et al. 2015). It lacks tracrRNA and uses its RNase activity to process pre-crRNA to form mature crRNA molecule (Fonfara et al. 2016). Cas12a has applications in therapeutics. As it produces overhangs instead of blunt ends DNA breaks, it may be useful in introduction of new sequences (Moreno-Mateos et al. 2017). Cas12b (C2c1) is dual-RNA-guided DNA nuclease which also requires tracrRNA for further processing. Cas12c, Cas12d (CasY), Cas12e (CasX), and Cas12f (Cas14) are other members of Cas12 family (Harrington et al. 2018).

Cas13 falls under type VI CRISPR-Cas system which is an RNA-guided ribonuclease. Cas13a lacks DNase domain but consists of two HEPN domains and linker domain and usually does not cut short RNA targets (Shmakov et al. 2015). Cas13a and Cas13b possess RNase domain for processing of crRNA (Smargon et al. 2017). To target RNA, the specific nucleotides are located in the upstream of duplex of double stranded RNA. Out of two HEPN domains, one turns into major groove of guided RNA duplex (Liu et al. 2017). RNA cleavage only proceeds in the presence of target RNA strand and HEPN domain. RNA molecule and HEPN domain interact with each other to form a crystal packing and this interaction is required for cleavage activity. crRNA maturation and RNA-guided RNA degradation are dual functionality of Cas13a (East-Seletsky et al. 2016).

The endonuclease deficient Cas9 (dCas9), also known as dead Cas9, functions as a base editor. It deaminates adenosine to inosine which is read as guanine by the translation machinery (Källman et al. 2003). Similarly, DNA base editors may generate point mutations from cytosine to thiamine (Komor et al. 2017). Fusion of dCas9 with different functional domains can mediate diverse modifications including transcriptional and epigenetic changes. Similar version of CRISPR-Cas13 along with RNA-guided ribonuclease can be used as base editor for editing mRNA sequences by editing adenosine to inosine. Another study reports base editing of mRNA in a programmable manner (Cox et al. 2017) known as RNA editing for

Table 8.1 Cas proteins which have been tested for gene editing in plants

Cas proteins	Sources	Plant	Sequence of PAM	Reference
SpCas9	*Streptococcus pyogenes*	*Arabidopsis*	NGG	Lin et al. (2018)
StCas9	*Streptococcus thermophilus*	*Arabidopsis*	NNGGAA	Steinert et al. (2015)
SaCas9	*Staphylococcus aureus*	*Arabidopsis*	NNGAA	Steinert et al. (2015)
LbCas12a	*Lachnospiraceae* bacterium ND2006	*Arabidopsis*	TTTV	Tang et al. (2017)
AsCas12a	*Acidaminococcus* sp. BV3L6	*Arabidopsis*	TTTV	Tang et al. (2017)
FnCas12a	*Francisella novicida*	*O. sativa*	TTV	Endo et al. (2016)
LwaCas13a	*Leptotrichia wadei*	*O. sativa*	PAM-independent	Abudayyeh et al. (2018)
LshCas13a	*Leptotrichia shahii*	*N. benthamiana*	PAM-independent	Aman et al. (2018)

programmable A to I replacement (REPAIR). A summary of different Cas proteins which have been tested in plants is given in Table 8.1.

8.2 Rapid Domestication of Wild Species to Harvest the Genetic Treasure Lost in Natural Domestication

Current crop agriculture owes to domestication of natural plant populations since the start of humanity. This human driven evolution has been, no doubt, a continuous source of food, feed, clothes, and shelter. However, the domestication pressure gradually took away a huge part of genetic diversity in terms of reduction in nutritional value and taste (Gruber 2017). With the advent of genetics in the late nineties, plant scientists, especially breeders, further narrowed the useful gene pool by their targeted breeding programs aiming for a few economical gains. The ongoing climate change scenario has further evidenced another genetic toll in the form of susceptibility to biotic and abiotic stresses. The mother nature, however, preserved the eroded genetic diversity in the form of wild species and relatives. Ranging from climate resilience to yield and quality, the wild plant species have all the traits which modern scientists wish to have in cultivated species (Hufford et al. 2013; Sawler et al. 2013). Due to incompatible modes of reproduction, scientists have only been able to bring back a fraction of genetic diversity through introgression or pre-breeding (Cooper et al. 2001; Dwivedi et al. 2008; Ogbonnaya et al. 2013).

Advancement in genome sequencing has greatly facilitated pinpointing the nucleotide changes shaping the domestication and evolution of crops. Comparative genomics based on complete genome sequences have shown that domestication and targeted breeding have been facilitated by traits due to loss of function or gain of function mutations in genes working independently or as a part of metabolic pathways. A comparison of genome sequence of model cereal maize and its proginitors best describes the genetic changes that resulted the current plant type, sink–source relationships and reduced tolerance to the environmental changes during its domestication from the closest progenitor teosinte. The hugely distinct morphology of maize from that of teosinte has been attributed to only six loci (Doebley et al. 1990). The first locus is teosinte *branched1* (*tb1*) which originated as a result of gain of function mutation in a transcription regulator gene which lowered side branching, altered sink–source interaction, and increased plant yield. The second mutation was the *grassy tillers1* (*gt1*) which was a mutation in the promoter sequence that reduced the number of ears per plant. The upright leaf morphology of maize was due to lower expression of squamosa promoter-binding protein and a bZIP transcription regulator due to *liguleless1* (*lg1*) and *liguleless2* (*lg2*) mutations, respectively. Another domestication trait was the increased number of kernel rows which was also due to point mutations causing amino acid substitutions in the Leucine Rich Repeat (LRR) domain converting *FAS-CIATED EAR2* (*FEA2*) gene of teosinte into maize *CLAVATA2*. Similarly, encased kernels of teosinte were converted into naked kernels in maize due to a single nucleotide change in the SBP-box transcriptional regulator resulting *teosinte glume architecture1* (*tga1*). Spread of maize to the temperate regions was due to photoperiod insensitive phenotype that resulted due to a transposon insertional mutation of a photoperiod associated CCT-containing protein causing loss of function. Likewise, domestication of rice and wheat has also been mostly due to simple mutations (Jiao et al. 2010; Kovach et al. 2007; Sang and Ge 2007; Hedden 2003; Livne et al. 2015). A large number of such domestication genes have been characterized in all major crops as shown in Table 8.2.

For a quite long time, there has been a debate among plant scientists to develop ways to enrich the exhausted genetic diversity of cultivated crops. Having this ability would permit plant scientists to design crops with the right phenotype for the right environment by harvesting the huge diversity present in the wild species. Increasing availability of genome sequence of crop plants especially model plants can be very helpful to identify orthologs in wild plants for targeted manipulation. CRISPR/Cas9 based genome editing has indeed opened new avenues to achieve precise mutations mimicking domestication (Meyer and Purugganan 2013). Multiplexing ability of CRISPR system to target several loci/targets simultaneously is particularly useful in de novo domestication of wild species. A proof of concept study (Fig. 8.2) has already been reported recently whereby the genome of *Solanum pimpinellifolium*, a wild species of tomato, was simultaneously edited at six loci in a single generation to produce domestication traits related to yield and productivity comparable to the cultivated *S. lycopersicum* but having all the traits from the wild species (Zsögön et al. 2018). Similarly, gene editing of promoters can be used to reverse the epistatic effects developed during domestication which have also been recently demonstrated (Soyk et al. 2017).

By deploying such a targeted editing, development of new and improved crops can be put on a fast track to effectively address the ongoing challenges of food

Table 8.2 Major field crops with already characterized domestication genes

Crops	Domestication genes	Function	Nature of mutation	Reference
Barley	*Btr1/2*	BTR1/2-LIKE proteins	Loss of function	Pourkheirandish (2015)
	Vrs1	Transcription factor	Gain of function	Komatsuda et al. (2007)
	Nud	Transcription factor	Gain of function	Taketa et al. (2008)
	INT-C (HvTB1)	Transcription factor	Gain of function	Ramsay et al. (2011)
Maize	*ZmSh1-1/1-5*	Transcription factor	Loss of function	Wang et al. (2005)
	Tb1	TCP-family transcription	Gain of function	Studer et al. (2011)
	ZmCCT	CCT-domain containing proteins	Loss of function	Yan et al. (2003)
	tga1	SBP-box transcription factor	Gain of function	Wang et al. (2014)
	lg1	Squamosa-promoter binding protein	Loss of function	Liu et al. (2017)
Rice	*Sh1*	Transcription factor	Loss of function	Lin et al. (2012a, b)
	GIF1	Cell wall invertase	Gain of function	Wang et al. (2008)
	PROG1	Transcription factor	Gain of function	Jin et al. (2008a, b)
	SD1	Gibberellin biosynthetic enzyme	Gain of function	Asano et al. (2011)
	Bh4	Amino acid transporter	Gain of function	Zhu et al. (2011)
Sorghum	*SbSh1*	Transcription factor	Loss of function	Lin et al. (2012a, b)
Sunflower	*HaFT1*	Transcription factor	Gain of function	Blackman et al. (2010)
Soybean	*SHAT1-5*	NAC-family transcription factor	Gain of function	Dong et al. (2014)
	DT1	CETS family of regulatory genes	Loss of function	Cai et al. (2015a, b)
	GA20ox	Gibberellin biosynthesis enzyme	Loss of function	Lu and Zhu (2017)
	G	Protease	Gain of function	Wang et al. (2018)
Tomato	*Style2.1*	Transcription factor	Loss of function	Chen et al. (2007)
	Fw2.2	Similar to human RAS	Gain of function	Frary et al. (2000)

(continued)

Table 8.2 (continued)

Crops	Domestication genes	Function	Nature of mutation	Reference
	SolyG	Protease	Gain of function	Wang et al. (2018)
Wheat	*Vrn1*	Transcription factor	Gain of function	Yan et al. (2003)
	Vrn2	Transcription factor	Gain of function	Yan et al. (2004)
	Q	Transcription factor	Gain of function	Simons et al. (2006)

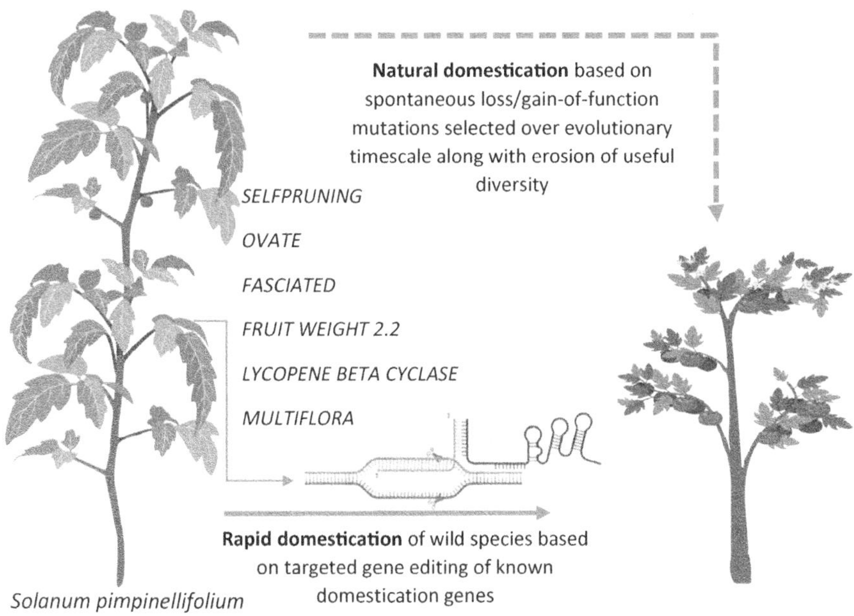

Fig. 8.2 Natural domestication vs. rapid domestication. Rapid domestication of wild tomato (*Solanum pimpinellifolium*) has been demonstrated by simultaneous editing of six loci corresponding to general plant growth habit (Self-pruning), fruit shape (Ovate), fruit size (Fasciated and Fruit Weight 2.2), fruit number (Multiflora), and nutritional quality (Lycopene Beta Cyclase) (Zsögön et al. 2018)

security. Table 8.3 summarizes major crops and their wild relatives having potential of rapid domestication using the latest gene editing technologies.

Table 8.3 Major wild species with a potential for rapid domestication

Crop Species	Candidate wild species for rapid domestication potential	Relevant traits	References
Triticum aestivum	*T.* turgidum L. ssp. Durum, *Triticum dicoccoides*	Improved grain quality, higher low molecular weight glutenins	Hoisington et al. (1999)
Zea mays	*Teosinte*, *Tripsacum*	Apomixis, tolerance to biotic and abiotic factors	
Oryza sativa	*Oryza longistaminata*, *O. rufipogon*	Metal and drought tolerance	Sanchez et al. (2013)
Glycine max	*Glycine soja*	Improved root biomass and architecture	Prince et al. (2015)
Manihot esculenta	*Manihot glaziovii*, *M. neosana*	Apomixis, larger roots	Nassar et al. (2008)
Potato (*Solanum tuberosum*)	*Solanum demissum*, *S. stoloniferum*	Disease resistance	Machida-Hirano et al. (2015)

8.3 Genome Editing to Improve Photosynthesis, the Ultimate Yield Trait

Meeting the increasing food demands due to increasing population is the greatest challenge for today's plant scientists. According to the U.S. Department of Commerce and Census Bureau, the current seven billion population is likely to touch the figure of 9.5 billion by 2050 reflecting a net increase of 35% in the next 20 years (USCB 2015). Due to a continuing trend in urbanization, demand for processed and fortified foods with additional meat and dairy products will rise as compared to direct consumption of staples. This trend will in turn require production of 85% more primary foodstuffs by the year 2050 (Ray et al. 2013). Furthermore, dwindling crop production environment as a result of global climate change requires plant scientists to work out new and more robust ways to keep balance between food supply and demand.

After ensuring the food security for over 50 years, the crop varieties of the green revolution have eventually reached their biological limits which is very much reflected by the ongoing stagnancy in yield increases over the past few decades. The yield potential of a plant is estimated how efficiently it intercepts solar radiation (εi) for transformation into biomass energy (εc) and partition it into the final plant harvest (εp). The green revolution made significant strides in the first (εi) and the third (εp) yield determinants leaving behind only marginal scope of improvement in the future (Zhu et al. 2008, 2010). It was easy for breeders to exploit large variation in traits responsible for solar interception and carbon partitioning like biomass reserves in leaves during vegetative stage and leaf growth, size, and longevity. On the other hand, efficiency of converting solar energy into biomass energy has remained evolutionary conserved not only within plant species but also across families. Therefore, this aspect of plant improvement has only been tapped a little (0.02) and will be the major focus area of plant genome engineering in future to keep yield increase pace at par with the increasing food demands (Long et al. 2015; Zhu et al. 2010).

Having a primordial importance in the food chain, photosynthesis is the most studied plant process in plants. From capturing light by chlorophyll to production of storage carbohydrates, the details of major photosynthetic processes are fully understood. The mechanistic differences in higher plants, algae, and photosynthetic prokaryotes have been deciphered and the corresponding genes and proteins involved in different pathways have been characterized. The scientists have now developed kinetic models for both C3 and C4 photosynthesis (Liu et al. 2013; Wang et al. 2014; Zhu et al. 2013). Being a conserved plant process, any improvement in one plant species can be reproduced in the others. Recent advancements in high-performance computing (HPC) have enabled simulation of kinetic models for photosynthesis to understand and predict clues to optimize the processes and resources from cellular to whole plant level (Drewry et al. 2014; Tholen et al. 2012; Tholen and Zhu 2011). HPC based in silico studies have shown what genes are needed to be upregulated or overexpressed and what gene products are negatively involved to be the candidates for downregulation or replacement in order to improve photosynthetic performance (McGrath and Long 2014; Xin et al. 2015). The pre- and postgenomic advancements have successfully set the stage for manipulation of this crucial process which could be a transformative strategy to regain the pace of yield increases (Simkin et al. 2017a, b; Kromdijk et al. 2016; Driever et al. 2017; Lopez-Calcagno et al. 2018; Barampuram and Zhang 2011).

Gene editing based approaches have a great potential to improve crop productivity by enhancing the photosynthetic efficiency of plants (Zhu et al. 2007a, b; Ort et al. 2015; Rae et al. 2017; South et al. 2018a, b). The immediate application of CRISPR/Cas system will be its application as a functional genomics tool to elucidate the less understood photosynthetic pathways to identify novel targets to fine-tune photosynthesis. These include C4, crassulacean acid metabolism, and photorespiration. Many protein components responsible for thylakoid assembly and degradation are still unknown. Likewise, factors regulating processes of electron transfer and carbon metabolism are not fully understood. In the past, functional studies of CAM pathway using chemical and physical mutagens (Cushman et al. 2008) and RNAi (Dever et al. 2015; Boxall et al. 2017) have been attempted. The former methodology produces random mutants requiring hefty screening and sometimes complicates developmental features owing to mutations in other parts of the genome. The RNAi based functional studies have also limitations like off target activities and incomplete degradation of RNA (Xu et al. 2006; Liu et al. 2016). Functional studies performed with CRISPR/Cas system in C_3 and C_4 plants are precise and easy to perform than the above conventional approaches (Feng et al. 2013; Shan et al. 2013; Xie et al. 2015; Liu et al. 2016; Shimatani et al. 2017; Kang et al. 2018; Salesse-Smith et al. 2018). With a small genome and easy to handle in tissue culture, *Kalanchoë fedtschenkoi* has recently emerged as a model plant for functional studies of CAM pathway (Boxall et al. 2017; Yan et al. 2004) and provides an excellent platform to

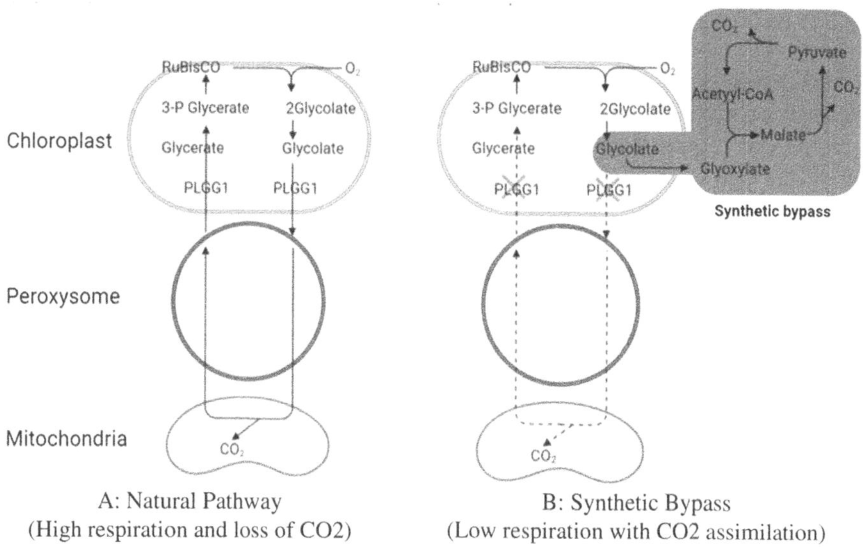

Fig. 8.3 Engineering of *Escherichia coli* glycolate oxidation pathway in tobacco plants (South et al. 2018a, b)

apply CRISPR/Cas based gene editing. A biallelic indel induced using CRISPR/Cas in the blue light receptor phototropin 2 (*KfePHOT2*) in Kalanchoë fedtschenkoi has revealed the importance of blue light signaling in the CAM plants. The mutants showed higher stomatal conductance and CO2 fixation during the night and lower levels of stomatal conductance and CO2 fixation in the late afternoon (Liu et al. 2019). In the long-term perspective, however, CRISPR/Cas will predominantly be focused on improving plant's capacity of reducing photorespiration, capturing more sunlight, and increasing flux through the Calvin cycle (South et al. 2018a, b).

In all organisms with oxygenic photosynthesis, photorespiration is an essential metabolic repair mechanism which is consisted of nine enzymatic reactions involving chloroplast, peroxisome, and mitochondrion (Fig. 8.3). In this process, 2-PGlycolate is converted into intermediate 3-PGlycerate which consumes 25% of previously fixed CO_2 in addition to energy. This process can reduce photosynthetic efficiency of C3 crops by 20–50%. This makes photorespiration a potentially wasteful process (Bauwe et al. 2010) and a prime target for genome engineering. To achieve complete metabolism of glycolate through a synthetic bypass, South et al. (2018a, b) engineered an alternative photorespiratory pathways in tobacco plants by transforming additional genes in tobacco chloroplasts from *Escherichia coli* glycolate oxidation pathway; glycolate oxidase and malate synthase from plants and catalase from *E. coli*; and the third pathway used plant malate synthase and a green algal glycolate dehydrogenase (Fig. 8.3). To limit the metabolite flux in the native pathway, they also knocked down chloroplastic glycolate transporter in the photorespiratory pathway using RNA interference (RNAi) approach. The resulting plants showed enhanced CO2 fixation activity of RuBisCO along with 40% yield gains.

Several steps involved in Calvin cycle can be manipulated for improving the process of photosynthesis. Functional studies have shown that decreasing the enzymes of this cycle resulted decrease in plant growth, decrease in assimilation of Co2, and ultimately decrease in photosynthesis (Lawson et al. 2006). Sedoheptulose-1,7-bisphosphatase (SBPase) and RuBisCO are the two major enzymes that control the flux of Calvin cycle (Poolman et al. 2000). Zhu et al. (2007a, b) proposed that increase in SBPase like enzymes along with the decrease in photorespiration will result in increased carbon assimilation and ultimate increase in the efficiency of photosynthesis. Moreover, by increasing the expression of any gene of Calvin cycle will also impart an increase in the efficiency of photosynthesis. For example, over-expression of SBPase enzyme in tobacco, Arabidopsis, and tomato resulted in an increase in biomass and carbon assimilation (Simkin et al. 2017a, b). Tamoi et al. (2006) demonstrated 20% increase in carbon assimilation in tobacco by over-expression of SBPase enzyme. By increasing the metabolites, i.e. increase of carbon assimilation will increase the efficiency of SBPase enzyme that ultimately increase in growth and development of plants. Driever et al. (2017) experimentally proved the increase of photosynthetic rate in wheat by increasing the activity of SBPase. SBPase enzyme and fructose-1,6-bisphosphate aldolase (FBPA) ultimately control the flow of carbon in Calvin cycle that runs process of photosynthesis. Manipulations in these enzymes may also increase photosynthesis. Uematsu et al. (2012) showed that over-expression of FBPA enzyme in tobacco resulted an increase in plant growth and process of photosynthesis. The over-expression of both SBPase and FBPA in tobacco has increased the biomass by 62% as compared to single gene over-expression (Simkin et al. 2015). Therefore, manipulation of the genes that are involved in the Calvin cycle can improve carbon assimilation, photosynthesis, and biomass of the plant.

Manipulations in the electron transport pathway can also result in improving the efficiency of photosynthesis by increasing the assimilation of carbon and increase of biomass. Manipulations in electron transport pathway and Calvin cycle have enabled the plants to response to the fluctuations of light in environment (Slattery et al. 2018). A summary of the prospects associated with modifications in different pathways to improve the efficiency of photosynthesis is given in Table 8.4.

8.4 Modulation of Gene Expression and Epigenetic Controls

CRISPR/Cas9 system offers a flexible, precise, and efficient tool to achieve a diverse range of genome modifications beyond gene editing (Gilbert et al. 2014). Utility of CRISPR/Cas system beyond gene editing emerged mainly because of unique platforms which the catalytically impaired or dead Cas9 (dCas) provides (Qi et al. 2013). Both HNH and RuvC catalytic domains are essential for creating DNA DSBs (Nishimasu et al. 2014) as each domain cleaves one strand of the double helix. A single point mutation in either of the domain converts it into a nickase, while

Table 8.4 Photosynthetic manipulations and expected outcomes

Manipulation	Efficiency gain (%)	References
Improving the usable spectrum of crop photosynthesis into NIR	10–30	Blankenship et al. (2011)
Improving rapid relaxation of heat dissipation at PSII	30	Zhu et al. (2007a)
Conversion of C3 plants to C4	30	Long et al. (2015) von Caemmerer et al. (2012)
Integration of cyanobacterial or microalgal CO_2/HCO_3 pumps in plants	5–10	McGrath and Long (2014)
Expression of cyanobacterial carboxysome system in plants	60	McGrath and Long (2014)
Expression of algal pyrenoid CO_2 concentrating system in plants	60	McGrath and Long (2014)
Improving RuBisCO's adaptability to CO_2	15–30	Zhu et al. (2008) Zhu et al. (2007b)
Expression of synthetic photorespiratory bypasses in plants	15	Long et al. (2015) Xin et al. (2015)
Optimization of regeneration of RubP	60	Zhu et al. (2007)
Improving light transmission to lower canopy leaves	15–60	Drewry et al. (2014) Long et al. (2015)

simultaneous mutations in both domains abolish its nuclease activity altogether (Jinek et al. 2012a, b). The dead Cas9 protein, though does not show nuclease activity, its ability of targeting to the specified location through gRNA offers an array of activities including manipulation of expression, editing of epigenome, and chromatin visualization (Adli 2018, Fig. 8.4). Binding of RNA-guided dCas9 on the target DNA sequence is strong enough to interfere with the regulatory proteins and RNA polymerase (Qi et al. 2013). This offers a unique platform to knockdown the expression of any gene. Addition of a repressor protein to dCas9 further improves repression as compared to dCas9 protein alone. This has been demonstrated by fusing Kruppel associated Box (KRAB, a zinc finger based transcriptional repressor protein) to dCas9 (Gilbert et al. 2013). KRAB proteins have been reported to attract additional co-repressors like KAP-1 (KRAB-box-associated protein-1) and HP1 (heterochromatin protein 1) which further suppress the expression of the target gene (Friedman et al. 1996). Similarly, the dCas9 platform can also be used to recruit transcriptional activators for upregulating gene expression (Fig. 8.4a) (Perez-Pinera et al. 2013a, b).

Epigenetics is a multi-layer regulatory mechanism operating to control gene expression patterns. These control mechanisms include DNA methylation, histone posttranslational modifications, and non-coding RNAs. DNA methylation is the most studied epigenetic mechanism which typically acts to repress gene transcription. DNA methylation is catalyzed by DNA methyltransferases (DNMTs). The nuclease deficient dCas9 fused with the catalytic domain of a DNA methyltransferases can specifically direct methylation of CpG islands. During this

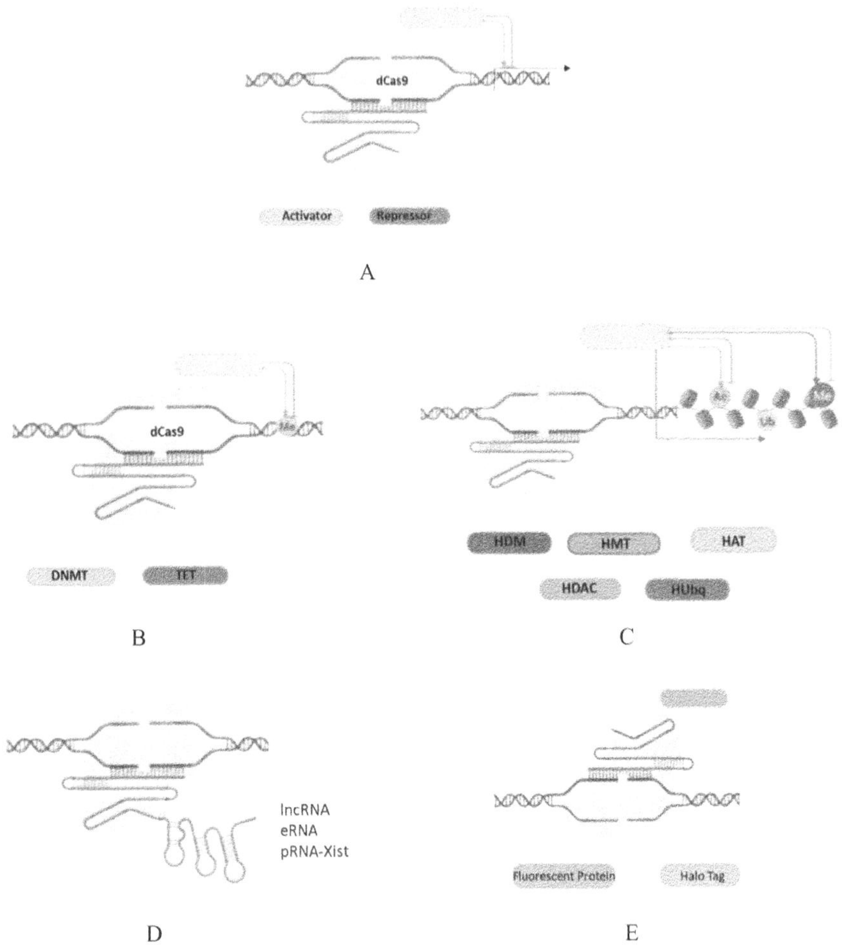

Fig. 8.4 CRISPR/dCas9 platforms to manipulate expression and epigenome modifications. (**a**) Transcriptional regulation. (**b**) DNA methylation. (**c**) Histone modifications. (**d**) RNA relocation. (**e**) Chromatin visualization. Colored boxes indicate different chromatin modifiers: *HDM* histone demethylase, *HMT* histone methyltransferase, *HAT* histone acetyltransferase, *HDAC* histone deacetylase, *HUbq* histone ubiquitin ligase, *DNMT* DNA methyltransferase, *TET* ten-eleven translocation enzymes. *Ac* acetylation, *Me* methylation, *Ub* ubiquitination

process, a methyl group is added to the 5th carbon of cysteine (5mC) or rarely to the 6th nitrogen of adenine (6 mA) that leads to transcriptional repression (Smith and Meissner 2013). Cysteine methylation is carried out by a set of DNMTs including DNMT3A and DNT3B (Chédin et al. 2002). To upregulate gene expression, a thymine DNA glycosylase (TDG) and ten-eleven translocation enzyme (TET1) can be fused with dCas9 to cause demethylation by removing methyl group from the methylated nucleotides (Fig. 8.4b) (Amabile et al. 2016a, b).

Posttranslational modification of histone tails is the major activity of epigenetic control mechanism to regulate gene expression both spatially and temporally. These modifications are catalyzed by a variety of epigenetic readers, writers, and erasers proteins (Strahl and Allis 2000). Chromatin modifying enzymes include histone demethylase (HDM), histone methyltransferase (HMT), histone acetyltransferase (HAT), histone deacetylase (HDAC), histone ubiquitin (HUbq), DNA methyltransferases (DNMT), and ten-eleven translocation enzymes (TET). Utilization of dCas9 protein in this area is mostly to either add histone methylation or acetylation or remove such epigenetic marks. While depending on position, histone methylation can be associated with both up- or downregulation of a gene, but histone acetylation always indicates active gene promoters and enhancers. Attachment of dCas9 with chromatin modifiers has shown epigenetic effects such as chromatin decondensation, histone tri-methylation of H3 at lysine 4, histone acetylation of H3 at lysine 27, and binding of Pol II to initiate mRNA transcription (Fig. 8.4c) (Chavez et al. 2016). These dCas9 mediated modifications can lead to robust gene expression.

Chromatin remodeling is also mediated by non-coding RNAs (ncRNAs) which act as cis-regulatory elements (Koerner et al. 2009a, b). Short ncRNAs, which are typically <30 nucleotides long, cause gene silencing and heterochromatin formation (Holoch and Moazed 2015) and are also involved in regulating gene imprinting process (Koerner et al. 2009a, b). On the other hand, long non-coding RNAs (lncRNAs) which are >200 nucleotides long are able to deploy chromatin-modifying effectors and result RNA–protein complexes to change chromatin topology (Panning et al. 1997; Pennisi 2013). Using dCas9, CRISPR/Cas9 system also allows incorporation of ncRNAs into gRNA sequence without lowering its efficiency to bind to specific target DNA (Shechner et al. 2015a, b) and mediate the required effect on chromatin structure (Fig. 8.4d). Some successful epigenetic medications made through CRISPR/dCas technology are listed in Table 8.5.

Chromatin is a dynamic structure and understanding its biology requires various visualization techniques (Essers et al. 2005). Most commonly used methods include chromosome conformation capture (3C) and DNA fluorescence in situ hybridization (FISH) (Croft et al. 1999; Dekker et al. 2013). Fluorescently tagged ZFPs and TALEs DNA-binding protein have been used to study repetitive sequences (Robinett et al. 1996; Hellwig et al. 2008). While in CRISPR/Cas9 system, dCas9 protein interacts with a fluorescent protein either as a part of fluorescent protein or via an RNA scaffold (Chen et al. 2013, 2014; Anton et al. 2014). CRISPR/Cas9 system is employed for non-repetitive sequences targeted by specific gRNA. For example, to visualize non-repetitive sequence of MUC4 locus, dCas9-eGFP fusion protein complex along with 26–36 gRNAs was used (Chen et al. 2013, 2014). The multiplex nature of CRISPR/dCas9 system is another benefit through which an array of gRNAs targeting the adjacent regions of non-repetitive sequences can be visualized (Fig. 8.4e). Using this approach, the non-repetitive region of the MUC4 locus was visualized by using a dCas9-eGFP fusion protein guided by 26–36 gRNAs (Chen et al. 2013).

Table 8.5 Successful reports of CRISPR/ddCas9 mediated epigenetic modifications

Objectives	Target	CRISPR/ dCas platform	Reference
Upregulation	Histone acetylation Histone acetylation DNA demethylation Recruitment of active transcription factors	dCas9-LSD1 dCas9-p300 dCas9-TET dCas9-VP64 or VPR	Thakore et al. (2015), Kearns et al. (2015), Hilton et al. (2015), Chavez et al. (2015)
Downregulation	Histone methylation DNA methylation	dCas9-KRAB dCas9-DNMT3A	Enriquez (2016), Hu et al. (2014)
Augmentation	Recruitment of multiple epieffector domains by the repeating peptide array SunTag Recruitment of multiple epieffector domains by the protruding guide RNA loops	dCas9-SunTag-epieffector dCas9-SAM-epieffector	Perez-Pinera et al. (2013a, b), McDonald et al. (2016), Amabile et al. (2016), Tanenbaum et al. (2014)

8.5 Genome Editing to Maximize Plant–Microbe Interactions: Nitrogen Fixation Under Compass

The green revolution-based interventions have been instrumental to address food security challenges since the mid-twenties. One of the major drivers behind this revolution was the improved crop responsiveness to soil fertilizers (Farmer 1986a, b). Increased use of fertilizers, one hand, is a continuous burden on economy, it is also causing soil and water pollution across the globe (Farmer 1986a, b). Compared to about 120 million tons of industrial nitrogen fertilizer produced annually via energy consuming Haber–Bosch process, the environment friendly biological nitrogen fixation (BNF) produces more than a double (413 million tons in 2010) (Fowler et al. 2013). This is achieved by nitrogen-fixing legumes crops who meet their nitrogen needs from BNF in association with Rhizobial bacteria by forming nodules on the roots of legumes crops. These bacteria get nutrients from plants for their survival and fix nitrogen for the plants in return for this symbiotic relationship (Oldroyd et al. 2011). With the advent of high-throughput genome sequencing, complete DNA sequence of several legume genomes is now publicly available including *G. max* (Schmutz et al. 2010), *L. japonicus* (Sato et al. 2008), *M. truncatula* (Young et al. 2011), and *Aeschynomene evenia* (Chaintreuil et al. 2016) and nearly 200 genes have been identified as essential genes for BNF.

CRISPR/Cas system offers a whole range of applications in legume research. Soybean is the first legume plant that has been edited using the CRISPR/Cas system with 54% editing efficiency. Several genes have been targeted in soybean including bar gene, GmSHR and GmFE1 (Cai et al. 2015a, b). Similar work has been reported for Lotus (Wang et al. 2016a, b) and Medicago truncatula (Meng et al. 2016). The CRISPR/Cas9 system based forward genetic screens will greatly speedup understanding of mechanistic details which have not been identified so far. Genome-wide sgRNAs libraries along with gene edits of BNF genes (Wang et al. 2017) will pave the way towards the ultimate aim, i.e., engineering BNF in non-leguminous crops. Four *Parasponia* species and the tropical tree *Parasponia andersonii* from Cannabis family (Cannabaceae) are the only known non-legume plant species which can develop nitrogen-fixing nodule symbiosis with rhizobium (Table 8.6). CRISPR/Cas9 based functional studies of legumes and *Parasponia* could lead to identify novel insight into evolution of this ancient process. Functional characterization of gene families and several small genes associated with BNF is another front to apply gene editing which were difficult to study with conventional approaches due to genetic redundancy and difficult to mutate (Roy et al. 2020). A summary of CRISPR/Cas based gene editing in major leguminous crops is given in Table 8.7.

8.6 Genome Editing to Modulate Plants Weeds Interaction

Owing to their environmental resilience and dynamic nature, weed infestation in agricultural crops remains a constant challenge for plant scientists despite technological advancements. These botanical pests cause substantial losses in crop yields by competing for light, water, and nutrients (Swanton et al. 2015; Ramesh et al. 2017). Weed management strategies not only need to minimize immediate yield losses due to crop-weed competition but should also prevent dispersal of weed seed or their vegetative parts into soils for their future perpetuation (Battle et al. 1996). Use of herbicides to control weeds has been the major approach to reduce crop losses (Johnson et al. 2009) and with availability of herbicide resistant GMO cultivars, there is significant rise in selective herbicides. Continuous and widespread use of chemical herbicides tends to develop herbicide resistance in weeds, change weed population dynamics, and pollute environment (Duary 2008; Johnson et al. 2009). In addition to continuous rise in the evolution of herbicide resistance in weeds due to continuous use of chemicals (Gould et al. 2018), sudden appearance of invasive alien weeds in cultivated areas is much difficult to control with routine management practices (Epanchin-Niell et al. 2010). For ecological sustainability of our cropping system, innovative approaches that yield environment friendly and economical weed management are direly needed.

Use of gene silencing and gene drive-based approaches to deter weed infestation holds a great promise to thwart growth and reduce fitness of weeds on a populations scale. Despite a broad application of gene editing in crop improvement, its use in weed control is lacking (Webber et al. 2015; Neve 2018).

Table 8.6 Summary of gene drive feasibility for major agricultural weeds (Adopted from Neve 2018)

Species	Distribution	Genome	Ploidy	Mating system	Vegetative	Fecundity	Seed bank persistence	Resistance risk	Potential
Alopecurus myosuroides	Europe, Asia	Large, no sequence	Diploid	Outcrossing	No	Medium	Medium	Very high	✓✓
Amaranthus palmeri	North and South America	Medium, sequenced	Diploid	Outcrossing (dioecious)	No	High	Low	Very high	✓✓
Amaranthus tuberculatus	North America	Medium, congener sequenced	Diploid	Outcrossing (dioecious)	No	High	Medium	Very high	✓✓
Avena fatua	America, Australia, Europe, Africa, Asia	Large, no sequenced	Hexaploid	Selfing	No	Low	Medium	Very high	×
Conyza canadensis	North America, Europe, Asia	Small, sequenced	Diploid	Selfing	No	Medium	Medium	High	×
Echinochloa crus-galli	America, Australia, Europe	Medium, sequenced	Hexaploid	Selfing	No	Medium	Medium	Very high	×
Kochia scoparia	North America	Medium, sequenced	Diploid	Outcrossing	No	Medium	Low	Intermediate	✓
Lolium rigidum	America, Australia, Europe, Africa, Asia	Large, congener sequenced	Diploid	Outcrossing	No	Medium	Medium	Very high	✓✓✓
Sorghum halepense	North and South America	Small, congener sequenced	Diploid and tetraploid	Selfing	Yes	Medium	Low	Intermediate	×

Table 8.7 CRISPR/Cas9 based gene editing reports in soybean, Lotus, and *Medicago truncatula*

Crops	Targeted genes	Transformation methods	References
Soybean	*Glyma06g14180* *Glyma08g02290* *Glyma12g37050*	Hairy root transformation mediated by K599	Sun et al. (2015)
	GmFE11, *GmFE12*, *bar gene*	Hairy root transformation mediated by K599	Cai et al. (2015a, b)
	DD43 and *DD20*	Transformation by particle bombardment	Li et al. (2015)
	GmPDS18 and *GmPDS11*	Hairy root transformation mediated by K599	Du et al. (2016)
	GFP, *miR1514*, *miR1509*	Hairy root transformation mediated by K599	Jacobs et al. (2015a, b)
Lotus	*LjLbs* and *LjSYMRK*	Hairy root transformation mediated by LBA1334	Wang et al. (2016a, b)
Medicago truncatula	*MtPDS*	Stable transformation mediated by AGL-1	Meng et al. (2016)

Exogenous application of gene silencing agent and self-perpetuating gene drives are the two major concepts which can be used for weed and pest management (Kumaran et al. 2019). RNAi based gene silencing using double stranded RNA (dsRNA) can suppress the target traits either by inhibiting translation or degrading the target mRNA post transcriptionally. Such an RNAi construct can be developed against specific genes of the target species affecting growth and development pathways or herbicide resistance. Topical application of RNAi agent on the plant will silence genes affecting the target traits like growth and fitness (Kumaran et al. 2019).

The idea of gene drives rests on the concept of selfish genes (Burt and Trivers 2009). Selfish genes are the genetic segments that can perpetuate at the cost of the rest of genome regardless whatever effect their transmission may cause to the organism's fitness (Lindholm et al. 2016). Gene drive is the mechanism of spreading a selfish genetic element to modulate population dynamics of the target species (Lindholm et al. 2016). Successful validation of gene drive concept in mosquitoes and the fungi *Candida albicans* (Kyrou et al. 2018; Shapiro et al. 2018) has motivated plant scientists to develop gene drives for controlling weeds (Esvelt et al. 2014). Developing a gene drive, for example, to spread herbicide susceptibility gene in wild population could be useful weed control strategy (Webber et al. 2015).

In CRISPR/Cas9 based gene drives, Cas9 facilitates copying of a gene of interest or induces required mutation in the targeted region on the homologous chromosome of the wild type. This results in segregation of the gene drive elements to all gametes at the time of meiosis. Further crossing of homozygous-edited with wild-type plants gives rise to heterozygotes which are once again converted to homozygous with the action Cas9. This way, transmission of the gene drive elements continues to all the offspring and spread throughout the population preferentially (Fig. 8.5) (Gantz and Bier 2015).

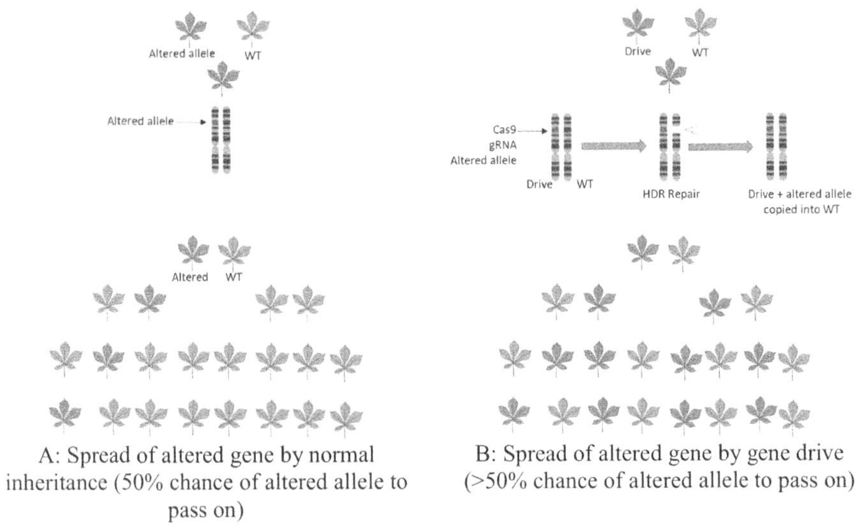

Fig. 8.5 Normal inheritance vs. gene drive inheritance patterns. (**a**) Spread of altered gene by normal inheritance (50% chance of altered allele to pass on). (**b**) Spread of altered gene by gene drive (>50% chance of altered allele to pass on)

Gene drive strategies for weeds can be based on population suppression or sensitization. In suppression drives, key viability traits like establishment, fecundity, and persistence are targeted to get rid of invasive weed species (Webber et al. 2015), while a sensitive trait like herbicide susceptibility is induced in weed population which easily amenable to a particular weed management practice (Kreiner et al. 2018). An endonuclease based selfish gene specifically designed to target the herbicide resistance gene of a weed species can be used to restore efficacy of an herbicide against which weed has developed resistance (Neve 2018). The same principle can be applied to rescue threatened or endangered plant species (Barrett et al. 2019). For such type of conservation work, it is important to identify the genes underlying the required traits like resistance to climate factors or the pathogens which can be used to design gene drives. The multiplex ability of CRISPR/Cas system to target multiple loci is very valuable aspect to handle polyploid species by targeting all homologous alleles within the genome. Table 8.6 describes details of major weed species found across the world and the potential of developing gene drive to minimize yield losses.

8.7 Future Perspective

The versatility of CRISPR/Cas system has empowered scientists with an unprecedented genomic toolbox enabling applications across the disciplines as well as biological kingdoms. Its remarkable flexibility, ease of use, and multiplex functionality have made it a method of choice in almost every biology lab. Its applications have broadened our vision how genetic material is organized in terms of topology and regulatory control on living processes. It has not only revolutionized basic biology but also overturned the applied fields with its immense applications. Applications of CRISPR/Cas in crop science are boundless. It not only helps to understand how plant genes evolved, but also helps to get evolved; it not only enables to add or delete genetic parts but also helps to spread them in the population; it not only facilitates to knockout genes, it can also be used to knockin or knockdown or knock-up target genes. Development of such a variety of tools just within a decade is stunning. Despite the unparallel progress and wide adoption, CRISPR/Cas also accompanies unique technical, social, and ethical challenges and concerns. Full potential of this technology will, no doubt, be realized when delivery of CRISPR/Cas reagents inside living cells is actualized for routine experiments.

References

Abudayyeh OO, Gootenberg JS, Essletzbichler P, Han S, Joung J, Belanto JJ, Verdine V, David BT, Kellner MJ, Regev A, Lander ES, Voytas DF, Alice Y (2018) RNA targeting with CRISPR-Cas13. Nature 550:280–284

Adli M (2018) The CRISPR tool kit for genome editing and beyond. Nat Commun 9:1911. https://doi.org/10.1038/s4467-018-04252-2

Amabile A, Migliara A, Capasso P et al (2016) Inheritable silencing of endogenous genes by hit-and-run targeted epigenetic editing. Cell 167(1):219–232

Amabile A, Migliara A, Capasso P, Biffi M, Cittaro D, Naldini L, Lombardo A (2016a) Inheritable silencing of endogenous genes by hit-and-run targeted epigenetic editing. Cell 167:219–232

Amabile A, Migliara A, Capasso P, Biffi M, Naldini L, Lombardo A (2016b) Inheritable silencing of endogenous gene by hit-and-run targeted epigenetic editing. Mol Ther 24:S287–S288

Aman R, Ali Z, Butt H, Mahas A, Aljedaani F, Khan MZ, Ding S, Mahfouz M (2018) RNA virus interference via CRISPR/Cas13a system in plants. Genome Biol 19:1–9

Anton T, Bultmann S, Leonhardt H, Markaki Y (2014) Visualization of specific DNA sequences in living mouse embryonic stem cells with a programmable fluorescent CRISPR/Cas system. Nucleus 5:163–172

Asano K, Yamasaki M, Takuno S, Miura K, Katagiri S, Ito T, Doi K, Wu J, Ebana K, Matsumoto T et al (2011) Artificial selection for a green revolution gene during japonica rice domestication. Proc Natl Acad Sci U S A 108:11034–11039

Barampuram S, Zhang ZJ (2011) Recent advances in plant transformation. In: Birchler JA (ed) Plant chromosome engineering. Springer Science, New York, pp 1–35

Barrett LG, Legros M, Kumaran N, Glassop D, Raghu S, Gardiner DM (2019) Gene drives in plants: opportunities and challenges for weed control and engineered resilience. Proc R Soc B 286:20191515. https://doi.org/10.1098/rspb.2019.1515

Battle M, Bender M, Sowers T, Tans PP, Butler JH, Elkins JW et al (1996) Atmospheric gas concentrations over the past century measured in air from fin at the south pole. Nature 383:231–235. https://doi.org/10.1038/383231a0

Bauwe H, Hagemann M, Fernie AR (2010) Photorespiration: players, partners and origin. Trends Plant Sci 15:330–336. https://doi.org/10.1016/j.tplants.2010.03.006

Blackman BK, Strasburg JL, Raduski AR, Michaels SD, Rieseberg LH (2010) The role of recently derived FT paralogs in sunflower domestication. Curr Biol 20:629–635

Blankenship RE, et al. (2011). Comparing photosynthetic and photovoltaic efficiencies and recognizing the potential for improvement. Science 332(6031):805–809

Boxall SF, Dever LV, Kneřová J, Gould PD, Hartwell J (2017) Phosphorylation of phosphoenolpyruvate carboxylase is essential for maximal and sustained dark CO2 fixation and core circadian clock operation in the obligate crassulacean acid metabolism species Kalanchoë fedtschenkoi. Plant Cell 29:2519–2536

Burstein D, Harrington LB, Strutt SC, Probst AJ, Anantharaman K, Thomas BC, Doudna JA, Banfield JF (2017) New CRISPR–Cas systems from uncultivated microbes. Nature 542 (7640):237–241

Burt A, Trivers R (2009) Genes in conflict: the biology of selfish genetic elements. Harvard University Press, Cambridge

Cai Y, Chen L, Liu X et al (2015a) CRISPR/Cas9-mediated genome editing in soybean hairy roots. PLoS One 10(8):e0136064

Cai Y, Chen L, Liu X, Sun S, Wu C, Jiang B, Han T, Hou W (2015b) CRISPR/Cas9-mediated genome editing in soybean hairy roots. PLoS One 10(8):e0136064

Carroll D (2012) A CRISPR approach to gene targeting. Mol Ther 10(9):1658–1660

Cass SD, Haas KA, Stoll B, Alkhnbashi OS, Sharma K, Urlaub H, Backofen R, Marchfelder A, Bolt EL (2015) The role of Cas8 in type I CRISPR interference. Biosci Rep 35(3):e00197

Chaintreuil C, Rivallan R, Bertioli DJ et al (2016) A gene-based map of the nod factor-independent Aeschynomene evenia sheds new light on the evolution of nodulation and legume genomes. DNA Res 23(4):365–376

Chavez A, Scheiman J, Vora S et al (2015) Highly efficient Cas9-mediated transcriptional programming. Nat Methods 12(4):326–328

Chavez A, Tuttle M, Pruitt BW, Ewen-Campen B, Chari R, Ter-Ovanesyan D, Haque SJ, Cecchi RJ, Kowal EJ, Buchthal J (2016) Comparison of Cas9 activators in multiple species. Nat Methods 13:563

Chédin F, Lieber MR, Hsieh C-L (2002) The DNA methyltransferase-like protein DNMT3L stimulates de novo methylation by Dnmt3a. Proc Natl Acad Sci 99:16916–16921

Chen KY, Cong B, Wing R, Vrebalov J, Tanksley SD (2007) Changes in regulation of a transcription factor lead to autogamy in cultivated tomatoes. Science 318:643–645

Chen B, Gilbert LA, Cimini BA, Schnitzbauer J, Zhang W, Li G-W, Park J, Blackburn EH, Weissman JS, Qi LS (2013) Dynamic imaging of genomic loci in living human cells by an optimized CRISPR/Cas system. Cell 155:1479–1491

Chen B, Gilbert LA, Cimini BA, Schnitzbauer J, Zhang W, Li G-W, Park J, Blackburn EH, Weissman JS, Qi LS (2014) Dynamic imaging of genomic loci in living human cells by an optimized CRISPR/Cas system. Cell 156:373

Cooper D et al (2001) Broadening the genetic base of crop production. CAB International, Wallingford

Cox DB, Gootenberg JS, Abudayyeh OO, Franklin B, Kellner MJ, Joung J, Zhang F (2017) RNA editing with CRISPR-Cas13. Science 358(6366):1019–1027

Croft JA, Bridger JM, Boyle S, Perry P, Teague P, Bickmore WA (1999) Differences in the localization and morphology of chromosomes in the human nucleus. J Cell Biol 145:1119–1131

Cushman JC, Agarie S, Albion RL, Elliot SM, Taybi T, Borland AM (2008) Isolation and characterization of mutants of common ice plant deficient in crassulacean acid metabolism. Plant Physiol 147:228–238

Dekker J, Marti-Renom MA, Mirny LA (2013) Exploring the three-dimensional organization of genomes: interpreting chromatin interaction data. Nat Rev Genet 14:390–403

Deltcheva E, Chylinski K, Sharma CM, Gonzales K, Chao Y, Pirzada ZA, Eckert MR, Vogel J, Charpentier E (2011) CRISPR RNA maturation by trans-encoded small RNA and host factor RNase III. Nature 471(7340):602–607

Dever LV, Boxall SF, Kneřová J, Hartwell J (2015) Transgenic perturbation of the decarboxylation phase of Crassulacean acid metabolism alters physiology and metabolism but has only a small effect on growth. Plant Physiol 167:44–59S

Doebley J, Stec A, Wendel J, Edwards M (1990) Genetic and morphological analysis of a maize-teosinte F2 population: implications for the origin of maize. Proc Natl Acad Sci U S A 87:9888–9892
Dong Y, Yang X, Liu J, Wang BH, Liu BL, Wang YZ (2014) Pod shattering resistance associated with domestication is mediated by a NAC gene in soybean. Nat Commun 5:3352
Drewry DT, Kumar P, Long SP (2014) Simultaneous improvement in productivity, water use and albedo through crop structural modification. Glob. Change Biol 20:1955–1967
Driever SM, Simkin AJ, Alotaibi S, Fisk SJ, Madgwick PJ, Sparks CA, Jones HD, Lawson T, Parry MAJ, Raines CA (2017) Increased SBPase activity improves photosynthesis and grain yield in wheat grown in greenhouse conditions. Philos Trans R Soc B Biol Sci 372:1730–1740
Du H, Zeng X, Zhao M et al (2016) Efficient targeted mutagenesis in soybean by TALENs and CRISPR/Cas9. J Biotechnol 217:90–97
Duary B (2008) Recent advances in herbicide resistance in weeds and its management. Indian J. Weed Sci 24:124–135
Dwivedi SL et al (2008) Enhancing crop gene pools with beneficial traits using wild relatives. Plant Breed Rev 30:179–230
East-Seletsky A, O'Connell MR, Knight SC, Burstein D, Cate JH, Tjian R, Doudna JA (2016) Two distinct RNase activities of CRISPR-C2c2 enable guide-RNA processing and RNA detection. Nature 538(7624):270–273
Endo A, Masafumi M, Kaya H, Toki S (2016) Efficient targeted mutagenesis of rice and tobacco genomes using Cpf1 from Francisella novicida. Sci Rep 6:1–9
Enriquez P (2016) CRISPR-mediated epigenome editing. Yale J Biol Med 89(4):471–486
Epanchin-Niell RS, Hufford MB, Aslan CE, Sexton JP, Port JD, Waring TM (2010) Controlling invasive species in complex social landscapes. Front Ecol Environ 8:210–216. https://doi.org/10.1890/090029
Essers J, van Cappellen WA, Theil AF, van Drunen E, Jaspers NG, Hoeijmakers JH, Wyman C, Vermeulen W, Kanaar R (2005) Dynamics of relative chromosome position during the cell cycle. Mol Biol Cell 16:769–775
Esvelt KM, Smidler AL, Catteruccia F, Church GM (2014) Concerning RNA-guided gene drives for the alteration of wild populations. eLife 5:S81–S97. https://doi.org/10.7554/eLife.03401
Farmer B (1986a) Perspectives on the 'green revolution' in South Asia. Mod Asian Stud 20 (1):175–199
Farmer BH (1986b) Perspectives on the 'green revolution' in South Asia. Mod Asian Stud 20 (1):175–199
Feng Z, Zhang B, Ding W et al (2013) Efficient genome editing in plants using a CRISPR/Cas system. Cell Res 23:1229–1232
Fonfara I, Richter H, Bratovič M, Le Rhun A, Charpentier E (2016) The CRISPR-associated DNA-cleaving enzyme Cpf1 also processes precursor CRISPR RNA. Nature 532 (7600):517–521
Fowler D, Coyle M, Skiba U, Sutton MA, Cape JN, Reis S, Sheppard LJ, Jenkins A, Grizzetti B, Galloway JN, Vitousek P (2013) The global nitrogen cycle in the twenty-first century. Philos Trans R Soc B 368(1621):20130164
Frary A, Nesbitt TC, Grandillo S, Knaap E, Cong B, Liu J, Meller J, Elber R, Alpert KB, Tanksley SD (2000) fw2.2: a quantitative trait locus key to the evolution of tomato fruit size. Science 289:85–88
Friedman JR et al (1996) A novel corepressor for the highly conserved KRAB repression domain. Genes Dev 10:2067–2078
Gantz VM, Bier E (2015) The mutagenic chain reaction: a method for converting heterozygous to homozygous mutations. Science 348:442–444
Garneau JE, Dupuis MÈ, Villion M, Romero DA, Barrangou R, Boyaval P, Fremaux C, Horvath P, Magadán AH, Moineau S (2010) The CRISPR/Cas bacterial immune system cleaves bacteriophage and plasmid DNA. Nature 468(7320):67–71

Gilbert LA et al (2013) CRISPR-mediated modular RNA-guided regulation of transcription in eukaryotes. Cell 154:442–451
Gilbert LA, Horlbeck MA, Adamson B, Villalta JE, Chen Y, Whitehead EH, Guimaraes C, Panning B, Hl P, Bassik MC (2014) Genome-scale CRISPR-mediated control of gene repression and activation. Cell 159:647–661
Gould F, Brown ZS, Kuzma J (2018) Wicked evolution: can we address the sociobiological dilemma of pesticide resistance? Science 360:728–732. https://doi.org/10.1126/science.aar3780
Gruber K (2017) Agrobiodiversity: the living library. Nature 544:S8–S10
Harrington LB, Burstein D, Chen JS, Paez-Espino D, Ma E, Witte IP, Cofsky JC, Kyrpides NC, Banfield JF, Doudna JA (2018) Programmed DNA destruction by miniature CRISPR-Cas14 enzymes. Science 362(6416):839–842
Hedden P (2003) The genes of the green revolution. Trends Genet 19:5–9
Hellwig D, Münch S, Orthaus S, Hoischen C, Hemmerich P, Diekmann S (2008) Live-cell imaging reveals sustained centromere binding of CENP-T via CENP-A and CENP-B. J Biophotonics 1:245–254
Hilton B, D'Ippolito AM, Vockley CM et al (2015) Epigenome editing by a CRISPR-Cas9-based acetyltransferase activates genes from promoters and enhancers. Nat Biotechnol 33(5):510–517
Hoisington D et al (1999) Plant genetic resources: what can they contribute toward increased crop productivity? Proc Natl Acad Sci U S A 96:5937–5943
Holoch D, Moazed D (2015) RNA-mediated epigenetic regulation of gene expression. Nat Rev Genet 16:71–84
Hu J, Lei Y, Wong WK et al (2014) Direct activation of human and mouse Oct4 genes using engineered TALE and Cas9 transcription factors. Nucleic Acids Res 42(7):4375–4390
Hufford MB et al (2013) The genomic signature of crop-wild introgression in maize. PLoS Genet 9 (5):e1003477
Ishino Y, Shinagawa H, Makino K, Amemura M, Nakata A (1987) Nucleotide sequence of the IAP gene responsible for alkaline phosphatase isozyme conversion in Escherichia coli and identification of the gene product. J Bacteriol 169(12):5429–5433
Jacobs TB, LaFayette PR, Schmitz RJ, Parrott WA (2015a) Targeted genome modifications in soybean with CRISPR/Cas9. BMC Biotechnol 15(1):16
Jacobs TB, LaFayette PR, Schmitz RJ, Parrott WA (2015b) Targeted genome modifications in soybean with CRISPR/Cas9. BMC Biotechnol 15(1):1–10
Jansen R, Embden JDV, Gaastra W, Schouls LM (2002a) Identification of genes that are associated with DNA repeats in prokaryotes. Mol Microbiol 43(6):1565–1575
Jansen R, Embden JDV, Gaastra Wand Schouls LM (2002b) Identification of genes that are associated with DNA repeats in prokaryotes. Mol Microbiol 43(6):1565–1575
Jiao Y et al (2010) Regulation of OsSPL14 by OsmiR156 defines ideal plant architecture in rice. Nat Genet 42:541–545
Jin J et al (2008a) Genetic control of rice plant architecture under domestication. Nat Genet 40:1365–1369
Jin J, Huang W, Gao JP, Yang J, Shi M, Zhu MZ, Luo D, Lin HX (2008b) Genetic control of rice plant architecture under domestication. Nat Genet 40:1365–1369
Jinek M, Chylinski K, Fonfara I, Hauer M, Doudna JA, Charpentier E (2012a) A programmable dual-RNA–guided DNA endonuclease in adaptive bacterial immunity. Science 337 (6096):816–821
Jinek M et al (2012b) A programmable dual-RNA-guided DNA endonuclease in adaptive bacterial immunity. Science 337:816–821
Johnson WG, Davis VM, Kruger GR, Weller SC (2009) Influence of glyphosate-resistant cropping systems on weed species shifts and glyphosate-resistant weed populations. Eur J Agron 31:162–172. https://doi.org/10.1016/j.eja.2009.03.008
Källman AM, Sahlin M, Öhman M (2003) ADAR2 A→ I editing: site selectivity and editing efficiency are separate events. Nucleic Acids Res 31(16):4874–4881

Kang BC, Yun JY, Kim ST, Shin Y, Ryu J, Choi M, Woo JW, Kim JS (2018) Precision genome engineering through adenine base editing in plants. Nat Plants 4:427–431
Kearns NA, Pham H, Tabak B et al (2015) Functional annotation of native enhancers with a Cas9–histone demethylase fusion. Nat Methods 12(5):401–403
Koerner MV, Pauler FM, Huang R, Barlow DP (2009a) The function of non-coding RNAs in genomic imprinting. Development 136:1771–1783
Koerner MV, Pauler FM, Huang R, Barlow DP (2009b) The function of non-coding RNAs in genomic imprinting. Development 136:1771–1783
Komatsuda T et al (2007) Six-rowed barley originated from a mutation in a homeodomain-leucine zipper I-class homeobox gene. Proc Natl Acad Sci U S A 104:1424–1429
Komor AC, Zhao KT, Packer MS, Gaudelli NM, Waterbury AL, Koblan LW, Kim YB, Badran AH, Liu DR (2017) Improved base excision repair inhibition and bacteriophage Mu Gam protein yields C: G-to-T: A base editors with higher efficiency and product purity. Sci Adv 3 (8):4774
Kovach MJ, Sweeney MT, McCouch SR (2007) New insights into the history of rice domestication. Trends Genet 23:578–587
Kreiner JM, Stinchcombe JR, Wright SI (2018) Population genomics of herbicide resistance: adaptation via evolutionary rescue. Annu Rev Plant Biol 69:611–635. https://doi.org/10.1146/annurev-arplant-042817-040038
Kromdijk J, Glowacka K, Leonelli L, Gabilly S, Iwai M, Niyogi K, Long SP (2016) Improving photosynthesis and crop productivity by accelerating recovery from photoprotection. Plant Sci 354:857–862
Krupovic M, Shmakov S, Makarova KS, Forterre P, Koonin EV (2016) Recent mobility of casposons, self-synthesizing transposons at the origin of the CRISPR-Cas immunity. Genome Biol Evol 8(2):375–386
Kumaran N, Choudhary A, Legros M, Sheppard AW, Barrett LG, Gardiner DM, Raghu S (2019) Gene technologies in weed management: a technical feasibility analysis. Curr Opin Insect Sci 38:6–14. https://doi.org/10.1016/j.cois.2019.12.003
Kyrou K, Hammond AM, Galizi R, Kranjc N, Burt A, Beaghton AK, Nolan T, Crisanti A (2018) A CRISPR–Cas9 gene drive targeting doublesex causes complete population suppression in caged Anopheles gambiae mosquitoes. Nat Biotechnol 36:1062–1066. https://doi.org/10.1038/nbt.4245
Lawson, T.; Bryant, B.; Lefebvre, S.; Lloyd, J.C.; Raines, C.A. (2006). Decreased SBPase activity alters growth and development in transgenic tobacco plants. Plant Cell Environ. 29, 48–58
Li Z, Liu ZB, Xing A et al (2015) Cas9-guide RNA directed genome editing in soybean. Plant Physiol 169(2):960–970
Lin Z et al (2012a) Parallel domestication of the Shattering1 genes in cereals. Nat Genet 44:720–724
Lin Z, Li X, Shannon LM, Yeh CT, Wang ML, Bai G, Peng Z, Li J, Trick HN, Clemente TE et al (2012b) Parallel domestication of the Shattering1 genes in cereals. Nat Genet 44:720–724
Lin CS, Hsu CT, Yang LH, Lee LY, Fu JY, Cheng QW, Wu FH, Hsiao HCW, Zhang Y, Zhang R, Chang WJ, Yu CT, Wang W, Liao LJ, Gelvin SB, Shih MC (2018) Application of protoplast technology to CRISPR/Cas9 mutagenesis: from single-cell mutation detection to mutant plant regeneration. Plant Biotechnol J 16(7):1295–1310
Lindholm AK et al (2016) The ecology and evolutionary dynamics of meiotic drive. Trends Ecol Evol 31:315–326. https://doi.org/10.1016/j.tree.2016.02.001
Liu H, Zhang H, Niedzwiedzki DM, Prado M, He G, Gross ML, Blankenship RE (2013) Phycobilisomes supply excitations to both photosystems in a megacomplex in cyanobacteria. Science 342:1104–1107
Liu D, Hu R, Palla KJ, Tuskan GA, Yang X (2016) Advances and perspectives on the use of CRISPR/Cas9 systems in plant genomics research. Curr Opin Plant Biol 30:70–77
Liu L, Li X, Ma J, Li Z, You L, Wang J, Wang M, Zhang X, Wang Y (2017) The molecular architecture for RNA-guided RNA cleavage by Cas13a. Cell 170(4):714–726

Liu D, Chen M, Mendoza B, Cheng H, Hu R, Li L, Trinh CT, Tuskan GA, Yang X (2019) CRISPR/Cas9-mediated targeted mutagenesis for functional genomics research of crassulacean acid metabolism plants. J Exp Bot 70(22):6621–6629

Livne S et al (2015) Uncovering DELLA-Independent gibberellin responses by characterizing new tomato procera mutants. Plant Cell 27:1579–1594

Long A et al (2015) Meeting the global food demand of the future by engineering crop photosynthesis and yield potential. Cell 161:56–66

Lopez-Calcagno PE, Fisk S, Brown KL, Bull SE, South PF, Raines CA (2018) Overexpressing the H-protein of the glycine cleavage system increases biomass yield in glasshouse and field grown transgenic tobacco plants. Plant Biotechnol J 17:141–151. https://doi.org/10.1111/pbi.12953

Lu Y, Zhu JK (2017) Precise editing of a target base in the rice genome using a modified CRISPR/Cas9 system. Mol Plant 10(3):523–525

Machida-Hirano R et al (2015) Diversity of potato genetic resources. Breed Sci 65:26–40

Makarova KS, Aravind L, Wolf YI, Koonin EV (2011a) Unification of Cas protein families and a simple scenario for the origin and evolution of CRISPR-Cas systems. Biol Direct 6(1):38

Makarova KS, Haft DH, Barrangou R, Brouns SJ, Charpentier E, Horvath P, Moineau S, Mojica FJ, Wolf YI, Yakunin AF, Van Der Oost J (2011b) Evolution and classification of the CRISPR–Cas systems. Nat Rev Microbiol 9(6):467–477

McDonald JI, Celik H, Rois LE et al (2016) Reprogrammable CRISPR/Cas9-based system for inducing site-specific DNA methylation. Biol Open 5(6):866–874

McGrath JM, Long SP (2014) Can the cyanobacterial carbon-concentrating mechanism increase photosynthesis in crop species? A theoretical analysis. Plant Physiol 164:2247–2261

Meng Y, Hou Y, Wang H et al (2016) Targeted mutagenesis by CRISPR/Cas9 system in the model legume Medicago truncatula. Plant Cell Rep 36(2):371–374

Meyer RS, Purugganan MD (2013) Evolution of crop species: genetics of domestication and diversification. Nat Rev Genet 14:840–852

Moreno-Mateos MA, Fernandez JP, Rouet R, Vejnar CE, Lane MA, Mis E, Khokha MK, Doudna JA, Giraldez AJ (2017) CRISPR-Cpf1 mediates efficient homology-directed repair and temperature-controlled genome editing. Nat Commun 8(1):1–9

Mulepati S, Bailey S (2011) Structural and biochemical analysis of nuclease domain of clustered regularly interspaced short palindromic repeat (CRISPR)-associated protein 3 (Cas3). J Biol Chem 286(36):31896–31903

Nam KH, Ding F, Haitjema C, Huang Q, DeLisa MP, Ke A (2012a) Double-stranded endonuclease activity in Bacillus halodurans clustered regularly interspaced short palindromic repeats (CRISPR)-associated Cas2 protein. J Biol Chem 287(43):35943–35952

Nam KH, Haitjema C, Liu X, Ding F, Wang H, DeLisa MP, Ke A (2012b) Cas5d protein processes pre-crRNA and assembles into a cascade-like interference complex in subtype IC/Dvulg CRISPR-Cas system. Structure 20(9):1574–1584

Nassar NMA, Hashimoto DYC, Fernandes SDC (2008) Wild Manihot species:botanical aspects, geographic distribution and economic value. Genet Mol Res 7:16–28

Neve P (2018) Gene drive systems: do they have a place in agricultural weed management? Pest Manag Sci 74:2671–2679

Nishimasu H et al (2014) Crystal structure of Cas9 in complex with guide RNA and target DNA. Cell 156:935–949

Nuñez JK, Kranzusch PJ, Noeske J, Wright AV, Davies CW, Doudna JA (2014) Cas1–Cas2 complex formation mediates spacer acquisition during CRISPR–Cas adaptive immunity. Nat Struct Mol Biol 21(6):528

Ogbonnaya FC et al (2013) Synthetic hexaploids: harnessing species of the primary gene pool for wheat improvement. Plant Breed Rev 37:35–122

Oldroyd GE, Murray JD, Poole PS, Downie JA (2011) The rules of engagement in the legume-rhizobial symbiosis. Annu Rev Genet 45:119–144

Ort DR, Merchange SS, Alric J, Barkan A, Blankenship RE, Bock R, Croce R, Hanson M, Hibberd JM, Long SP, Moore T, Moroney J, Niyogi KK, Parry MA, Peralta-Yahya P, Prince RC,

Redding KE, Spalding MH, van Wijk K, WFJ V, von Caemmerer S, APM W, Yeates TO, Yuan JS, Zhu XG (2015) Redesigning photosynthesis to sustainably meet global food and bioenergy demand. Proc Natl Acad Sci U S A 112:8529–8536

Panning B, Dausman J, Jaenisch R (1997) X chromosome inactivation is mediated by Xist RNA stabilization. Cell 90:907–916

Pennisi E (2013) Long noncoding RNAs may alter chromosome's 3D structure. Science 340:910

Perez-Pinera P, Kocak DD, Vockley CM et al (2013a) RNA guided gene activation by CRISPR-Cas9-based transcription factors. Nat Methods 10(10):973–976

Perez-Pinera P et al (2013b) RNA-guided gene activation by CRISPR-Cas9-based transcription factors. Nat Methods 10:973–976

Poolman MG, Fell DA, Thomas S. (2000). Modelling photosynthesis and its control. Journal of Experimental Botany 51, 319–328

Pourkheirandish M (2015) Evolution of the grain dispersal system in barley. Cell 162:527–539

Prince SJ et al (2015) Genetic variants in root architecture-related genes in a Glycine soja accession, a potential resource to improve cultivated soybean. BMC Genomics 16:132

Qi LS et al (2013) Repurposing CRISPR as an RNA-guided platform for sequence specific control of gene expression. Cell 152:1173–1183

Rae BD, Long BM, Forster B, Nguyen ND, Velanis CN, Atkinson N, Hee WY, Mukherjee B, Price GD, McCormick AJ (2017) Progress and challenges of engineering a biophysical CO2-concentrating mechanism into higher plants. J Exp Bot 68:3717–3737

Ramesh K, Matloob A, Aslam F, Florentine S, Chauhan BS (2017) Weeds in a changing climate: vulnerabilities, consequences and implications for future weed management. Front Plant Sci 8:95. https://doi.org/10.3389/fpls.2017.00095

Ramsay L, Comadran J, Druka A, Marshall DF, Thomas WT, Macaulay M, MacKenzie K, Simpson C, Fuller J, Bonar N et al (2011) Intermedium-C, a modifier of lateral spikelet fertility in barley, is an ortholog of the maize domestication gene teosinte branched 1. Nat Genet 43:169–172

Ray DK, Mueller ND, West PC, Foley JA (2013) Yield trends are insufficient to double global crop production by 2050. PLoS ONE 8:e66428

Robinett CC, Straight A, Willhelm C, Sudlow G, Murray A, Belmont AS (1996) In vivo localization of DNA sequences and visualization of large-scale chromatin organization using lac operator/repressor recognition. J Cell Biol 135:1685–1700

Roy S, Liu W, Nandety RS, Crook A, Mysore KS, Pislariu CI, Frugoli J, Dickstein R, Udvardi MK (2020) Celebrating 20 years of genetic discoveries in legume nodulation and symbiotic nitrogen fixation. Plant Cell 32(1):15–41. https://doi.org/10.1105/tpc.19.00279

Salesse Smith CE, Sharwood RE, Busch FA, Kromdijk J, Bardal V, Stern DB (2018) Overexpression of Rubisco subunits with RAF1 increases Rubisco content in maize. Nat Plants 4:802–810

Samai P, Pyenson N, Jiang W, Goldberg GW, Hatoum-Aslan A, Marraffini LA (2015) Co-transcriptional DNA and RNA cleavage during type III CRISPR-Cas immunity. Cell 161 (5):1164–1174

Sanchez PL, Wing RA, Brar DS (2013) The wild relatives of rice: genomes and genomics. In: Zhang Q, Wing RA (eds) Genetics and genomics of rice. Springer, New York, pp 9–25

Sang T, Ge S (2007) Genetics and phylogenetics of rice domestication. Curr Opin Genet Dev 17:533–538

Sato S, Nakamura Y, Kaneko T et al (2008) Genome structure of the legume, Lotus japonicus. DNA Res 15(4):227–239

Sawler J et al (2013) Genomics assisted ancestry deconvolution in grape. PLoS ONE 8(11):e80791

Schmutz J, Cannon SB, Schlueter J et al (2010) Genome sequence of the palaeopolyploid soybean. Nature 463(7278):178–183

Schunder E, Rydzewski K, Grunow R, Heuner K (2013) First indication for a functional CRISPR/Cas system in Francisella tularensis. Int J Med Microbiol 303(2):51–60

Shan Q, Wang Y, Li J et al (2013) Targeted genome modification of crop plants using a CRISPR–Cas system. Nat Biotechnol 31:686–688

Shao Y, Cocozaki AI, Ramia NF, Terns RM, Terns MP, Li H (2013) Structure of the Cmr2-Cmr3 subcomplex of the Cmr RNA silencing complex. Structure 21(3):376–384

Shapiro RS et al (2018) A CRISPR–Cas9-based gene drive platform for genetic interaction analysis in Candida albicans desirable genetic changes in wild populations. Nat Microbiol 3:73. https://doi.org/10.1038/s41564-017-0043-0

Shechner DM, Hacisuleyman E, Younger ST, Rinn JL (2015a) Multiplexable, locus-specific targeting of long RNAs with CRISPR-Display. Nat Methods 12:664–670

Shechner DM, Hacisüleyman E, Younger ST, Rinn JL (2015b) CRISPR display: a modular method for locus-specific targeting of long noncoding RNAs and synthetic RNA devices in vivo. Nat Methods 12:664

Shimatani Z, Kashojiya S, Takayama M et al (2017) Targeted base editing in rice and tomato using a CRISPR–Cas9 cytidine deaminase fusion. Nat Biotechnol 35:441–443

Shmakov S, Abudayyeh OO, Makarova KS, Wolf YI, Gootenberg JS, Semenova E, Minakhin L, Joung J, Konermann S, Severinov K, Zhang F (2015) Discovery and functional characterization of diverse class 2 CRISPR-Cas systems. Mol Cell 60(3):385–397

Shmakov SA, Makarova KS, Wolf YI, Severinov KV, Koonin EV (2018) Systematic prediction of genes functionally linked to CRISPR-Cas systems by gene neighborhood analysis. Proc Natl Acad Sci 115(23):E5307–E5316

Simkin AJ, McAusland L, Headland LR, Lawson T, Raines CA (2015) Multigene manipulation of photosynthetic carbon assimilation increases CO2 fixation and biomass yield in tobacco. J Exp Bot 66(13):4075–4090

Simkin AJ, Lopez-Calcagno PE, Davey PA, Headland LR, Lawson T, Timm S, Bauwe H, Raines CA (2017a) Simultaneous stimulation of sedoheptulose 1,7-bisphosphatase, fructose 1,6-bisphophate aldolase and the photorespiratory glycine decarboxylase-H protein increases CO2 assimilation, vegetative biomass and seed yield in Arabidopsis. Plant Biotechnol J 15:805–816

Simkin AJ, Lopez-Calcagno PE, Davey PA, Headland LR, Lawson T, Timm S, Bauwe H, Raines CA (2017b) Simultaneous stimulation of sedoheptulose 1,7-bisphosphatase, fructose 1, 6-bisphophate aldolase and the photorespiratory glycine decarboxylase-H protein increases CO 2 assimilation, vegetative biomass and seed yield in Arabidopsis. Plant Biotechnol J 15 (7):805–816

Simons KJ, Fellers JP, Trick HN, Zhang Z, Tai YS, Gill BS, Faris JD (2006) Molecular characterization of the major wheat domestication gene Q. Genetics 172:547–555

Sinkunas T, Gasiunas G, Fremaux C, Barrangou R, Horvath P, Siksnys V (2011) Cas3 is a single-stranded DNA nuclease and ATP-dependent helicase in the CRISPR/Cas immune system. EMBO J 30(7):1335–1342

Slattery RA, Walker BJ, Weber APM, Ort DR. (2018). The impacts of fluctuating light on crop performance. Plant Physiology 176, 990–1003

Smargon AA, Cox DB, Pyzocha NK, Zheng K, Slaymaker IM, Gootenberg JS, Abudayyeh OA, Essletzbichler P, Shmakov S, Makarova KS, Koonin EV (2017) Cas13b is a type VI-B CRISPR-associated RNA-guided RNase differentially regulated by accessory proteins Csx27 and Csx28. Mol Cell 65(4):618–630

Smith ZD, Meissner A (2013) DNA methylation: roles in mammalian development. Nat Rev Genet 14:204–220

South PF, Cavanagh AP, Lopez-Calcagno PE, Raines CA, Ort DR (2018a) Optimizing photorespiration for improved crop productivity. J Integr Plant Biol 60(12):1217–1230

South PF, Cavanagh AP, Lopez-Calcagno PE, Raines CA, Ort DR (2018b) Optimizing photorespiration for improved crop productivity. J Integr Plant Biol 60:1217–1230. https://doi.org/10.1111/jipb.12709

Soyk S et al (2017) Bypassing negative epistasis on yield in tomato imposed by a domestication gene. Cell 169:1142–1155

Steinert J, Schiml S, Fauser F, Puchta H (2015) Highly efficient heritable plant genome engineering using Cas9 orthologues from Streptococcus thermophilus and Staphylococcus aureus. Plant J 84:1295–1305

Strahl BD, Allis CD (2000) The language of covalent histone modifications. Nature 403:41–45

Studer A, Zhao Q, Ross-Ibarra J, Doebley J (2011) Identification of a functional transposon insertion in the maize domestication gene tb1. Nat Genet 43:1160–1163

Sun X, Hu Z, Chen R et al (2015) Targeted mutagenesis in soybean using the CRISPR-Cas9 system. Sci Rep 5:10342

Swanton CJ, Nkoa R, Blackshaw RE (2015) Experimental methods for crop–weed competition studies. Weed Sci 63:2–11. https://doi.org/10.1614/WS-D-13-00062.1

Taketa S, Amano S, Tsujino Y, Sato T, Saisho D, Kakeda K, Nomura M, Suzuki T, Matsumoto T, Sato K et al (2008) Barley grain with adhering hulls is controlled by an ERF family transcription factor gene regulating a lipid biosynthesis pathway. Proc Natl Acad Sci U S A 105:4062–4067

Tanenbaum ME, Gilbert LA, Qi LS, Weissman JS, Vale RD (2014) A protein-tagging system for signal amplification in gene expression and fluorescence imaging. Cell 159(3):635–646

Tang X, Lowder LG, Zhang T, Malzahn AA, Zheng X, Voytas DF, Zhong Z, Chen Y, Ren Q, Li Q, Kirkland ER, Zhang Y, Qi Y (2017) A CRISPR-Cpf1 system for efficient genome editing and transcriptional repression in plants. Nat Plants 3:1–5

Thakore PI, D'Ippolito AM, Song L et al (2015) Highly specific epigenome editing by CRISPR-Cas9 repressors for silencing of distal regulatory elements. Nat Methods 12(12):1143–1149

Tamoi M, Nagaoka M, Miyagawa Y, Shigeoka S. (2006). Contribution of fructose-1,6-bisphosphatase and sedoheptulose-1,7-bisphosphatase to the photosynthetic rate and carbon flow in the Calvin cycle in transgenic plants. Plant & Cell Physiology 47, 380–390

Tholen D, Zhu X-G (2011) The mechanistic basis of internal conductance: a theoretical analysis of mesophyll cell photosynthesis and CO2 diffusion. Plant Physiol 156:90–105

Tholen D, Boom C, Zhu X-G (2012) Opinion: prospects for improving photosynthesis by altering leaf anatomy. Plant Sci 197:92–101

Uematsu K, Suzuki N, Iwamae T, Inui M, Yukawa H. (2012). Increased fructose 1,6-bisphosphate aldolase in plastids enhances growth and photosynthesis of tobacco plants. Journal of Experimental Botany 63, 3001–3009

USCB (2015) World Population 1950-2050. International database. U.S. Department of Commerce, Census Bureau, Washington, DC

von Caemmerer S, Quick WP, Furbank RT (2012). The development of C4 rice: Current progress and future challenges. Science 336(6089):1671–1672

Wang H et al (2005) The origin of the naked grains of maize. Nature 436:714–719

Wang E, Wang J, Zhu X, Hao W, Wang L, Li Q, Zhang L, He W, Lu B, Lin H et al (2008) Control of rice grain-filling and yield by a gene with a potential signature of domestication. Nat Genet 40:1370–1374

Wang Y, Long SP, Zhu X-G (2014) Elements required for an efficient NADP-malic enzyme type C4 photosynthesis. Plant Physiol 164:2231–2246

Wang L, Wang L, Tan Q et al (2016a) Efficient inactivation of symbiotic nitrogen fixation related genes in Lotus japonicus using CRISPR-Cas9. Front Plant Sci 7:1333

Wang L, Wang L, Tan Q, Fan Q, Zhu H, Hong Z, Zhang Z, Duanmu D (2016b) Efficient inactivation of symbiotic nitrogen fixation related genes in Lotus japonicus using CRISPR-Cas9. Front Plant Sci 7:1333

Wang L, Wang L, Zhou Y, Duanmu D (2017) Use of CRISPR/Cas9 for symbiotic nitrogen fixation research in legumes. In: Progress in molecular biology and translational science. Elsevier, London, pp 187–213. https://doi.org/10.1016/bs.pmbts.2017.03.010

Wang M, Li W, Fang C, Xu F, Liu Y, Wang Z, Yang R, Zhang M, Liu S, Lu S et al (2018) Parallel selection on a dormancy gene during domestication of crops from multiple families. Nat Genet 50:1435–1441

Webber BL, Raghu S, Edwards OR (2015) Opinion: is CRISPR-based gene drive a biocontrol silver bulletor global conservation threat? Proc Natl Acad Sci 112:10565–10567

Xie K, Minkenberg B, Yang Y (2015) Boosting CRISPR/Cas9 multiplex editing capability with the endogenous tRNA-processing system. Proc Natl Acad Sci U S A 112:3570–3575

Xin CP, Tholen D, Devloo V, Zhu XG (2015) The benefits of photorespiratory bypasses: how can they work? Plant Physiol 167:574–585

Xu P, Zhang Y, Kang L, Roossinck MJ, Mysore KS (2006) Computational estimation and experimental verification of off-target silencing during posttranscriptional gene silencing in plants. Plant Physiol 142:429–440

Yan L, Loukoianov A, Tranquilli G, Helguera M, Fahima T, Dubcovsky J (2003) Positional cloning of the wheat vernalization gene VRN1. Proc Natl Acad Sci U S A 100:6263–6268

Yan L, Loukoianov A, Blechl A, Tranquilli G, Ramakrishna W, SanMiguel P, Bennetzen JL, Echenique V, Dubcovsky J (2004) The wheat VRN2 gene is a flowering repressor downregulated by vernalization. Science 303:1640–1644

Young ND, Debelle F, Oldroyd GE et al (2011) The Medicago genome provides insight into the evolution of rhizobial symbioses. Nature 480(7378):520–524

Zetsche B, Gootenberg JS, Abudayyeh OO, Slaymaker IM, Makarova KS, Essletzbichler P, Volz SE, Joung J, Van Der Oost J, Regev A, Koonin EV (2015) Cpf1 is a single RNA-guided endonuclease of a class 2 CRISPR-Cas system. Cell 163(3):759–771

Zhu XG, De Sturler E, Long SP (2007a) Optimizing the distribution of resources between enzymes of carbon metabolism can dramatically increase photosynthetic rate: a numerical simulation using an evolutionary algorithm. Plant Physiol 145(2):513–526

Zhu X-G, de Sturler E, Long SP (2007b) Optimizing the distribution of resources between enzymes of carbon metabolism can dramatically increase photosynthetic rate: a numerical simulation using an evolutionary algorithm. Plant Physiol 145:513–526

Zhu XG, Long SP, Ort DR (2008) What is the maximum efficiency with which photosynthesis can convert solar energy into biomass? Curr Opin Biotechnol 19:153–159

Zhu XG, Long SP, Ort DR (2010) Improving photosynthetic efficiency for greater yield. Annu Rev Plant Biol 61:235–261

Zhu BF, Si L, Wang Z, Zhou Y, Zhu J, Shangguan Y, Lu D, Fan D, Li C, Lin H (2011) Genetic control of a transition from black to straw-white seed hull in rice domestication. Plant Physiol 155:1301–1311

Zhu X-G, Wang Y, Ort DR, Long SP (2013) e-Photosynthesis: a comprehensive dynamic mechanistic model of C3 photosynthesis: from light capture to sucrose synthesis. Plant Cell Environ 36:1711–1727

Zsögön A, Čermák T, Naves E et al (2018) De novo domestication of wild tomato using genome editing. Nat Biotechnol 36:1211–1216. https://doi.org/10.1038/nbt.4272

Regulatory, Ethical, and Social Aspects of CRISPR Crops

9

Aftab Ahmad, Muhammad Zubair Ghouri, Nayla Munawar, Muhammad Ismail, Sidra Ashraf, and Syed Ovais Aftab

Abstract

CRISPR/Cas has become the new face of genome editing with rapidly expanding applications in academic research (reverse genetics and functional genomics), as well as applied research to develop new crop varieties with improved traits. The policies and regulations of CRISPR-edited crops are debated worldwide. For example, the USA has been deregulating CRISPR-edited organisms (CEOs) especially crops under existing regulatory laws, whereas the EU regulates all types of CRISPR-edited crops as genetically modified crops. These crops hold tremendous potential for global agriculture and food security. Therefore, the question of how these crops will be regulated determines the future of CRISPR. In this chapter, we highlight product- and process-based regulations of genetically modified (GM) crops because genome editing regulations are also based on these existing regulations. Also, we discuss that the outcomes of site-directed nucleases (ZFNs, TALENs, and CRISPR/Cas) are sometimes similar to conventional mutagenesis techniques, although the worldwide community is divided over their regulation. The USA has been deregulating site-directed nuclease (SDN1 and SDN2) crops, whereas the EU has been following process-based

A. Ahmad (✉) · M. Z. Ghouri · S. O. Aftab
Center for Advanced Studies in Agriculture and Food Security (CASAFS), University of Agriculture, Faisalabad, Pakistan
e-mail: aftab.ahmad@uaf.edu.pk

N. Munawar
Department of Chemistry, United Arab Emirates University, Al-Ain, UAE

M. Ismail
Department of Forestry and Range Management, University of Agriculture, Faisalabad, Punjab, Pakistan

S. Ashraf
Department of Biochemistry, Faculty of Sciences, University of Agriculture, Faisalabad, Pakistan

A. Ahmad et al. (eds.), *CRISPR Crops*,
https://doi.org/10.1007/978-981-15-7142-8_9

regulation for genome-edited crops and has classified the outcomes of SDN1, SDN2, and SDN3 as GMOs. Furthermore, we emphasize that there is a need for a new product based-scalable regulatory system for genome-edited crops to face the global challenges of food security, food safety, and sustainable agriculture. We summarize that although CRISPR-edited crops are transgene-free, public acceptance of these crops remains a challenge. We also highlight the ethical issues of CRISPR-edited crops and the importance of governance because scientists have cautioned about CRISPR-based gene drives. We also briefly discuss the regulatory cost and time required for the approval of these crops. Finally, we discuss the future prospects of CEOs and challenges related to their regulation.

9.1 Introduction

Agriculture is facing global challenges because of the increasing world population, climate change, and increasing biotic and abiotic stresses. Food prices continue to increase, as the world population is expected to reach 9 billion by 2050 (Raza et al. 2019; Godfray and Garnett 2014). Agricultural production must be increased to keep the prices down and feed people while meeting the demands for better nutrition (Sasson 2012). However, we cannot afford to create new arable land by deforestation, as we need forests and wilderness to decrease the level of carbon dioxide (CO_2) (Carrington 2019). Developing countries have been facing huge problems due to the shortage of fresh water, and climate change is leading to more frequent droughts and floods that affect agricultural production (Misra 2014). Therefore, new crop varieties are needed to increase agricultural production to face these environmental challenges. Genome editing tools, such as the clustered regularly interspaced short palindromic repeats (CRISPR)/CRISPR-associated system9 (Cas9), offer promising solutions for agriculture with rapid breeding and trait development, multiple and precise modifications, and transgene-free crops.

Recently, genome editing techniques such as zinc-finger nucleases (ZFNs), transcription activator-like effector nucleases (TALENs), and CRISPR/Cas are enabling previously unachievable, precise, and targeted modifications in the genome. The discovery of the CRISPR/Cas9 system is viewed as a game changer, given its efficiency, simplicity, cost-effectiveness, and continuously expanding applications across species. CRISPR/Cas creates a double-stranded break (DSB) in the genomic DNA at the target site, which is subsequently repaired by nonhomologous end joining (NHEJ) or homology-directed repair (HDR) (Gaj et al. 2013; Nemudryi et al. 2014; Javed et al. 2018).

NHEJ is the preferred pathway of DSB repair in eukaryotes and can be used to create insertion–deletion (indel) mutations of various lengths, resulting in gene disruption (Lieber 2010). Relying on the NHEJ repair pathway, CRISPR/Cas9 has been widely used to engineer major crops such as rice, wheat, barley, potato, cotton, and maize (Jaganathan et al. 2018; Arora and Narula 2017; Zaidi et al. 2019).

CRISPR-edited crops created through the NHEJ pathway are transgene-free and consequently more socially acceptable (Zhang et al. 2019; Araki and Ishii 2015). Such crops could bypass the existing regulations of genetically engineered crops, as they are not different from conventional crops (Resources 2016). However, like any other new technology, CRISPR/Cas system also has some limitations, and concerns of unintended genomic modifications may lead to the imposition of regulations (Omodamilola and Ibrahim 2018; Cribbs and Perera 2017). Therefore, despite the tremendous potential of CRISPR/Cas system in agriculture and human health, the question of how the CRISPR-edited organisms (CEOs) will be regulated will determine the future of CRISPR. Rapid developments in the CRISPR/Cas suite have also been accompanied by scientific and legal discussions within and across borders to determine the regulatory policies potentially governing CRISPR-edited crops.

The status of CEOs is currently debated worldwide, and countries have been divided on whether to regulate CEOs according to process-based or final product-based regulation (Wolt et al. 2016). Similar to the regulation of genetically engineered crops, the United States of America (USA) has been following product-based regulations for CEOs, whereas the European Union (EU) has been regulating CEOs under process-based regulation (Wolt et al. 2016; Jones 2015). In addition, there is also a discussion on whether the point mutations created by CRISPR will be regulated as conventionally induced mutations or spontaneous mutations (Duensing et al. 2018). Moreover, it is difficult to trace the origin of or the technique used to create these mutations, if no foreign DNA is integrated into the host genome.

Finally, this mosaicism in the legal status of CRISPR-edited crops among countries may lead to public distrust and clash with standards of the World Trade Organization (WTO) (Mehravar et al. 2019). By contrast, if a conducive and scalable regulatory system is established, it may increase the potential impact of the CRISPR/Cas system to address the challenges of food security and sustainable agriculture. In the first section of this chapter, we describe the existing process- and product-based regulation of genetically engineered crops (Sects. 9.2.1 and 9.2.2). We also describe the current regulation of CRISPR-edited crops in different countries (Sects. 9.4, 9.5, 9.6, 9.7 and 9.8). Finally, we discuss the ethical issues, cost of regulation, public perception, and social concerns regarding CRISPR-edited crops (Sects. 9.11, 9.12, 9.13 and 9.14).

9.2 Biosafety Regulations and Benefits of GMOs

Traditional breeding is an integral and accepted part of global agriculture. Therefore, products developed through traditional breeding are considered safe and are not subjected to regulation in any country (Dubock 2017). By contrast, genetically modified (GM) crops, which are grown worldwide on a consistently increasing area every year, are under strict regulations in many countries because of the potential impact of foreign DNA and genetic modifications on food safety and the environment (Nollet et al. 2012). Therefore, before discussing the regulatory

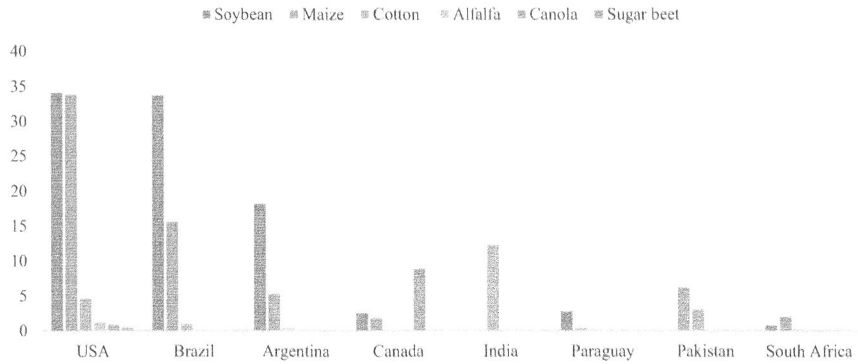

Fig. 9.1 Transgenic crops grown worldwide. The USA and Brazil are at the top, with 34.05 and 33.7 Mha of genetically modified (GM) soybean and 33.84 and 15.6 Mha of GM maize, respectively. India is the leading producer of GM cotton (12.24 Mha), followed by the USA (4.58 Mha), and Pakistan (3.0 Mha). Source: ISAAA (2017)

concepts of CRISPR-edited crops, it is worth understanding the existing regulatory framework of GMOs globally because these regulatory concepts may impact the legal status and regulation of CRISPR-edited crops (Wolt et al. 2016).

Today, GMOs including crops are a reality in our lives whether as feed for livestock, food for humans, or raw material for the industry. Most of the GM crops are generated by adding one or more genes derived from plants or unrelated incompatible species using *Agrobacterium*-mediated transformation or biolistics (Priyadarshan 2017). Flavr Savr tomato was the first GM crop approved in the USA for human consumption in 1996, which was followed by the approval of GM cotton, soybean, and maize for commercial cultivation (Wunderlich and Gatto 2015). To date, 29 different crops and fruit trees have been successfully modified in more than 40 countries for improving traits such as quality and insect resistance and stacking multiple traits in a single genetic background (Borém and Fritsche-Neto 2014).

The USA, Brazil, India, China, Argentina, and Pakistan are the leading countries in the commercial cultivation of GMOs (Fig. 9.1) (Ishii and Araki 2016). GM crops have been adopted globally, especially in developing countries, for various benefits such as higher yield, lower pesticide application, greater disease resistance, and enhanced quality (Phillips 2008). Despite the global adoption and potential benefits of GMOs, there are multiple concerns related to genes derived from incompatible species, development of antibiotic resistance, and horizontal gene transfer, which trigger strict regulation of GMOs in some countries (Bawa and Anilakumar 2013). Globally, GMOs are regulated in three ways: (1) product-based regulation (depending on the type of trait expressed), (2) a process-based regulation

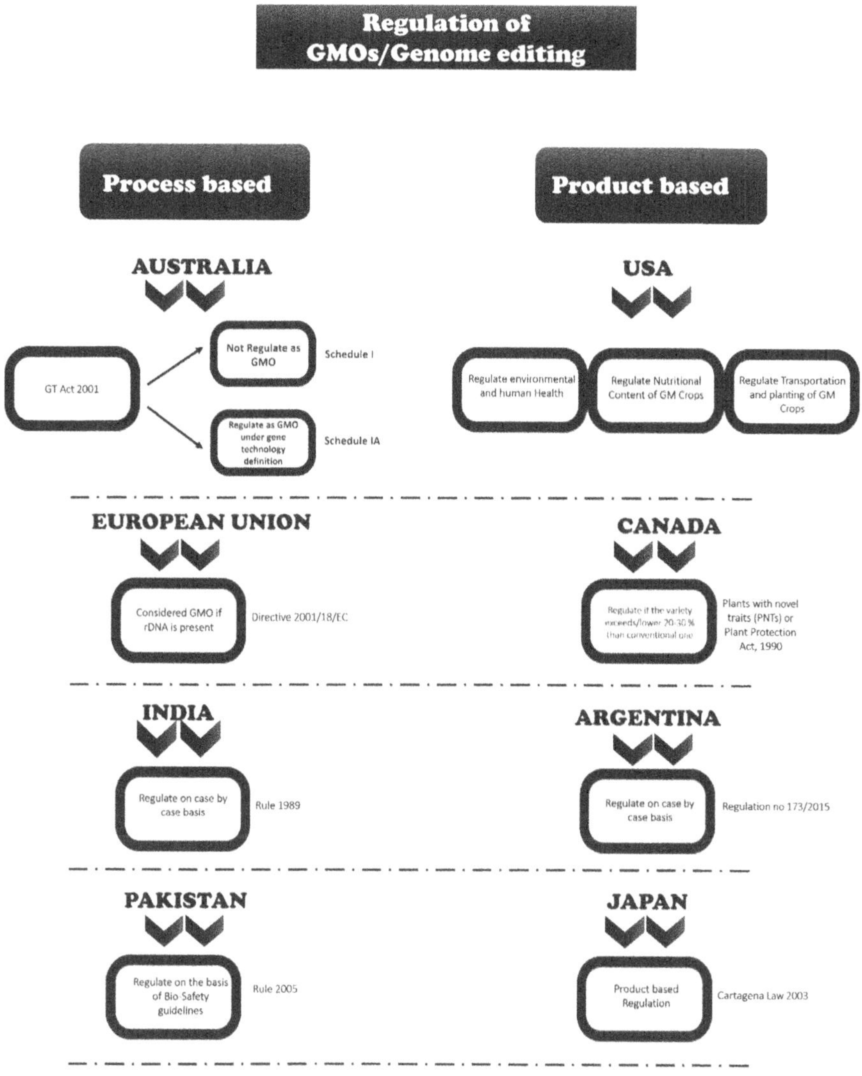

Fig. 9.2 Regulation of genetically modified organisms (GMOs) or genome-edited crops in different countries of the world. Australia, the European Union (EU), India, and Pakistan follow process-based regulation, whereas the USA, Canada, Japan, and Argentina follow product-based regulation for GM or genome-edited products

(genetically modified or not), and (3) plants with novel trait (PNT)-based regulation (see Fig. 9.2 for the product, process, and PNT-based regulations), and there is a strong division in the world regarding the type of regulation of GMOs (Sprink et al. 2016). This division probably stems from the fact that not all countries have ratified the Cartagena biosafety protocol (CBP). Among the top 10 GM crop-growing countries, the USA, Argentina, Russia, Canada, Australia, and China have not signed the CBP and have adopted product-based regulation of GMOs (Smyth 2017).

9.2.1 Process-Based Regulation

Recombinant DNA (rDNA) technology, also known as genetic engineering or modern biotechnology, is one of the common triggers to invoke regulation and has been used as a simple dichotomy between conventionally bred and GM plants (McHughen 2016). In the 1980s, genetic engineering was used to transfer a gene from an incompatible species to plants for the first time (Gupta et al. 1984). Before genetic engineering, it was a common belief that nature does not transfer genes across incompatible species. Therefore, any violation of this barrier among species indicated that the product must have been generated using the rDNA technology (McHughen 2016). Moreover, it was also believed that the rDNA technology was risky, and consequently the product was considered unsafe and was required to pass through strict safety assessments. Therefore, with each development in the field of genetic engineering, concerns were raised in the scientific and legal/regulatory communities, and at least, the legal community was convinced that GM plants containing bacterial genes were derived using the rDNA technology (McHughen 2016).

The EU has been following a process-based regulation for GMOs. Under this type of regulation, importance is given to the process by which the GMOs are generated rather than the trait they express. In the EU, GM crops are subjected to process-based regulatory oversights through detailed procedures under EU Directive 2001/18EC (2001) and EC Regulation 258/97 (1997) (Araki and Ishii 2015; Sprink et al. 2016; Ishii and Araki 2016). Currently, GM maize is the only GM crop cultivated on a limited area in some of the 28 member states of EU (BBC News 2015).

9.2.2 Product-Based Regulation

In process-based regulation, it is possible to reveal some risks but not to determine the safety of the end products (Dederer and Hamburger 2019). For example, in two plants carrying the same mutation or deletion, one of which is generated through chemical mutagenesis and the other by genome editing, both traits will be indistinguishable (Van et al. 2017). Under product-based regulation, both plants will need to pass through regulation; however, if the two identical traits are treated differently only because different methods were used to create these plants, the consumers could be misled, if one product is considered safer than the other. Also, product-based regulation does not leave any regulatory gap between GMOs and non-GMOs. For example, process-based regulation differentiates between GMOs and non-GMOs based on the process involved but not on the actual risk associated with the product;

therefore, the same risk arises from the particular trait produced by conventional breeding, which may not be strictly regulated (Sprink et al. 2016).

Compared with process-based regulation, the product-based approach considers the risks associated with a particular product, not with a process. Scientifically, risks are not associated with the genetic modification technique or method used but with a particular expressed trait (McHughen 2016; Sprink et al. 2016; Smyth 2017). Moreover, product-based regulation is also compatible with free trade agreement or the WTO law. Product-based regulation helps avoid unnecessary trade barriers more easily than process-based regulation (Prakash et al. 2011; Sprink et al. 2016).

In contrast to the EU, the USA endorses product-based regulation of GMOs under the 7CFR part 340 and part 360. In addition, the USA is one of the largest producers of GM crops in the world (Araki and Ishii 2015; Ishii and Araki 2016). The regulatory frame for GMOs in the USA is implemented through three federal agencies: United States Department of Agriculture (USDA), Environmental Protection Agency (EPA), and Food and Drug Administration (FDA). The difference between the regulatory framework of the USA and the EU suggests that product-based regulation encourages the cultivation of GM crops, whereas process-based system discourages it.

9.2.3 PNT-Based Regulation

A PNT is defined as a plant that contains or expresses a specific trait, which is not present in any of the previously approved/cultivated products, or one that displays a trait at a level significantly outside the range of the already existing trait in approved cultivars (Government of Canada 2018). The PNT-based regulation does not differentiate between processes, and a PNT can be produced with any of the several known techniques such as conventional mutagenesis, genetic engineering, and genome editing (Government of Canada 2018). Unlike the EU and the USA, the Canadian regulatory framework for GMOs is based on the PNT system, and all PNTs must pass through risk assessments. The Canadian Food Inspection Agency (CFIA) and the Health Canada are responsible for the regulation of food derived from GMOs (Seeds Act, R.S.C. 1985) and for the regulation of food and drugs (Food and drugs Act, R.S.C. 1985) (Canada Health 2018). All other countries use one of the three regulation systems for GMOs (Table 9.1).

9.3 New Plant Breeding Technologies (NPBTs)

NPBTs has been the method of choice for many years to develop crops with superior traits and consequently to address the issue of food security. Plant breeding involves trial and error that involves crossing between parental lines and selfing to produce new verities (Osei et al. 2014; Witcombe et al. 2013). Plant breeding involves the identification of a plant with a superior trait and crossing it with another plant to transfer the desired trait into the offspring (Glenn et al. 2017). However, some

Table 9.1 Genome-edited crops approved in different countries

Crops	Technique	Gene target	Regulator	Regulatory response	Reference
Grapevine	Grafting	GFLV coat protein expressed in GM rootstock	French Ministry of Agriculture, Biomolecular Engineering Commission	Approved	Lieber (2010)
Grapevine	Intragenesis or cisgenesis	VvMybA1	USDA APHIS	Deregulated	https://www.aphis.usda.gov/biotechnology/downloads/reg/loi/Grapevine/Inquiry_BR_response_042412.pdf
Apple	Cisgenesis	Apple scab resistance gene (Vf)	USDA APHIS	Might be regulated	https://www.aphis.usda.gov/biotechnology/downloads/reg_loi/aphis_response_schouten.pdf
Maize	ZFN	IPKI knockout via NHEJ	USDA APHIS	Deregulate	http://www.aphis.usda.gov/biotechnology/downloads/reg_loi/DOW_ZFN_IPKI_052610.pdf
Maize	OMM	AHAS108 and AHAS 109			Zhu et al. (1999)
Maize	ZFN	IPK1 (insertional disruption)	USDA APHIS	Deregulated	Shukla et al. (2009)
Potato	RNAi	Low polyphenols Reduced acrylamide Reducing sugars	Health Canada	Deregulated	https://www.isaaa.org/gmapprovaldatabase/event/default.asp?EventID=381
Apple	RNAi	PPO (polyphenol oxidase genes)	CFIA	Declined regulation	https://www.inspection.gc.ca/plant-health/plants-with-novel-traits/approved-under-review/decision-documents/dd2015-110/eng/1498237204300/1498237333410
Potato	TALEN	PPO5	USDA APHIS	Declined regulation	https://www.aphis.usda.gov/brs/aphisdocs/16_06401p_pdet_pprsa.pdf
Maize	EMN	MS26	N/A	–	D'Halluin et al. (2013)

Soybean	TALEN	FAD2-IA and FAD2-IB	USDA APHIS	Deregulated	http://www.aphis.usda.gov/biotechnology/downloads/reg_loi/brs_response_cellectis_air_fad2k0_soy_cbidel.pdf
Rice	TALEN	OsSWEET11 and OsSWEET14	USDA APHIS	Deregulated	http://www.aphis.usda.gov/biotechnology/downloads/reg_loi/aphis_resp_isu_ting_rice.pdf

Source: Wolt et al. (2016)

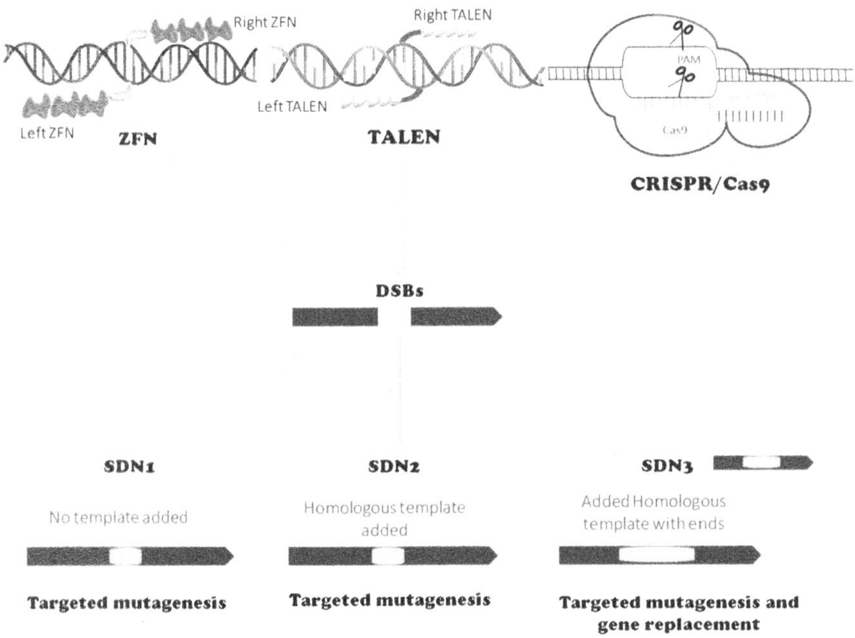

Fig. 9.3 Overview of site-directed nuclease (SDN) technologies. DSBs created by SDN1 can be repaired through the NHEJ pathway, which leads to gene knockout, silencing, or deletion. In SDN2, DSBs are repaired by the addition of small template, thereby adding a few bases. Repair pathway for SDN3 needs the inclusion of a long stretch of DNA with homology to a region downstream or upstream of the DSB

unrelated traits may also be transferred to the offspring during this process, and additional breeding cycles are required to eliminate these traits. Therefore, breeding is laborious, is random, and has off-target effects. This led to the development of genetic engineering in the 1980s. Subsequently, plant breeding moved from cisgenic to transgenic with introgression of genes from distant noncrossable species (Schouten et al. 2006). However, transgenic plants were produced by the random integration of the transgene in the plant genome.

Recent years have witnessed rapid development and increased use of NBPTs, also known as genome editing, to induce previously impossible, precise, and targeted genome modifications (Zaidi et al. 2019). Genome editing has been used for several applications in plants such as to improve disease resistance, yield, and abiotic stress tolerance (Sedeek et al. 2019). Genome editing has enabled a range of precise genomic modifications, including insertions, deletions, base editing, and epigenome editing. NBPTs are a set of new innovative methods based on oligonucleotide-mediated genome modification (ODM), site-directed nucleases (SDNs; e.g., ZFNs and TALENs), or a combination of both (CRISPR/Cas) (Saha et al. 2019; Aglawe et al. 2018). The molecular mechanism of SDNs involves the induction of a DSB at a specific site in the genome, which is subsequently repaired through the NHEJ or HDR pathway. Repair of DSBs by SDNs may result in three different outcomes, designated as SDN1, SDN2, and SDN3 (Fig. 9.3). SDNs provide unprecedented control over genomic modifications and consequently over agronomically important traits efficiently and cost-effectively.

9.3.1 SDN1

SDN1 repairs a DSB via NHEJ without any donor or template. The repair of DSB in SDN1 may result in mutation or deletion, leading to altered gene activity, gene silencing, or gene knockout. Also, the outcome of SDN-mediated repair will be indistinguishable from conventional and natural mutagenesis (Lassoued et al. 2019).

9.3.2 SDN2

SDN2 repairs DSB using a small donor template, homologous to the DSB site. The template contains one or a few bases different from the original sequence in the gene, resulting in specifically altered sequence (mutation) in the targeted gene. Similar to SDN1, it is difficult to differentiate SDN2 from conventional or natural mutations (Sprink et al. 2016; Wolt 2017).

9.3.3 SDN3

SDN3 repairs DSB using a long stretch of DNA, with a homologous arm of 500 nt or greater to the upstream and downstream sequence of DSB and between their homology arms any sequence like cisgene, transgene, synthetic gene, promoter sequence, or regulatory sequence can be provided. Applications of SDN3 in plants are complicated and inefficient. The use of SDN3 for DSB repair could result in two possible outcomes. If the donor sequence is derived from a noncrossable species, the final plants will be transgenic, which will need to be regulated (Sprink et al. 2016; Wolt 2017). However, if the donor sequence is from a crossable partner of the same species, the final product could theoretically result from conventional crossing,

Table 9.2 Classification of products (GMOs) obtained through oligonucleotide-mediated genome modification (ODM), site-directed nuclease 1 (SDN1), SDN2, and SDN3 in different countries

Countries	ODM	SDN-1	SDN-2	SDN-3
Argentina	−ve	−ve	−ve	+ve
Australia	+ve	−ve	+ve	+ve
The EU	+ve	+ve	+ve	+ve
Japan	−ve	−ve	−ve	+ve/−ve
The USA	−ve	−ve	−ve	+ve/−ve

+ve = GMO, −ve = non-GMO
Source: Hamburger (2019)

which will need to be subjected to several backcrosses. In this case, GMO regulation will be inconsistent, as the same product could be derived from natural crossing. The classification of products produced through ODM, SDN1, SDN2, and SDN3 as GMOs is summarized in Table 9.2.

9.3.4 Concerns Regarding CRISPR/Cas-Based Genome Editing

Genome editing with engineered nucleases is increasingly being used for a wide range of applications in animals, plants, and microorganisms (Osakabe and Osakabe 2014). In addition, rapid developments in the CRISPR/Cas suite have enabled scientists to gain dramatic control over precise modifications in the genome, including deletions, insertions, base editing, and directed evolution. CRISPR/Cas9 and CRISPR/Cas12 (Cpf1) has been widely used in plants for various applications such as rapid trait development, disease resistance, and improved quality (Phillips 2008). Moreover, with the use of CRISPR/Cas, scientists can develop plants that are more socially acceptable than GMOs (Morrison and de Saille 2019). CRISPR/Cas has been proposed as a revolutionary tool with published proof of concept to change the future landscape of agriculturally important crops. Although there are many studies published on the precise and intended modifications in plants using CRISPR, very few studies have evaluated off-target and unintended effects. Although CRISPR has been proposed as a precise technology, the outcomes are not always precise, and intended changes may result in environmental risks and food safety issues (Duensing et al. 2018). In addition, most of the genome-edited plants developed using CRISPR/Cas must go through the process of transformation either using *Agrobacterium* or biolistics, followed by tissue culture, to introduce the desired changes. Here, we discuss concerns about CRISPR-mediated genome editing, which may impose the regulation of CRISPR-edited crops.

9.3.5 Choice of Reagents and Delivery

CRISPR/Cas has been adopted widely, and most of the CRISPR-edited plants created to date have been passed through the same procedure (Gaj et al. 2013; Nemudryi et al. 2014). Moreover, most of the studies on CRISPR/Cas have been focused on creating gene knockout plants using NHEJ to repair DSBs. Only a limited number of studies have used the HDR repair mechanism. Also, most of the studies used *Agrobacterium* as a preferred tool for the delivery of plasmids, which trigger the regulation of CRISPR-edited plants (Rozov et al. 2019). The choice of delivery method and CRISPR/Cas reagents may trigger the regulation of CRISPR-edited plants. Therefore, appropriate off-target prediction, choice of modifications, source of nucleases, and method of delivery may contribute to the deregulation of CRISPR-edited plants.

9.3.6 Reagents

Genome editing using CRISPR/Cas relies on the presence of a functional Cas9–guide RNA (sgRNA) complex in the nucleus to create DSBs, which are subsequently repaired through specific DNA repair systems. Functional Cas9 and sgRNA have been successfully used in the form of a plasmid, in-vitro transcribed mRNA, or ribonucleoproteins (RNPs) for successful genome editing in plants (Farboud et al. 2018; Razzaq et al. 2019). Plasmids are the most common reagents used for plant genome editing; however, any stable and random integration in the genome may trigger the regulation of the CRISPR-edited plants. However, the transgene can be removed through crossing to bypass regulations (Yang et al. 2017). The use of in-vitro transcribed mRNA for genome editing will lead to the expression of Cas nuclease in the cell and will not result in stable integration of the *Cas* gene in the plant genome. The use of RNA as an editing reagent will also reduce off-target editing. Delivery of purified RNPs has been successfully used for DNA-free genome editing and generation of transgene-free genome-edited plants with the lowest off-targets (Razzaq et al. 2019). Although the use of purified RNPs as genome editing reagents will help avoid potential regulations, the delivery of purified RNPs is challenging.

9.3.7 Delivery Methods

Agrobacterium is currently the most dominant delivery method for CRISPR reagents in plants; however, it results in random and stable integration of the transfer DNA (T-DNA) in the plant genome. Genome-edited plants produced by *Agrobacterium*-mediated transformation carry the *Cas9* gene or transgene in their genome and therefore must be regulated as transgenic plants (Rojas-Vásquez and Gatica-Arias 2019). Although CRISPR/Cas cassette can be removed from the plant genome by crossing to avoid regulation (Yang et al. 2017), however, researchers have to make sure that the *Cas* gene has been removed. Similarly, the biolistic delivery method also leads to stable integration of the CRISPR-Cas cassette at a random location in the plant genome (Banakar et al. 2019). However, the delivery of mRNA through the biolistic approach produces transgene-free genome-edited plants, which will be able to bypass regulation. Transient transformation of protoplasts is also a good option to avoid stable integration; however, protoplast transformation has not been standardized for all crops.

9.3.8 Off-targets

Off-target editing is one of the main concerns of the CRISPR/Cas technology, which may trigger the regulation of genome-edited plants. Because Cas nuclease can tolerate a few mismatches in the sgRNA, the sgRNA sometimes binds to multiple sites, in addition to the intended site, and the Cas nuclease may then cleave some of those unintended sites, leading to off-target editing and consequently unexpected changes (Wolt 2017; Hajiahmadi et al. 2019). Therefore, characterization of the off-targets through sequencing, metabolomics, transcriptomics, and phenomics is important for the commercialization of CRISPR-edited plants.

9.3.9 Types of Modifications

Modifications induced by SDN1 and SDN2 are indistinguishable from those caused by conventional or natural mutagenesis. Therefore, genome-edited plants produced using these nucleases are deregulated. However, plants carrying modifications induced by SDN3 are considered the same as GMOs (Troadec and Pagès 2019; Wolt et al. 2016).

9.3.10 Gene Drives

Gene drives are always classified under SDN3 and are strictly regulated because of biosafety issues. CRISPR-based gene drive is a system that involves forced inheritance of traits through generations of species bypassing the natural selection of that trait (Collins 2018). The basic criteria for a functional gene drive include (a) sexual reproduction, (b) short generation time, (c) drive stability, and (d) appropriate population structure. In a CRISPR-based gene drive, Cas9 and sgRNA must present within flanking arms homologous to the insertion site in target germ cells (Oberhofer et al. 2018). In CRISPR drives, a few engineered individuals may spread the intended trait to the entire population (Esvelt et al. 2014; Noble et al. 2018). Because CRISPR-based gene drive always contains foreign genetic elements, plants generated using this technology could be regulated under the existing regulatory framework for GMOs.

Gene drives have been demonstrated to eradicate vector-borne diseases such as malaria, zika virus, and dengue fever; however, a number of gene drives released in the environment may raise biosafety issues (CGDRO 2016;Min et al. 2018;The Economist 2018). Moreover, once released, gene derives are impossible to recall. Therefore, genetic changes in a population induced by a gene drive are permanent. Scientists have cautioned about the severe consequences of any off-targets in gene drives (ServickJul et al. 2017;CGDRO 2016). In addition, the use of gene drives for agricultural purposes could have adverse effects on human health (Brossard et al. 2019). Moreover, transfer of a suppression drive to nontarget species could have harmful effects on the environment as well as the food chain. Currently, there exists no regulatory mechanism worldwide to control the environmental release of gene drives. Moreover, CRISPR-based gene drives could also be used as biocrime (Mueller 2019). Therefore, scientists have been calling for an international moratorium on gene drive research.

9.4 Current Regulation of Genome-Edited Organisms in Different Countries

Genome editing differs from classical genetic engineering in a number of ways. Genome editing leads to a discrete change, as small as a single base pair, at a very precise location in the genome, with relative ease and in a short time. Moreover, the change can be introduced without the presence of a transgene in the final product, and sometimes the resulting change could be indistinguishable from that induced by conventional breeding (Eckerstorfer et al. 2019). Therefore, scientists argue that the risk imposed by such genome-edited crops could be as little as that imposed by conventionally bred crops. Therefore, the emergence of genome editing techniques has been accompanied by significant legal, scientific, and policy discussions over the regulation of these crops under the existing regulation of GMOs.

Because changes introduced using the CRISPR/Cas system could be similar to natural mutations, scientists advocate that CRISPR-edited crops should not be regulated under current GMO regulation. It has been stated that CRISPR-edited crops should be regulated under only product-based regulations, relying on the scientific risks or benefits of each trait. Because it is obvious that some CRISPR crops do not contain any foreign DNA, the existing regulations of GMOs are considered ambiguous and vague for application to CRISPR-edited crops worldwide (McHughen 2016; Prakash et al. 2011; Eckerstorfer et al. 2019). Regulators also support the scientific debate that regulation and biosafety of CRISPR-edited crops should be focused on the phenotype of the plant arising from the application. It has been suggested that transgene-free and NHEJ repair pathway-based crops should not be regulated (Friedrichs et al. 2019; Wolt et al. 2016).

The legal status of genome-edited crops, especially those generated using the CRISPR/Cas technology, is currently uncertain, as there is considerable debate whether CRISPR/Cas-edited crops should be deregulated or regulated under the current GMO laws. In the following sections, we discuss how different countries are currently regulating CRISPR-edited crops.

9.5 Regulation of Genome-Edited Crops in Argentina

According to Argentinian regulation, GMOs are defined as "*any vegetable or organism that possesses a novel combination of genetic material derived through modern biotechnology*." This definition follows the description of GMOs under the CBP. Argentina's regulation could be classified as product based as it defines crops as possessing a novel combination of genetics or as process based because the novel combination must be derived through the process of modern biotechnology (Friedrichs et al. 2019; Whelan and Lema 2019). Argentina was the first country to adopt new regulations for genome-edited crops (Resolution no. 173/2015) (Friedrichs et al. 2019). It clearly indicates that genome-edited crops without a transgene will bypass regulations under product-based regulation. Crops generated using SDN1 and SDN2 will not be regulated as GMOs, but those created using

SDN3 will be regarded as GMOs, with one exception of perfect allele replacement. The risk assessment and regulation will be on a case-by-case basis, implying that each transformation event will be assessed and regulated, if needed (Whelan and Lema 2019).

Regulation involves a process that determines whether the product was derived by genome editing, and this process precedes the application of the regulatory framework to the plant in question. Farmers could therefore ask for preliminary classification of a plant while it is in development. In addition, it provides an opportunity for developers to receive the regulatory classification of the envisaged product at an earlier stage of research and development. If the final plant corresponds to the earlier predictions, then preliminary assessments retain their validity. However, if a competent regulatory authority identifies some risk associated with genome-edited plants, they can issue safety-related recommendations to breeders.

9.6 Regulation of Genome-Edited Crops in Australia

A bination regulatory authority, Food Standards Australia and New Zealand (FSANZ), regulates GMOs in Australia and New Zealand. FSANZ issues a single regulatory approval for both countries. However, it is interesting that GMOs are cultivated only in Australia and not in New Zealand. This is because, in addition to FSANZ, Australia has a separate regulatory framework for GMO cultivation (Brent et al. 2003; Ludlow 2019).

Both the regulatory frameworks of Australia are responsible for the cultivation of GMOs and their products following process (gene technology)-based trigger; however, both define gene technology. Because of this difference, the regulatory framework for genome-edited organisms in Australia depends on whether they will be used for cultivation or food production. It indicates that genome-edited organisms will be regulated under general GMO regulations but not by GMO food framework (Brent et al. 2003; Ludlow 2019). No genome-edited crops are cultivated in Australia; however, Australian authorities have been reviewing whether some of the genome-edited crops fall under the gene technology framework. In the meantime, Australia has adopted interim regulation and considers crops generated using SDN1 as conventional crops but those generated using ODM, SDN2, and SDN3 as GMOs.

9.7 Regulation of Genome-Edited Crops in the USA

The regulation of genome-edited organisms in the USA is controlled by the FDA, EPA, and USDA. The approach of the USA to regulate GMOs is quite different. GMOs can escape regulation, if they do not trigger the regulatory framework. A unique feature of the US regulatory framework is product-based trigger for the regulation of food (Wolt 2017). Regarding the cultivation of plants, the trigger is plant pest, plant pest risk, or plant-containing protectant, which needs regulatory requirements. Here, we discuss the role of each US agency in the regulation of genome-edited crops.

9.7.1 FDA

FDA regulates food and feed derived using the rDNA technology, with a strong focus on product-based regulation. FDA regulates the safety of food and feed derived from genetically engineered plants, on the basis of compositional equivalence to their non-GMO counterparts related to allergens, antinutrients, and toxins (Wolt 2017).

9.7.2 EPA

EPA regulates genetically engineered crops for their pesticidal properties to detect pesticide residues in food. Few products of new genome editing techniques, such as knockout plants with resistance to viruses and pests, fall under the regulatory authority of EPA.

9.7.3 USDA

Regulation of genome-edited crops by USDA depends on whether genome-edited plants fulfill the corresponding definition of regulated articles. To address the uncertainty about the regulation of genome-edited crops, USDA published the following statement: "*Plants that could otherwise have been developed through traditional breeding techniques as long as they are developed without the use of a plant pest as the donor or a vector and they are not themselves plant pests*" (USDA APHIS 2018). It comprises single base substitutions or insertions from compatible plant relatives and deletions or complete null segregants. Therefore, plants generated via ODM, SDN1, and SDN2 will be considered as deregulated genome-edited plants, as these changes could also be produced by conventional breeding (Wolt et al. 2016). However, plants generated using SDN3 will be regulated on a case-by-case basis to determine whether individual applications of SDN3 fall within the definition of regulated articles.

To date, USDA has declined to regulate numerous genome-edited crops; for example, ZFN12 maize produced using the ZFN technology for reduced production of phytate was not regulated by USDA (Grossman 2019). In addition, Animal and Plant Health Inspection Service (APHIS) working under USDA declined to regulate PPO5 potato edited using TALENs to knock out the *PPO5* gene (Sciences 2020). Similarly, APHIS did not regulate CRISPR/Cas-edited mushroom with an antibrowning phenotype (Stoye 2015). However, the import of mushrooms is subjected to plant protection and quarantine measures, and mushrooms could be subjected to FDA and EPA regulations (Grossman 2019). Moreover, APHIS deregulated waxy corn developed by DuPont Pioneer using the CRISPR/Cas technology (Unglesbee 2016). Similarly, waxy corn might be then subjected to regulation by the FDA and EPA. In May 2018, APHIS also declined to regulate a CRISPR-edited tomato variety (Grossman 2019). Therefore, CEOs may escape USDA regulations, if they are indistinguishable from conventionally bred crops.

9.8 Regulation of Genome-Edited Crops in the EU

Under the EU legislation, GM crops are defined as "*an organism—in which the genetic material has been altered in a way that does not occur naturally through mating and or natural recombination*" (Custers et al. 2018). In addition, techniques of genetic engineering are subject to regulation, whereas mutagenesis is excluded from regulation. According to the ruling of the European Court of Justice (ECJ) as part of case C-520/1650, "*all organisms obtained from site-directed mutagenesis will be subjected to GMO regulatory regime of EU*" (Hajiahmadi et al. 2019). In the proceedings, SDN1 and ODM were mentioned as an example of site-directed nucleases. Therefore, all types of genome-edited organisms created using SDN1, SDN2, and SDN3 are regulated under the EU regulatory framework for GMOs. Because of process-based interpretation of the ECJ, all types of genome editing techniques result in GMOs, and the same applies to food derived from them (Hajiahmadi et al. 2019). Among all examined regulatory frameworks in the world, the regulatory framework of the EU is unique, as it regulates the cultivation of CEOs as well as food derived from them, without considering their product-based characteristics.

9.9 Regulation of Genome-Edited Crops in Canada

Since the Canadian regulatory framework is PNT based, no conclusion can be made on the use of a genome editing technique. The type of regulation applied to the trait is not dependent on the specific technique used to develop that trait (Hajiahmadi et al. 2019). For example, a plant generated through conventional breeding could also be considered as a PNT and consequently will be regulated under PNT-based regulation (Hajiahmadi et al. 2019).

PNT is solely a product-based trigger for the regulatory framework. Although, there is no standard definition of "Novel," Canadian breeders assume that any trait they have been selected should be 20–30% altered (higher or lower) than conventional varieties. Therefore, it would be predictable that some genome-edited crops will not be PNTs. To date, no CRISPR-edited crops have been cultivated in Canada; however, this is expected to change in 2020.

9.10 A New Dynamically Scalable System for Genome-Edited Crops

One of the most important developments in the modern history of life sciences is the invention and proliferation of genome editing technologies such as ZFNs, TALENs, and most notably, CRISPR/Cas (Ricroch et al. 2016). Developments in CRISPR/Cas offer new opportunities to address the global issue of food security and safety. Therefore, how genome editing and products derived from genome editing will be regulated are important aspects for the future development of this technology (Georges and Ray 2017). In addition to other technologies, RNAi, cisgenics, marker-assisted selection, and transgenics are also available, putting more pressure on the existing challenges of the regulation of CEOs.

Agencies, governments, and non-governmental organizations are struggling to establish new regulations under the current regulatory framework in the world. There are inconsistencies and controversies among the current regulatory frameworks in different countries. The USA and the EU have responded differently to the regulation of genome editing technologies. In addition, new genetic modifications such as multiplex genome-editing, gene drives and germline gene editing in humans have further complicated social, political, and regulatory issues (Greely 2019). Experts believe that there should be a dynamically scalable, consistent, reliable, and predictable system for accommodating all these modifications and addressing multiple challenges in the current regulations (Marchant and Stevens 2015; Wolt et al. 2016; Davison and Ammann 2017). In addition, there should be a coherent strategy to ensure that all risky products are given appropriate attention, and less risky products are not deliberately delayed. Regulation should be product or phenotype based, and modern technologies should pass through risk assessment (Hamburger 2019; Wolt 2017; Wolt et al. 2016). Moreover, risk assessment should be adopted for each trait and should be flexible and reliable.

Evolving modifications and emerging new technologies require continuous consideration, and it is critical for the regulatory systems to keep pace. If the same product has been already approved, repetitive risk assessment should be avoided and/or replaced by considerable monitoring and implementation of precision agriculture. Moreover, policy makers should not use a strict definition of GMOs because the concept of genetic modification is always evolving.

9.11 Public Perceptions of CEOs

GM crops like cotton, soybean, and corn have been in cultivation since the 1990s, and the area under these GM crops is continuously increasing each year (Raman 2017; Royal Society 2020). More than 28 countries have adopted GMOs for commercial cultivation (Royal Society 2020). However, even in countries where the cultivation of GMOs is legally allowed, not all people accept them as food products. In addition, certain countries like the EU and New Zealand have not allowed commercial cultivation of GMOs (Graciela et al. 2014). Many factors such as the lack of trust in developers, lack of knowledge, poor risk assessment, and poor risk–benefit communication contribute to the negative attitude of the public toward GMOs (Ishii and Araki 2016). Therefore, these factors must be considered, if we believe that GMOs can contribute to food security and safety.

Genome editing technologies like ZFNs, TALENs, and CRISPR/Cas, which are subsequently repaired via the NHEJ and HDR pathways, rely on DSBs in the genome. These technologies have resulted in myriads of benefits to the agriculture community. Studies suggest that NHEJ is the preferred pathway for DSB repair; however, the NHEJ repair pathway was a major concern about GMOs. Similar to GMOs, consumer acceptance of genome-edited crops is not optimistic. Although transgene-free plants have been produced by genome editing, they lack explanations because the method appears to be the same as that used for the production of GMOs. Despite the significant benefits of genome-edited crops, public acceptance is a major challenge for these crops worldwide (Bawa and Anilakumar 2013;Ishii and Araki 2016). Moreover, scientists and legislators need to engage with the public and listen to their concerns regarding genome-edited crops and must be transparent about each product (Betizeau and Betizeau 2017). Scientists need to clarify how genome-edited crops are different from GMOs and explain the advantages and disadvantages of GMOs. However, scientists and regulatory authorities need to perform extensive research and monitor the potential risks of genome-edited crops to develop the public trust in CEOs (Phillips 2008). The government should foster communication between the public, developers, and regulators to increase public acceptance of genome-edited crops (Ishii and Araki 2016). With better regulation, clear information, sincere consideration of the right to know, and careful risk–benefit communication, genome-edited crops can be better integrated into the society.

9.12 Ethics Issues of CRISPR-Edited Crops

Ethical concerns regarding genome editing, especially using CRISPR/Cas, need careful consideration both in health and agricultural sciences (Mulvihill et al. 2017; Rossant 2018; Brokowski and Adli 2019; Morrison and de Saille 2019; Cribbs and Perera 2017). For example, recently a Chinese scientist claimed the first CRISPR-edited infant with potential immunity to the human immunodeficiency virus (HIV) (Li et al. 2019). This claim resulted in widespread criticism, which highlighted the importance of ethical and governance issues regarding CRISPR/Cas technologies. The USA and the EU have banned germline genome editing in humans (Sykora and Caplan 2017). Similarly, in agricultural products, some countries protect the rights of their consumers and impose mandatory labeling of GM foods (Oh and Ezezika 2014). In these countries, people may demand clear traceability of genome-edited crops in food products to avoid purchasing or consuming them because of ethical issues (Ishii and Araki 2016; Maaß et al. 2019). It is important to consider whether to label all transgene-free genome-edited crops and food derived from them. In France, activists repeatedly interrupt the grafting of grapevine trait, although it was approved by a competent authority (Wolt 2017). Similarly, the

European Plant Science Organization (EPSO) intends to uncouple the questions of labeling from those of environment and safety assessment (Ishii and Araki 2016). Therefore, it is important to consider labeling while developing genome-edited crops in society. Similarly, introducing a tag in the plant genome could help in the identification of genome-edited crops; however, this might be opposed by scientists, as it would require an additional HDR step, thus increasing the burden on researchers and the cost of development (Duensing et al. 2018). Nonetheless, tagging may be considered as an option for cultivation and marketing, as the primary objective of genome editing is food security and food safety and not environmental disturbance or confusion among the public (Ishii and Araki 2016). Moreover, it is important to consider isolated trials, if genome-edited crops cause environmental concerns. However, it would be difficult to distinguish between transgene-free and conventional crops. Conversely, if risk assessment proved that genome-edited crops have no environmental implications, transgene-free crops may be cultivated openly without any tagging (Duensing et al. 2018; Friedrichs et al. 2019; Maaß et al. 2019). Therefore, it has been highlighted that genome-edited crops should be evaluated from the viewpoint of novel traits before field traits. Despite considerable awareness about genome editing technologies, some people resist genome-edited crops because of their religious beliefs (Carroll and Charo 2015). CEOs and GMOs are considered unethical since it is considered immoral and unethical to transfer genes across species. Those who think that CEOs are against religion may ask for clear labeling.

9.13 Social Concerns Related to CRISPR-Edited Crops

CRISPR/Cas has given scientists much control over genome modification and rapid trait development in agricultural crops. CRISPR has enabled the development of customized crops with superior features, which was previously impossible (Sedeek et al. 2019). CRISPR/Cas is on the leading edge of massive innovations and will produce crops with novel and diverse traits, which were never possible by conventional genetic engineering. However, CRISPR-edited agricultural crops may not be considered within the formal regulatory framework. The regulatory standards and framework for genome-edited crops will set an example for future innovations. Increasing the cultivation and release of complex and novel genome-edited crops in the environment would be difficult to monitor and recall. Therefore, the intended phenotype of genome-edited crops is of greater concern than the process used to create them because most genome-edited cops would be indistinguishable from conventional crops (Duensing et al. 2018). Similarly, there is a need to increase collaboration among regulatory authorities to focus on the risk assessment of complex products and to adopt extra regulatory measures, if required. In addition, there is a need to enhance the capacity and capabilities of scientists regarding regulatory science education with active public engagement (Carroll and Charo 2015; Morrison and de Saille 2019). Extra research governance and risk management are also required at the developmental stage of genome-edited crops (Wolt

2017). The USA has already started the process by establishing an Institutional Biosafety Committee (IBC) for risk management and confining gene drives.

Moreover, despite the differences among legal definitions, there is a need to find an adequate regulatory framework for avoiding sociotechnical resistance and for promoting international trade and agricultural innovations (Whelan and Lema 2017). Therefore, when a scientific advisory committee decides that gene/genome edited organisms (GEOs) are equivalent to conventional crops, political authorities may decide otherwise. One example of this is the EU ban on commercial cultivation of GMOs, when political authorities did not follow scientific advice. Thus, it is possible that because of political decisions, some genome-edited crops may be overregulated.

9.14 Cost of Regulation

Advances in genome editing have enabled the development of customizable crops by adding favorable traits or deleting undesirable traits. Engineered nucleases are considered superior to traditional genetic engineering techniques because of their specificity, scope, and accuracy (Lassoued et al. 2019). To date, more than 50 crops have been engineered by genome editing as a proof of concept, and genome-edited crops have recently started appearing in markets. The scientific community agrees that genome editing has enabled crops with precise and targeted modifications at a lower cost and within a shorter time frame than GMOs. Only a few studies have been focused on the cost of development and commercialization of genome-edited crops. Results from a study in 2017 indicate that genome-edited crops may reach the market within a time span of 5 years at a cost 10.5 M because of the deregulation of transgene-free genome-edited crops by APHIS (Lassoued et al. 2019).

By contrast, genetically engineered crops may cost 24.5 M and take 14 years to reach the market. In genome-edited crops, majority of the cost is associated with research and development, whereas in genetically engineered crops, the regulation cost is also high. Genome editing has significantly reduced the cost and time involved in engineering a trait in crops compared with GMOs. Regional differences, however, may exist because of the differences in regulatory and commercialization procedures.

9.15 Future Outlook

Genome editing technologies are being widely used to engineer agriculturally important crops for various applications ranging from disease resistance to functional genomics. Genome editing techniques have been described as precise; however, they are prone to off-target editing, on-target effects, and interference with the regulation of genes. The engineered traits could have environmental implications and unexpected interactions. In addition, gene drives could have biosafety issues, as they cannot be recalled once released, and any change induced by a gene drive will be permanent in a given population. Since scientific studies have shown off-target editing in CRISPR-edited plants, government agencies should adopt precautionary measures to approve and market CEOs. Genome-edited crops and food derived from them should be independently evaluated for environmental and food safety before approving for consumption. If scientific, regulatory, and political authorities do not establish a well-defined regulatory system, progress in the genome editing crops will be seriously hampered. The regulatory system should have the capacity to accommodate future developments, if complex traits and multiplex genome-edited crops are being used. Furthermore, there is a need to adopt a global regulatory system that could also address social concerns of individual countries about genome-edited crops. Developers need to clearly explain genome editing to the public and inform them about the advantages and disadvantages of genome-edited crops. There is also a need for actively engaging the public in the development, regulatory aspects, and political issues of genome-edited organisms to better integrate the CEOs in the society. Different definitions and regulatory systems in different countries increase confusion and distrust in the public. Therefore, if we believe that genome editing crops could contribute to food security and sustainable agriculture, we need a better universal regulatory system to build public trust in genome-edited crops.

References

Aglawe SB, Barbadikar KM, Mangrauthia SK, Madhav MS (2018) New breeding technique "genome editing" for crop improvement: applications, potentials and challenges. Biotech 8 (8):336

Araki M, Ishii T (2015) Towards social acceptance of plant breeding by genome editing. Trends Plant Sci 20(3):145–149

Arora L, Narula A (2017) Gene editing and crop improvement using CRISPR-Cas9 system. Front Plant Sci 8(8):1932

Banakar R, Eggenberger AL, Lee K, Wright DA, Murugan K, Zarecor S, Lawrence-Dill CJ, Sashital DG, Wang K (2019) High-frequency random DNA insertions upon co-delivery of CRISPR-Cas9 ribonucleoprotein and selectable marker plasmid in rice. Sci Rep 9(1):1–3

Bawa AS, Anilakumar KR (2013) Genetically modified foods: safety, risks and public concerns—a review. J Food Sci Technol 50(6):1035–1046

BBC News (2015) EU changes rules on GM crop cultivation. BBC News, BBC. https://www.bbc.com/news/world-europe-30794256

BetizeauM, BetizeauM (2017) Ecole Normale Superieure of Lyon, & ETH/University of Zurich. Regulation of Genetically Modified Organisms: Present and Future. Retrieved fromhttps://www.reatch.ch/de/content/regulation-genetically-modified-organisms-present-and-future

Borém A, Fritsche-Neto R (2014) Omics in plant breeding. Wiley Blackwell, Hoboken

Brent P, Bittisnich D, Brooke-Taylor S, Galway N, Graf L, Healy M, Kelly L (2003) Regulation of genetically modified foods in Australia and New Zealand. Food Control 14(6):409–416

Brokowski C, Adli M (2019) CRISPR ethics: moral considerations for applications of a powerful tool. J Mol Biol 431(1):88–101

Brossard D, Belluck P, Gould F, Wirz CD (2019) Promises and perils of gene drives: navigating the communication of complex, post-normal science. Proc Natl Acad Sci 116(16):7692–7697

Canada Health (2018) Government of Canada. Canada.ca, Government of Canada.https://www.canada.ca/en/health-canada/services/food-nutrition/genetically-modified-foods-other-novel-foods/factsheets-frequently-asked-questions/part-1-regulation-novel-foods.html

Carrington D (2019) Tree planting has mind-blowing potential to tackle climate crisis. The Guardian 2019:4

Carroll D, Charo RA (2015) The societal opportunities and challenges of genome editing. Genome Biol 16(1):242

CGDRO (2016) Committee on gene drive research in non-human organisms: recommendations for responsible conduct; board on life sciences; division on earth and life studies; national academies of sciences, engineering, and medicine. In: Gene drives on the horizon: advancing science, navigating uncertainty, and aligning research with public values. National Academies Press, Washington, DC

Collins JP (2018) Gene drives in our future: challenges of and opportunities for using a self-sustaining technology in pest and vector management. BMC Proc 12(8):9

Cribbs AP, Perera SM (2017) Focus: genome editing: science and bioethics of CRISPR-Cas9 gene editing: an analysis towards separating facts and fiction. Yale J Biol Med 90(4):625

Custers R, Casacuberta J, Eriksson D, Sági L, Schiemann JH (2018) Genetic alterations that do or do not occur naturally; consequences for genome edited organisms in the context of regulatory oversight. Front Bioeng Biotechnol 6:213

Davison J, Ammann K (2017) New GMO regulations for old: determining a new future for EU crop biotechnology. GM Crops Food 8(1):13–34

Dederer HG, Hamburger D (2019) Regulation of genome editing in plant biotechnology: a comparative analysis of regulatory frameworks of selected countries and the EU. Springer, New York

D'Halluin K, Vanderstraeten C, Van Hulle J, Rosolowska J, Van Den Brande I, Pennewaert A, D'Hont K, Bossut M, Jantz D, Ruiter R, Broadhvest J (2013) Targeted molecular trait stacking in cotton through targeted double-strand break induction. Plant Biotechnol J 11(8):933–941

Dubock A (2017) An overview of agriculture, nutrition and fortification, supplementation and biofortification: golden rice as an example for enhancing micronutrient intake. Agric Food Secur 6(1):59

Duensing N, Sprink T, Parrott WA, Fedorova M, Lema MA, Wolt JD, Bartsch D (2018) Novel features and considerations for ERA and regulation of crops produced by genome editing. Front Bioeng Biotechnol 2018:6

Eckerstorfer MF, Engelhard M, Heissenberger A, Simon S, Teichmann H (2019) Plants developed by new genetic modification techniques-comparison of existing regulatory frameworks in the EU and non-EU countries. Front Bioeng Biotechnol 7:26

Esvelt KM, Smidler AL, Catteruccia F, Church GM (2014) Emerging technology: concerning RNA-guided gene drives for the alteration of wild populations. elife 17(3):e03401

Farboud B, Jarvis E, Roth TL, Shin J, Corn JE, Marson A, Meyer BJ, Patel NH, Hochstrasser ML (2018) Enhanced genome editing with Cas9 ribonucleoprotein in diverse cells and organisms. JoveJ Vis Exp 135:e57350

FriedrichsS, TakasuY, KearnsP, DagallierB, OshimaR, SchofieldJ, MoredduC (2019) Meeting report of the OECD conference on genome editing: applications in agriculture—implications for health, environment and regulation

Gaj T, Gersbach CA, Barbas CF, Talen ZF (2013) CRISPR/Cas-based methods for genome engineering. Trends Biotechnol 31(7):397–405

Georges F, Ray H (2017) Genome editing of crops: a renewed opportunity for food security. GM Crops Food 8(1):1–2

Glenn KC, Alsop B, Bell E, Goley M, Jenkinson J, Liu B, Martin C, Parrott W, Souder C, Sparks O, Urquhart W (2017) Bringing new plant varieties to market: plant breeding and selection practices advance beneficial characteristics while minimizing unintended changes. Crop Sci 57(6):2906–2921

Godfray HC, Garnett T (2014) Food security and sustainable intensification. Philos Trans R Soc B Sci 369(1639):20120273

Greely HT (2019) CRISPR'd babies: human germline genome editing in the 'He Jiankui affair'. J Law Biosci 6(1):111–183

Government of Canada (2018) Government of Canada/Gouvernement Du Canada. *Canadian Food Inspection Agency*, Government of Canada/Government Du Canada. https://www.inspection.gc.ca/plant-health/plants-with-novel-traits/applicants/directive-94-08/eng/1512588596097/1512588596818

GracielaN, SoaresE, AhmadT, WendyA (2014) Restrictions on genetically modified organisms. Retrieved from https://www.loc.gov/law/help/restrictions-on-gmos/index.php

Grossman MR (2019) Genetic engineering in the United States: Regulation of crops and their food products. In: Regulation of genome editing in plant biotechnology. Springer, Cham, pp 263–312

Gupta PP, Schieder O, Gupta M (1984) Intergeneric nuclear gene transfer between somatically and sexually incompatible plants through asymmetric protoplast fusion. Mol Gen Genet 197 (1):30–35

Hajiahmadi Z, Movahedi A, Wei H, Li D, Orooji Y, Ruan H, Zhuge Q (2019) Strategies to increase on-target and reduce off-target effects of the CRISPR/Cas9 system in plants. Int J Mol Sci 20 (15):3719

Hamburger D (2019) Comparative analysis: the regulation of plants derived from genome editing in Argentina, Australia, Canada, the European Union, Japan and the United States. In: Regulation of genome editing in plant biotechnology. Springer, Cham, pp 313–363

ISAAA (2017) Global status of commercialized Biotech/GM crops in 2017: biotech crop adoption surges as economic benefits accumulate in 22 years. ISAAA brief no. 53. ISAAA, Ithaca, NY

Ishii T, Araki M (2016) Consumer acceptance of food crops developed by genome editing. Plant Cell Rep 35(7):1507–1518

Jaganathan D, Ramasamy K, Sellamuthu G, Jayabalan S, Venkataraman G (2018) CRISPR for crop improvement: an update review. Front Plant Sci 2018:9

Javed MR, Sadaf M, Ahmed T, Jamil A, Nawaz M, Abbas H, Ijaz A (2018) CRISPR-Cas system: history and prospects as a genome editing tool in microorganisms. Curr Microbiol 75 (12):1675–1683

Jones HD (2015) Regulatory uncertainty over genome editing. Nat Plants 1(14011):10–38

Lassoued R, Macall DM, Hesseln H, Phillips PW, Smyth SJ (2019) Benefits of genome-edited crops: expert opinion. Transgenic Res 28(2):247–256

Li JR, Walker S, Nie JB, Zhang XQ (2019) Experiments that led to the first gene-edited babies: the ethical failings and the urgent need for better governance. J Zhejiang Univ 20(1):32–38

Lieber MR (2010) The mechanism of double-strand DNA break repair by the nonhomologous DNA end-joining pathway. Annu Rev Biochem 79:181–211

Ludlow K (2019) Regulation of genome editing in plant biotechnology: Australia. In: Regulation of genome editing in plant biotechnology. Springer, Cham, pp 63–110

Maaß O, Consmüller N, Kehlenbeck H (2019) Socioeconomic impact of genome editing on agricultural value chains: the case of fungal-resistant and coeliac-safe wheat. Sustainability 11 (22):6421

Marchant GE, Stevens YA (2015) A new window of opportunity to reject process-based biotechnology regulation. GM Crops Food 6(4):233–242

McHughen A (2016) A critical assessment of regulatory triggers for products of biotechnology: product vs. process. GM Crops Food 7(3-4):125–158

Mehravar M, Shirazi A, Nazari M, Banan M (2019) Mosaicism in CRISPR/Cas9-mediated genome editing. Dev Biol 445(2):156–162

Min J, Smidler AL, Najjar D, Esvelt KM (2018) Harnessing gene drive. J Responsible Innov 5 (suppl 1):S40–S65

Misra AK (2014) Climate change and challenges of water and food security. Int J Sustain 3 (1):153–165

Morrison M, de Saille S (2019) CRISPR in context: towards a socially responsible debate on embryo editing. Palgrave Commun 5(1):1–9

Mueller S (2019) Are Market GM plants an unrecognized platform for bioterrorism and biocrime? Front Bioeng Biotechnol 7:121

Mulvihill JJ, Capps B, Joly Y, Lysaght T, Zwart HA, Chadwick R (2017) Ethical issues of CRISPR technology and gene editing through the lens of solidarity. Br Med 122(1):17–29
Nemudryi AA, Valetdinova KR, Medvedev SP, Zakian SM (2014) TALEN and CRISPR/Cas genome editing systems: tools of discovery. Acta Nat 6:22
Noble C, Adlam B, Church GM, Esvelt KM, Nowak MA (2018) Current CRISPR gene drive systems are likely to be highly invasive in wild populations. elife 7:e33423
Nollet LM, Toldrá F, Benjakul S, Paliyath G, Hui YH (2012) Food biochemistry and food processing. Wiley, Hoboken
Oberhofer G, Ivy T, Hay BA (2018) Behavior of homing endonuclease gene drives targeting genes required for viability or female fertility with multiplexed guide RNAs. Proc Natl Acad Sci 115 (40):E9343–E9352
Oh J, Ezezika OC (2014) To label or not to label: balancing the risks, benefits and costs of mandatory labelling of GM food in Africa. Agric Food Secur 3(1):8
Omodamilola OI, Ibrahim AU (2018) CRISPR technology; advantages, limitations and future direction. J Biomed Pharm Sci 1(115):2
Osakabe Y, Osakabe K (2014) Genome editing with engineered nucleases in plants. Plant Cell Physiol 56(3):389–400
Osei MK, Asante MD, Agyeman A, Adebayo MA, Adu-Dapaah H (2014) Plant breeding: a tool for achieving food sufficiency. In: Sustainable horticultural systems. Springer, Cham, pp 253–274
Phillips T (2008) Genetically modified organisms (GMOs): transgenic crops and recombinant DNA technology. Nat Educ 1(1):213
Prakash D, Verma S, Bhatia R, Tiwary BN (2011) Risks and precautions of genetically modified organisms. ISRN Ecol 2011:13
Priyadarshan PM (2017) Refinements to Hevea rubber breeding. Tree Genet Genomes 13:20. https://doi.org/10.1007/s11295-017-1101-8
Raman R (2017) The impact of genetically modified (GM) crops in modern agriculture: a review. GM Crops Food 8(4):195–208
Raza A, Razzaq A, Mehmood SS, Zou X, Zhang X, Lv Y, Xu J (2019) Impact of climate change on crops adaptation and strategies to tackle its outcome: a review. Plants 8(2):34
Razzaq A, Saleem F, Kanwal M, Mustafa G, Yousaf S, Arshad I, Muhammad H, Hameed MK, Khan MS, Joyia FA (2019) Modern trends in plant genome editing: an inclusive review of the CRISPR/Cas9 toolbox. Int J Mol Sci 20(16):4045
Ricroch AE, Ammann K, Kuntz M (2016) Editing EU legislation to fit plant genome editing. EMBO Rep 17(10):1365–1369
Rojas-Vásquez R, Gatica-Arias A (2019) Use of genome editing technologies for genetic improvement of crops of tropical origin. Plant Cell Tiss Org 11:1–30
Rossant J (2018) Gene editing in human development: ethical concerns and practical applications. Development 145(16):dev150888
Resources N (2016) National academies of sciences, engineering, and medicine. Regulation of current and future genetically engineered crops. In: Genetically engineered crops: experiences and prospects, 2016, May 17. National Academies Press (US)
Royal Society (2020) What GM crops are being grown and where? Royal Society. Retrieved fromhttps://royalsociety.org/topics-policy/projects/gm-plants/what-gm-crops-are-currently-being-grown-and-where/
Rozov SM, Permyakova NV, Deineko EV (2019) The problem of the low rates of CRISPR/Cas9-mediated knock-ins in plants: approaches and solutions. Int J Mol 20(13):3371
Saha SK, Saikot FK, Rahman MS, Jamal MA, Rahman SK, Islam SR, Kim KH (2019) Programmable molecular scissors: applications of a new tool for genome editing in biotech. Mol Ther NuclAcids 14:212–238
Sasson A (2012) Food security for Africa: an urgent global challenge. Agric Food Secur 1(1):2
Schouten HJ, Krens FA, Jacobsen E (2006) Cisgenic plants are similar to traditionally bred plants. EMBO Rep 7(8):750–753

SciencesSP (2020) Confirmation that low PPO5 potatoes are not a regulated article. https://www.aphis.usda.gov/biotechnology/downloads/reg_loi/16-320-01_air_response_signed. Accessed January 4, 2020
Sedeek KE, Mahas A, Mahfouz M (2019) Plant genome engineering for targeted improvement of crop traits. Front Plant Sci 10
ServickJulKet al (2017) How will we keep controversial gene drive technology in check? Science
Shukla VK, Doyon Y, Miller JC, DeKelver RC, Moehle EA, Worden SE, Mitchell JC, Arnold NL, Gopalan S, Meng X (2009) Precise genome modification in the crop species Zea mays using zinc-finger nucleases. Nature 459:437–441
Smyth SJ (2017) Canadian regulatory perspectives on genome engineered crops. GM Crops Food 8 (1):35–43
Sprink T, Eriksson D, Schiemann J, Hartung F (2016) Regulatory hurdles for genome editing: process-vs. product-based approaches in different regulatory contexts. Plant Cell Rep 35 (7):1493–1506
StoyeE (2015) Crispr-edited mushroom dodges regulation. Retrieved fromhttps://www.chemistryworld.com/news/crispr-edited-mushroom-dodges-regulation/1010298.article
Sykora P, Caplan A (2017) The Council of Europe should not reaffirm the ban on germline genome editing in humans. EMBO Rep 18(11):1871–1872
The Economist (2018) The promise and peril of gene drives. The Economist, the Economist Newspaper. https://www.economist.com/briefing/2018/11/08/the-promise-and-peril-of-gene-drives
Troadec MB, Pagès JC (2019) Where are we with unintended effects in genome editing applications from DNA to phenotype: focus on plant applications. In: Transgenic research, vol 28. Springer, New York, pp 125–133
UnglesbeeE(2016) Gene editing comes to agriculture. Retrieved fromhttps://www.dtnpf.com/agriculture/web/ag/news/article/2016/06/17/gene-editing-comes-agriculture
USDA APHIS (2018) Details on USDA Plant Breeding Innovations. Retrieved fromhttps://www.aphis.usda.gov/aphis/ourfocus/biotechnology/brs-news-and-information/2018_brs_news/pbi-details.
Van de Wiel CC, Schaart JG, Lotz LA, Smulders MJ (2017) New traits in crops produced by genome editing techniques based on deletions. Plant Biotechnol Rep 11(1):1–8
Whelan AI, Lema MA (2017) A research program for the socioeconomic impacts of gene editing regulation. GM Crops Food 8:74–83. https://doi.org/10.1080/21645698.2016.1271856
Whelan AI, Lema MA (2019) Regulation of genome editing in plant biotechnology: Argentina. In: Regulation of genome editing in plant biotechnology. Springer, Cham, pp 19–62
Witcombe JR, Gyawali S, Subedi M, Virk DS, Joshi KD (2013) Plant breeding can be made more efficient by having fewer, better crosses. BMC Plant Biol 13(1):22
Wolt JD (2017) Safety, security, and policy considerations for plant genome editing. In: Progress in molecular biology and translational science, vol 149. Academic, New York, pp 215–241
Wolt JD, Wang K, Yang B (2016) The regulatory status of genome-edited crops. Plant Biotechnol J 14(2):510–518
Wunderlich S, Gatto KA (2015) Consumer perception of genetically modified organisms and sources of information. Adv Nutr 6(6):842–851
Yang H, Wu JJ, Tang T, Liu KD, Dai C (2017) CRISPR/Cas9-mediated genome editing efficiently creates specific mutations at multiple loci using one sgRNA in Brassica napus. Sci Rep 7 (1):7489
Zaidi SS, Vanderschuren H, Qaim M, Mahfouz MM, Kohli A, Mansoor S, Tester M (2019) New plant breeding technologies for food security. Science 363(6434):1390–1391
Zhang Y, Malzahn AA, Sretenovic S, Qi Y (2019) The emerging and uncultivated potential of CRISPR technology in plant science. Nat Plants 5(8):778–794
Zhu T, Peterson DJ, Tagliani L, St. Clair G, Baszczynski CL, Bowen B (1999) Targeted manipulation of maize genes in vivo using chimeric RNA/DNA oligonucleotides. Proc Natl Acad Sci USA 96:8768–8773

Challenges and Future Perspective of CRISPR/Cas Technology for Crop Improvement

10

Zulqurnain Khan, Sultan Habibullah Khan, and Aftab Ahmad

Abstract

Since the inception of CRISPR technology, an unprecedented progress has been observed in the field of genome editing (GE) in animals and plants with equal success. Undoubtedly, the technology is very simple, cost-effective, time-saving, and precise compared to old techniques such as MNs, ZFNs, and TALENs. Although, off-targeting has been a serious issue with CRISPR technology but it is the most promising technique hitherto. A number of approaches have been used to address the issue of off-targeting such as use of modified gRNA and other variants of CRISPR/Cas. In this chapter, current scenario of CRISPR technology will be discussed along with its pros and cons. Moreover, the strategies to address the issues associated with CRISPR technology will be discussed with future perspectives of the technology.

Z. Khan (✉)
Institute of Plant Breeding and Biotechnology, Muhammad Nawaz Sharif University of Agriculture, Multan, Pakistan
e-mail: zulqurnain.khan@mansuam.edu.pk

S. H. Khan
Centre of Agricultural Biochemistry and Biotechnology (CABB), University of Agriculture, Faisalabad, Pakistan

Center for Advanced Studies in Agriculture and Food Security (CASAFS), University of Agriculture, Faisalabad, Pakistan
e-mail: sultan@uaf.edu.pk

A. Ahmad
Center for Advanced Studies in Agriculture and Food Security (CASAFS), University of Agriculture, Faisalabad, Pakistan
e-mail: aftab.ahmad@uaf.edu.pk

A. Ahmad et al. (eds.), *CRISPR Crops*,
https://doi.org/10.1007/978-981-15-7142-8_10

10.1 Introduction

Food security is a major challenge for rapidly growing world population. For the improvement of food security there is a need to improve the crop yield and tolerance to biotic and abiotic stresses. Conventional breeding takes many years to introduce desirable traits. On the other hand, modern tools for GE (GE) exhibit the capability of integrating a foreign gene into a predetermined site of the genome precisely, allowing for accurate substitution of an existing allele with an alternative one. The latest developments in GE tools have paved the way for targeted mutagenesis, opening new horizons for precise genome engineering. Various GE tools such as transcription activator-like effector nucleases (TALENs), zinc finger nucleases (ZFNs) have enabled plant scientists to manipulate desired genes in crop plants. Conversely, CRISPR/Cas9 is an entrancing, easy-to-design, cost-effective, and versatile tool for precise and efficient plant GE. CRISPR/Cas9-based GE has demonstrated great potential for crop improvement. GE and targeted genome modifications with engineered nucleases and artificial DNA-binding proteins have been frequently reported in the last decade. This era has witnessed a number of successful reports of the subject in plants and animals. Genetic engineering, cell biology, synthetic biology, cytogenetic, drug delivery, and other fields of life sciences have been revolutionized with the use of GE tools. Exponentially growing world population in the changing climate has been facing chronic food shortage issues. To address the issues of food safety and food security, there is a dire need to adopt new techniques such as GE tools which guarantee precise targeted genome modifications. The era of GE starts from the late 1990s with the discovery of meganucleases which recognize a long DNA sequence to create DSBs (Silva et al. 2011). Following meganucleases, other popular engineered nucleases such as zinc finger nucleases (ZFNs), transcription activator-like effector-based nucleases (TALEN), and the clustered regularly interspaced short palindromic repeats (CRISPR/Cas9) system have been extensively used for GE in plants as well as animals (Weeks et al. 2016). Apart from creating DSBs at predefined target sites (Fig. 10.1), these tools have been used extensively for gene activation, gene

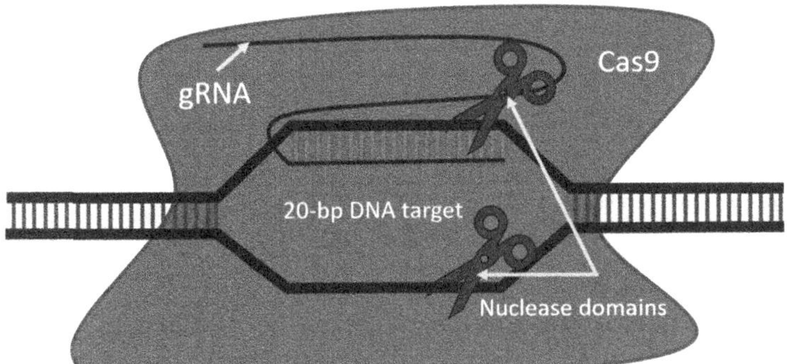

Fig. 10.1 A CRISPR/Cas system targeting dsDNA. The figure is showing Cas9 protein complex with gRNA targeting a dsDNA with 20 bp target site. The pair of scissors is indicating two nuclease domains HNH and RuvC of Cas9 protein which create a DSB at the target site

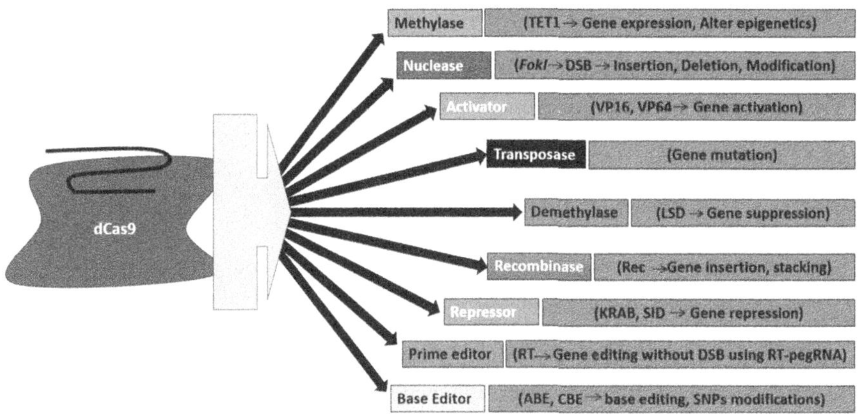

Fig. 10.2 Engineering of deactivated CRISPR/Cas (dCas) with fusion of different effector domains for multiple purposes

repression, epigenome modifications, etc., by fusion with a number of effector domains (Fig. 10.2). Bacterial CRISPR/Cas systems have been used to target DNA artificially (i.e., targeted modification) since 2013. Now, different forms of nucleases in addition to Cas 9 are reported with their novel functions. Cas9 and Cas12 are widely used as DNA nucleases for inducing site-specific DNA breaks for different kinds of genome engineering applications, and in their mutated forms as DNA-binding proteins to modify gene expression. Moreover, histone modifications, as well as cytosine methylation or base editing, were achieved with these systems in plants. Recently, with the discovery of the nuclease Cas13a (previously called C2c2), molecular biologists have obtained a system that enables sequence-specific cleavage of single-stranded RNA molecules. The latest experiments with this and also the alternative Cas13b system demonstrated that these proteins can be used in a similar manner in eukaryotes for RNA manipulation as Cas9 and Cas12 for DNA manipulations. The first application of Cas13a for post-transcriptional regulation of gene expression in plants has been reported (Terns 2018; Basso et al. 2019). Recent results show that the system is also applicable for combating viral infection in plants. As single-stranded RNA viruses are by far the most abundant class of viruses in plants, the application of this system is of special promise for crops. Cas9 can be transformed into a DNA-binding protein by mutating its two nuclease domains. This way, its DNA cleavage activity is deactivated but its DNA-binding potential remains intact. By fusing this nuclease "dead" Cas9 (dCas9) directly or indirectly to an effector domain, it can then be used to guide diverse enzymatic functions to any specific site in the genome. This enables the use of CRISPR/Cas for many diverse site-specific manipulations, such as regulation of gene expression, epigenetic modifications, imaging of genomic loci in live cells, or base editing without DSB induction enabled by fusion of deaminases (Tong et al. 2012). Adopting CRISPR/Cas technology with new additional tools will be very helpful for solving prevailing issues of food shortage. Changing climate has been posing threat to crop production by challenging crops with a number of biotic and abiotic stresses resulting in low crop growth and yield. Moreover, using advanced sophisticated techniques is very crucial to get acceptance from regulatory authorities. Customer's acceptance is also important to solve the ethical and societal issues. Biotechnologists are hopeful that the CRISPR technology will be proved as an elixir for solving the prevailing problems in all fields of life sciences.

10.2 CRISPR/Cas: A Revolution in GE

Soon after the discovery of CRISPR-Cas9 GE system, a number of multi-disciplinary approaches were started in a variety of fields adapting this marvelous tool. In agriculture sector, CRISPR technology has been considered as a celestial tool which can be used efficiently to ensure the food security for world masses. Researchers showed ultimate interest in using this technology for multiple purposes. The first major crop edited using CRISPR technology was rice (Feng et al. 2013). In this study, Cas9 and sgRNA were controlled by CaMV 35S and OsU6-2 promoters, respectively, which resulted in efficient modification of the rice genome. The first study demonstrated the practicability of the system and stimulated further researches into adapting CRISPR technology to other potential food crops. In the same year, five different target genes of Arabidopsis and *Nicotiana benthamiana* were evaluated for mutagenesis using sgRNA:pcoCas9 (Li et al. 2013). It was observed that the binding strength of sgRNA to target sequences was critical to define mutagenesis. Subsequently, the introduction of multiple sgRNAs targeting multiple genes simultaneously enabled an improved success rate of mutagenesis. The CRISPR system with engineered sgRNA successfully directed DNA breaks at specifically defined chromosomal loci in *N. benthamiana* (Nekrasov et al. 2013). These studies demonstrated the simplicity of this system in plant genome engineering comparing contemporary methods. Furthermore, studies regarding designing of sgRNA to improve the efficiency of CRISPR-based GE in crops with high nutritional value, such as maize (Liang et al. 2014), rice (Miao et al. 2013), wheat (Wang et al. 2014), sweet orange (Jia and Wang 2014), potato (Wang et al. 2015), and tomato (Brooks et al. 2014) provided hopeful results for a wide range of possible applications of the technology.

Many plants with increased biotic stress resistance, including resistance to fungal, bacterial, and viral diseases and insects, have been obtained via CRISPR/Cas9 knockout. For example, powdery mildew is a devastating fungal disease in crops. Using TALEN and CRISPR/Cas9, Wang's laboratory knocked out all six *TaMLO* alleles in wheat and obtained plants with increased resistance to powdery mildew (Wang et al. 2014). Similarly, Nekrasov et al. (2013) showed that CRISPR/Cas9-mediated knockout of *MLO* confers resistance to powdery mildew in tomato. Rice blast is a destructive fungal disease; blast-resistant rice was obtained via knock out of OsERF922, an ethylene responsive factor transcription factor gene (Wang et al. 2015). Bacterial blight in rice is caused by *Xanthomonas oryzae* pv. oryzae. Deletion of the *OsSWEET13* promoter led to the production of plants resistant to this disease (Zou et al. 2019). With regard to viral diseases, use of CRISPR/Cas9 has also produced tungro disease-resistant eif4g rice (Macovei et al. 2018), broad potyvirus-resistant eif4e cucumber (Chandrasekaran et al. 2016), and cotton leaf curl disease-resistant CLCuD cotton (Iqbal et al. 2016). Recently, Liu et al. (2017) found that disrupting OsCYP71A1 blocked serotonin biosynthesis and greatly increased salicylic acid levels, thereby conferring resistance to plant hoppers and stem borers, the two most destructive pests of rice. Among abiotic stresses, contamination of arable lands has created the need to prevent the accumulation of toxic heavy metals in crops. By knocking out OsARM1, OsNramp5, and OsHAK1, breeders have developed rice strains with low levels of cadmium, radioactive cesium, and arsenic, respectively (Nieves-Cordones et al. 2017; Tang et al. 2017; Dahan-Meir et al. 2018). In 2018, research on the *OsPYL* abscisic acid receptor gene family revealed that pyl1/4/6 triple knockout rice created by CRISPR/Cas9 editing had increased grain yield, greater high temperature tolerance, and reduced preharvest sprouting compared with wild type (Miao et al. 2013).

The need for improved food security makes yield the primary target of gene editing for crop improvement. Yield is a complex trait that is dependent on many factors. Knocking out negative regulators known to affect yield-determining factors such as grain number (*OsGn1a*), grain size (*OsGS3*), grain weight (*TaGW2, OsGW5, OsGLW2, orTaGASR7*), panicle size (*OsDEP1, TaDEP1*), and tiller number (*OsAAP3*) created the expected phenotypes in plants with loss-of-function mutations in these genes, demonstrating that CRISPR/Cas9 is an effective tool for improving yield-related traits (Li et al. 2013; Liu et al. 2017; Lu et al. 2018; Zhang et al. 2016, 2018). Simultaneous knockout of three grain weight-related genes (*GW2, GW5*, and *TGW6*) in rice led to trait pyramiding, which greatly increased grain weight (Xu et al. 2016).

However, because most yield-related traits are quantitative and controlled by quantitative trait loci, simply knocking out individual factors may not be sufficient to increase yield in the field. Huang et al. (2018) recently developed a method for the large-scale identification of genes that contribute to complex quantitative traits, such as yield, by combining pedigree analysis, whole-genome sequencing, and CRISPR/Cas9 technology. The authors sequenced 30 cultivars of the parents and descendants of the Green Revolution miracle rice variety IR8 and selected 57genes retained in all high-yielding lines for gene editing via knockout or knockdown using Cas9 or dCas9. Phenotypic analysis revealed that many of these genes are important for rice yield. This work provided insight into the mechanism of yield development and may facilitate the molecular breeding of improved rice.

Quality traits vary depending on the specific breeding requirements. To date, quality improvements by GE have impacted starch content, fragrance, nutritional value, and storage quality in crops. Rice with low amylose content, and thus improved rice eating and cooking quality, was generated by knockout of Waxy via CRISPR/Cas9 (Ma et al. 2015; Zhang et al. 2018). DuPont Pioneer produced a CRISPR/Cas9 knockout waxy corn line with high yields for commercial use (Waltz 2016). CRISPR/Cas9 was also used to produce high-amylose and resistant starch rice by mutating the starch branching enzyme gene SBE IIb; consuming high-amylose foods should benefit patients with diet-related noninfectious chronic diseases (Sun et al. 2018). Fragrance is an important quality trait in rice, and rice varieties with desirable fragrances when cooked have increased commercial value. A defect in the betaine aldehyde dehydrogenase 2 (*BADH2*) gene results in the biosynthesis of 2-acetyl-1-pyrroline, the major fragrance compound in fragrant rice. Using TALEN-targeted disruption of *OsBADH2*, our group has created a fragrant rice line with a similar 2-acetyl-1-pyrroline content (0.35–0.75 mg/kg) to that of the natural mutant fragrant rice variety (Shan et al. 2015). Gluten proteins from cereal crops trigger celiac disease in more than 7% of individuals in Western countries. The α-gliadin gene family, the major gluten-encoding gene family in wheat, consists of nearly100 genes or pseudogenes. CRISPR /Cas9 editing offers a new way to alter traits controlled by large gene families with redundant functions. Indeed, by simultaneously knocking out most conserved domains of α-gliadin family members, researchers have created low-gluten wheat (Sanchez-Leon). Other high-quality crops produced by CRISPR/Cas9 editing include seeds with high oleic acid oil in *Camelina sativa* (Jiang et al. 2017; Morineau et al. 2017) and *Brassica napus* (Okuzaki et al. 2018), tomatoes with a long shelf life (Ito et al. 2015; Li et al. 2018), high-value tomato with enhanced lycopene (Li et al. 2018) or γ-aminobutyric acid content (Li et al. 2018; Nonaka et al. 2017), and potato (hairy roots) with reduced levels of toxic steroidal glycoalkaloids (Nakayasu et al. 2018).

10.2.1 Base Editing and Prime Editing: Fresh Arrival in CRISPR Toolbox

In plants, the GE through CRISPR system (Woo et al. 2015) holds promise as a revolutionary tool for basic and applied research. CRISPR cleaves a double-stranded DNA at a target sequence and these double strand breaks (DSBs) are repaired by endogenous error-prone non-homologous end-joining events which create small deletions (indels) and/or insertions at a specific site by homologous recombinations. However, inducing point mutations rather than indels in plants remains a challenge for agricultural applications, although, point mutagenesis is one of the key strategies in plants for crop improvement (Yin et al. 2017).

Base-editing technology ("Base editor") is a more advanced and emerging gene modification technology which is based on CRISPR system. This technique is more efficient and simple compared to other variants (Komor et al. 2016). It does not demand the DSBs or DNA templates for replacing the specific bases. To achieve specific amino acid changes, single base-editing technique can be applied by precisely changing a single base and can also create a gene knockout by introducing a premature stop codon. The current system of base editing contains three major components: Cas9 nickase (nCas9) or deactivated Cas9 (dCas9), uracil glycosylase inhibitor (UGI), and cytidine deaminase. The enzymatic inactivation of RuvC or HNH domains in Cas9 nuclease generates nCas9 and dCas9, responsible for cleaving double strands of DNA. The nCas9 keeps the ability to be programmed with sgRNA and then targets specific sequences of DNA to cut single-stranded (ssDNA) regions. Afterwards, through cytidine deaminase the cytidine (C) is converted to uracil (U) contained by a ssDNA and it is then replaced with thymine (T) during replication of a DNA or repair mechanism. In mammalian cells the serial cytidine base editors like BE1, BE2, and BE3 were subsequently developed (Komor et al.

2016). Many successful applications of base editing were then reported in many species of plants like in wheat, rice, maize, and also in tomato (Liu et al. 2017; Zong et al. 2017; Ren et al. 2018; Li et al. 2017; Shimatani et al. 2017). Currently, the third-generation base editor, BE3, is one of the most commonly used systems armed with a UGI, the function of UGI is to inhibit the endogenous activity of base excision. So, the resulting base editor changes a cytidine to a thymine on non-target sequence and completes a C•G conversion to a T•A (Standage-Beier et al. 2015). Recently, in mammalian cells, the adenine base editors (ABEs) have been developed which can convert A•T into G •C (Gaudelli et al. 2017), and this novel system has also been applied in many plant species (Kang et al. 2018; Hua et al. 2018; Yan et al. 2018). Prime editing, a versatile and precise GE method, directly writes new genetic information into a specified DNA site using a catalytically impaired Cas9 endonuclease fused to an engineered reverse transcriptase, programmed with a prime editing guide RNA (pegRNA) that both specifies the target site and encodes the desired edit. More than 175 edits in human cells have been performed including targeted insertions, deletions, and all 12 types of point mutations were observed without requiring DSBs or provision of donor DNA templates (Anzalone et al. 2019).

10.3 GE at the Cost of Off-Targeting

Undoubtedly, CRISPR technology is much sophisticated, specific, and precise but, still there are a number of concerns associated with the use of this technology, particularly off-targeting. However, the high frequency of off-target activity ($\geq$50%) instigate changes at sites other than the intended on-target site is a significant concern, particularly for restorative and clinical applications. Here, we review the basic mechanisms underlying off-target cutting and strategies for minimizing off-target cleavage in the CRISPR/Cas9 system. The methods for detecting off-target mutations and improved off-target specificity in the CRISPR/Cas9 system will provide solid genotype–phenotype correlations and consequently, enable devoted interpretation of genome editing data that allows you to simply facilitate the basic and clinical utility of this technology.

To address the off-targeting, a number of CRISPR tools have been developed. Owing to the short target site in case of CRISPR comparative to TALENs, the chances for getting off-targets increase. Cas9 nickases and dCas9 with FokI fusion have been used (Fig. 10.3) increasing the target site. In both cases, a pair of Cas protein is required which decreases the chances of off-targeting in the genome (Mikami et al. 2016; Guilinger et al. 2014; Havlicek et al. 2017). Moreover, targeting with dCas9 as DNA-binding protein to transcriptional blockage is another promising way to decrease the off-targets in the genome (Vigouroux et al. 2018; Widom et al. 2019). A number of effector domains may be used which work on epigenome, transcriptional, and post-transcriptional levels for managing off-targeting (Fig. 10.2).

Off-targeting should be managed at different levels from designing to delivery of the GE reagents into the cells. A number of designing tools have been developed for

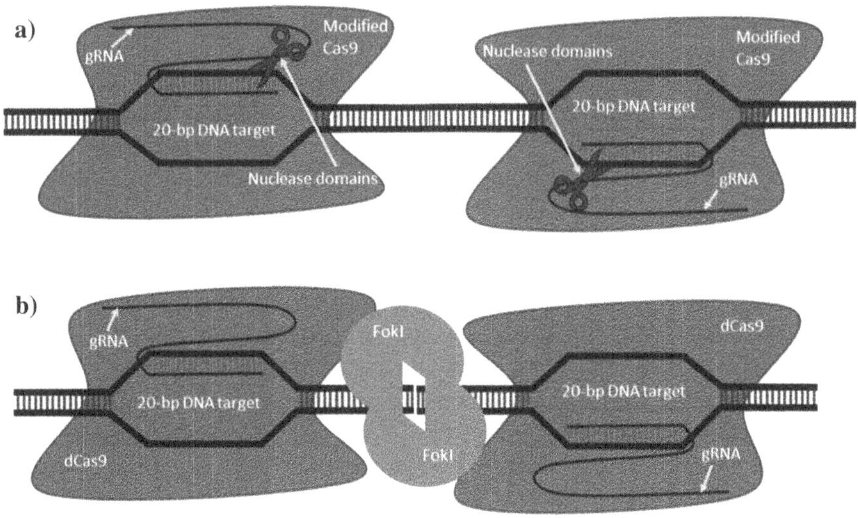

Fig. 10.3 Modified CRISPR/Cas system to manage off-targeting. (**a**) A modified Cas9 has been shown with one of the inactivated nuclease domains. This tool is called paired Cas9 nickases. (**b**) In this figure, a pair of Cas9 with deactivated nuclease (dCas9) has been used along with a dimer of FokI nuclease domain to create DSB at the target site. Using these approaches, off-targeting can be managed increasing the possibility of safe application of CRISPR/Cas technology

genome-wide screening of on-targets and off-targets such as CHOPCHOP (Montague et al. 2014), CRISPR-P (Liu et al. 2017), ZiFi Targetter (Sander et al. 2010), E-CRISPR (Heigwer et al. 2014), Fly CRISPR (Gratz et al. 2014), CRISPRdirect (Naito et al. 2015), etc. These tools are very useful for designing sgRNA. Moreover, some tools such as Cas OFF Finder (Bae et al. 2014) can be employed for efficient screening of off-targets in the genome. It is further suggested that, progress in bioinformatics is needed for genome-wide screening of potential gRNA sites to be considered for targeting in the genome. Another important technology such as machine learning or deep learning can be used for further precision and accuracy in the designing of sgRNA. These tools based on neural networks such as CNN (Convolutional Neural Network) have powerful algorithms for high throughput analysis of the data with specificity. Use of CNN models and machine learning in genomics has been successfully reported by different researchers (Eraslan et al. 2019; Zou et al. 2019). Use of these new tools coupled with previous conventional tools would be fruitful in addressing the off-targeting concerns regarding CRISPR technology.

Obtaining successful and efficient editing of the gene has been a prime objective of the researchers working in the field of GE. For this purpose, delivery of the GE reagents to the target cells is very crucial. GE tools can be delivered to plants as DNA, mRNA, and proteins (Tong et al. 2012; Tesson et al. 2011; Wefers et al. 2013; Chen et al. 2013; Liu et al. 2014; Jia et al. 2014). In case of CRIPSR/Cas specifically,

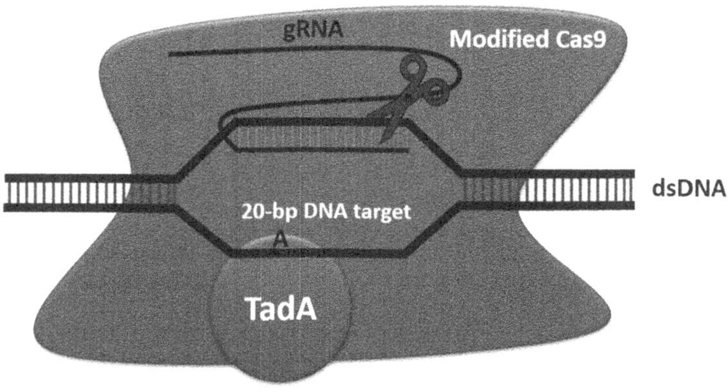

Fig. 10.4 Base editors are based on modified Cas9 with ABE (Adenosine base editor)/CBE (cytosine base editor) which have the ability to replace the corresponding base in the target sequence. TadA base editor converts the adenine to guanine on the complementary strand

being an RNA-guided nuclease system, the delivery is possible as DNA and ribonucleoproteins (RNPs) which have been reported previously (Liang et al. 2017). Application of the CRISPR as DNA may have more implications regarding integration of the transgene in the target organisms. Application of the GE tools as proteins has many advantages including DNA-free editing of genomes which have further opportunities for less regulation of the produced genome editing organism by the regulatory authorities (Malnoy et al. 2016; Kim et al. 2017).

To mitigate off-targeting in the target organism, other important mechanisms to be considered are method of transformation and type of vector. Researchers have used different types of bacterial and viral vectors to obtaine efficient GE in plants. In case of viral vectors, geminivirus-based vectors have been efficiently used to deliver GE reagents and donor templates (Baltes et al. 2014). In addition, nanoparticles have been used for efficient delivery of the RNPs in the cells to get GE with more ease and success. Nanoparticle-mediated delivery of RNPs and donor DNA has increased the efficiency of producing genome edited cells without integration of the foreign gene in the target genome (Lee et al. 2017; Mout et al. 2017; Glass et al. 2018). Almost, all transformation methods have been reported to transform the organisms with GE tools from PEG-mediated transformation to Agrobacterium and biolistic method. All the factors, from designing, cloning, selection of vector to transformation method, are important to make targeted and site-specific modifications in the genome with higher efficiency and low off-targeting. Moreover, recent tools in the CRISPR toolbox, such as prime editing and base editing (Figs. 10.4 and 10.5), are the most favorite for avoiding off-targets due to their ability to make targeted modifications without creating a DSB at the target site.

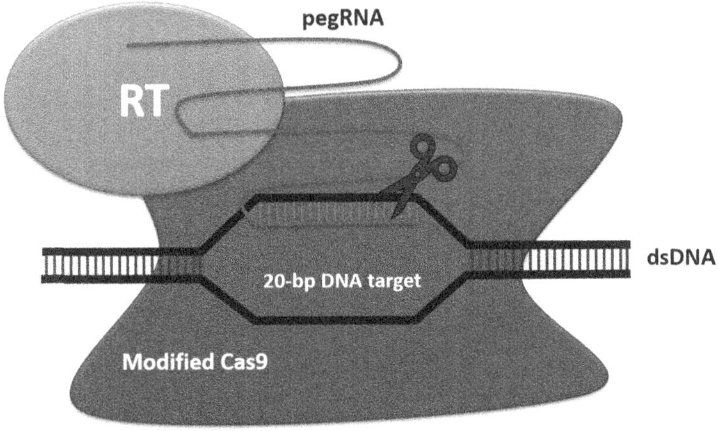

Fig. 10.5 Prime editors are based on modified Cas9, reverse transcriptase (RT) enzyme, and pegRNA (prime editing RNA) complex. Prime editing uses pegRNA to target the sequence while RT makes DNA from the pegRNA having the donor sequence. Cas9 produced nick in the single strand to edit the desired sequence accordingly

10.4 Regulation of Genome Edited Crops

There is a continuous discussion about whether an organism genetically edited with CRISPR-Cas is a genetically modified organism (GMO). Recently, a choice from the EU Court of Justice decided that such items are considered GMOs, at any rate in the EU (Anonymous 2018; EU-Commission 2001). The GMO is an organism to which genes of known feature are transferred, that are incorporated into random locations alongside the chromosome. On the other hand, a GMO that is inserted with specific alterations of gene functions target in unique locations of the genome, to silence or improve its expression, consequently fending off non-unique random modifications (Georges and Ray 2017).

At the point when another innovation of these attributes shows up, it is inescapable that questions emerge among individuals about its protected use, its guideline, and its morals (Araki and Ishii 2015). In this sense, we could feature two key issues the off-target edition and the technology of the gene drive. Utilization of CRISPR-Cas9 for altering the genome in plants shows that the recurrence of a change outside the target is quite low. In a study carried out on transgenic soybean, where transgenes were incorporated by Agrobacterium, it was discovered that the genomic variety was less (one order of magnitude) contrasted with soybean plants transformed by radiation, and a much less (two orders of magnitude) in regard to the diverse soybean crops (Anderson et al. 2016). The opinion among the scientific community is also remarkable, since 88% of scientists from the AAAS (American Association for the Advancement of Science) consider genetically edited or modified foods safe for intake, in line with a examine executed with the aid of Pew trust and the AAAS in

2015 (Funk et al. 2015). In spite of this, and to diminish the issue of mutation, it is valuable to chip away at improving the structure of reagents, computational tools, and trial procedure, just as a more noteworthy thought of the genome investigation of the lines of plants altered by CRISPR, since they can give extra certifications to forestall this marvel. Another perspective to consider with respect to the safety of crops and food is the creation of gene drives. A gene drive is a system permitting the biased inheritance of a genetic element so that within a population their posterity has over half likelihood to acquire a specific attribute. Specially, drive genes in agriculture are being investigated inside the pests as a control. The critical research exertion has concentrated on gene drives, and a few fruitful effects have been accounted for on the elimination of mosquitoes (Anstead et al. 2016; Leftwich et al. 2018; Roitman et al. 2017). Different applications include the control of weed for farming (Neve 2018) and the resensitization of life forms that are impervious to pesticides or herbicides (Collins 2018; Zentner and Wade 2017). In nature, these systems are monitored, however, with the utilization of CRISPR-Cas9 coordinated to germ cells or undeveloped organisms, profoundly effective components can be made (Akbari et al. 2015). This is less significant in residential plants, because of the trouble in scattering the gene. In any case, it is a more prominent worry in wild plants, as it is the situation when used to expel obtrusive species (Esvelt et al. 2014). To maintain a strategic distance from this, a reversible gene drive can be structured, in which the unique characteristic of the target gene is restored by using second gene drive (DiCarlo et al. 2015). In addition to the above, to avoid these undesirable changes, it is critical to analyze the sequence of the genome acquired after editing. This portrayal will uncover the automatic physical impacts, just as the impacts of intrigue, and this method is centered around the resulting assessment of the danger of potential impacts that transgenic crops and their items can have, considering applicable parts of human and ecological security (Schnell et al. 2015). The development of such a novel and incredible technology as CRISPR-Cas has made it important to direct its utilization and the items acquired with this technology. In this sense, the guideline of genetically altered foods and plants is diverse relying upon the nation wherein we are found. In the EU, the order of GMOs depends in transit by which the adjustment has happened. GMO is considered if the hereditary change happens "in a way that doesn't happen normally through mating as well as characteristic recombination," and utilizing any of the recorded methods. Recently, living beings altered by the CRISPR-Cas method have been remembered for this arrangement (Anonymous 2018). In any case, in different nations, for example, Canada, all foods that are hereditarily changed, including traditional rearing, are considered "new foods." These new items require a notice to the Health Department of Canada before promoting. After this notification, a whole protection assessment is made taking into consideration the characteristics of the product itself, instead of the manufacturing system. Along these lines, with this accentuation on the item contrasted with the procedure, the new qualities of the plants will be dependent upon a comparable guideline paying little heed to the strategy utilized (Anonymous 2005). On the other way, in the United States, there are three entities accountable for directing GMOs. Genetically changed foods are directed by the FDA. They are directed by their segments, so that on the off chance that they are substances "by and large perceived as safe," like those presently found in unmodified foods, they do not require earlier endorsement to be put available. Be that as it may, when a GMO item "contrasts fundamentally in the structure, function or arrangement of the substances as of now found in food," its endorsement before commercialization is essential. The regulation of genetically altered plants and crops relates to the United States Department of Agriculture (USDA 2018). It is answerable for controlling the planting, transportation, or importation of these vegetables. This organization gave an announcement reporting that they would not direct harvests adjusted by CRISPR or other genome altering methods (USDA 2018). The third organization included, the EPA (Environmental Protection Agency), is answerable for the regulation of genetically altered microorganisms and pesticides (Acosta 2014). Concerning property, it ought to be noticed that the primary patent enrollment to cover the utilization of CRISPR was completed by Jennifer Doudna (UC Berkeley) and Emmanuelle Charpentier (University of Umeå, Sweden), at the time of their commitment to revelation on May 25, 2012, which was distributed universally in November 2013 (Jinek et al. 2013). On the other hand, Feng Zhang (Broad Institute) additionally propelled a patent enlistment use of CRISPR in eukaryotic cells, however, utilized a quicker procedure than Doudna and Charpentier, so their application was affirmed before in spite of being introduced later (Sherkow 2017). This has prompted a lawful question in the United States over the misuse privileges of CRISPR that has gone on for quite a long time and has demonstrated the privilege to Doudna and Charpentier (Anonymous 2018). Around the world, there are 2052 enlisted licenses identified with CRISPR (US Patent Collection 2018), of which 344 are enrolled in Canada (Government of Canada 2015) and 1239 in the United States (US Patent Collection 2018).

10.5 Future of CRISPR-Based GE

CRISPR is a versatile gene editing technology that has just discovered various applications over the most recent couple of years. Utilizing CRISPR in farming or food frameworks is quicker, simpler, more exact, and presumably less expensive than conventional genetic engineering techniques. The research led so far, demonstrated that an effective CRISPR-Cas9-interceded genome altering requires the presentation of sgRNA containing a roughly 20-base sequence precise to the target DNA. With the improvement of accuracy in quality focusing on framework, there might be no more requirement for a long nucleotide sgRNA, which, therefore, will rearrange the structure of specific sgRNA. Other than crop improvement, fermentation procedures likewise have profited by genetic engineering by creating strains with improved useful properties. Strains of *Lactobacillus* spp. are microorganisms with prominent importance in the food industry, principally as probiotics and psychobiotics (Misra and Mohanty 2019). In this way, utilizing CRISPR-Cas9 innovation in these microorganisms could essentially improve their opposition against a few pressure conditions and thusly, expanding the profitability of the maturation procedure, which upgrades the nature of food items. There are still a few spaces to fill in our insight into the CRISPR-Cas9 genome altering framework. The mechanism of separation of Cas9 from planned sgRNA and its subsequent reuse are still hazy and need further investigation. One of the rests of the difficulties in plant altering is the adequacy of the transmission pace of the mutation to the next generation. New, bigger scale investigation should address the off-target mutation transformations rate in plant genome and how to minimize them (Zischewski et al. 2017). Besides, auxiliary investigations on Cas9 will empower the structure of new variations with novel PAM specificities. The utilization of various sgRNAs gathered into a conveyance framework can likewise open up open doors for the erasure of gene clusters. All the more critically, making an all-around accessible database that contains all the data on CRISPR-Cas9 structure with the guide of bioinformatics will facilitate revelations. As of now, most investigations concentrated on the use of CRISPR-Cas9 tool to improve agricultural traits of crops. New disclosures in Cas protein will make ready for an increasingly differing application in a progressively proficient manner. Ongoing examinations with Cas13 proteins uncovered that the effectiveness and accuracy of this framework could be altogether expanded. However, the improvement of these frameworks is profoundly required for additional application in food crops. Defeating these constraints will expand the pace of creation of new and more beneficial crop. Recently, the choice of the Court of Justice of the European Union (ECJ) in Luxembourg caused to notice the utilization of CRISPR on crops (Callaway 2018). As per the ECJ, the crops made with this tool are dependent upon a 2001 order, which is a law produced for more seasoned reproducing systems and required a long endorsement process. This disillusioning choice was censured by established researchers, expressing that this circumstance could altogether block a large portion of the interests in crop explores utilizing CRISPR device in the EU. Then again, this choice was cheered by the natural association "Friends of the Earth" in Amsterdam since they unequivocally accepted that the exacting guidelines for the support of these harvests are important. The choice of ECJ and consequent responses by established researchers show that, other than specialized restrictions, there are political issues to be defeated to progress in this incredible technology.

References

Acosta L (2014) Restrictions on genetically modified organisms: The Law Library of Congress of United States. https://www.loc.gov/law/help/restrictionson-gmos/usa.php#Foodstuffs. Accessed 24 Nov 2018

Akbari OS, Bellen HJ, Bier E, Bullock SL, Burt A, Church GM, Cook KR, Duchek P, Edwards OR, Esvelt KM (2015) Safeguarding gene drive experiments in the laboratory. Science 349 (6251):927–929

Anderson JE, Michno J-M, Kono TJ, Stec AO, Campbell BW, Curtin SJ, Stupar RM (2016) Genomic variation and DNA repair associated with soybean transgenesis: a comparison to cultivars and mutagenized plants. BMC biotechnology 16(1):41

Anonymous (2005) The regulation of genetically modified food. https://www.canada.ca/en/healthcanada/services/scienceresearch/reportspublications/biotechnology/regulation-genetically-modified-foods.html. Accessed 24 Nov 2018

Anonymous (2018). Doudna-Charpentier team awarded U.S. patent for CRISPR-Cas9. Berkeley News. http://news.berkeley.edu/2018/06/19/doudnacharpentier-team-awarded-u-s-patent-for-crispr-cas9/. Accessed 25 Nov 2018

Anstead CA, Batterham P, Korhonen PK (2016) A blow to the fly—Lucilia cuprina draft genome and transcriptome to support advances in biology and biotechnology. Biotechnol Adv 34 (5):605–620

Anzalone AV, Randolph PB, Davis JR, Sousa AA, Koblan LW, Levy JM, Chen PJ, Wilson C, Newby GA, Raguram A (2019) Search-and-replace genome editing without double-strand breaks or donor DNA. Nature 576(7785):149–157

Araki M, Ishii T (2015) Towards social acceptance of plant breeding by genome editing. Trends in Plant Science 20(3):145–149

Bae S, Park J, Kim J-S (2014) Cas-OFFinder: a fast and versatile algorithm that searches for potential off-target sites of Cas9 RNA-guided endonucleases. Bioinformatics 30 (10):1473–1475

Baltes NJ, Gil-Humanes J, Cermak T, Atkins PA, Voytas DF (2014) DNA replicons for plant genome engineering. Plant Cell 26(1):151–163

Basso MF, Ferreira PCG, Kobayashi AK, Harmon FG, Nepomuceno AL, Molinari HBC, Grossi-de-Sa MF (2019) Micro RNA s and new biotechnological tools for its modulation and improving stress tolerance in plants. Plant Biotechnol J 17(8):1482–1500

Brooks C, Nekrasov V, Lippman ZB, Van Eck J (2014) Efficient gene editing in tomato in the first generation using the clustered regularly interspaced short palindromic repeats/CRISPR-associated9 system. Plant Physiology 166(3):1292–1297

Callaway E (2018) EU law deals blow to CRISPR crops. Nature 560(7716):16–16

Chandrasekaran J, Brumin M, Wolf D, Leibman D, Klap C, Pearlsman M, Sherman A, Arazi T, Gal-On A (2016) Development of broad virus resistance in non-transgenic cucumber using CRISPR/Cas9 technology. Mol Plant Pathol 17(7):1140–1153

Chen S, Oikonomou G, Chiu CN, Niles BJ, Liu J, Lee DA, Antoshechkin I, Prober DA (2013) A large-scale in vivo analysis reveals that TALENs are significantly more mutagenic than ZFNs generated using context-dependent assembly. Nucleic acids research 41(4):2769–2778

Collins JP (2018) Gene drives in our future: challenges of and opportunities for using a self-sustaining technology in pest and vector management. BMC Proceedings 8:9

Dahan-Meir T, Filler-Hayut S, Melamed-Bessudo C, Bocobza S, Czosnek H, Aharoni A, Levy AA (2018) Efficient in planta gene targeting in tomato using geminiviral replicons and the CRISPR/Cas9 system. Plant J 95(1):5–16

DiCarlo JE, Chavez A, Dietz SL, Esvelt KM, Church GM (2015) Safeguarding CRISPR-Cas9 gene drives in yeast. Nature Biotechnology 33(12):1250–1255

Eraslan G, Avsec Ž, Gagneur J, Theis FJ (2019) Deep learning: new computational modelling techniques for genomics. Nature Reviews Genetics 20(7):389–403

Esvelt KM, Smidler AL, Catteruccia F, Church GM (2014) Emerging technology: concerning RNA-guided gene drives for the alteration of wild populations. Elife 3:e03401

EU-Commission (2001) Directive 2001/18/EC of the European Parliament and of the Council of 12 March 2001 on the deliberate release into the environment of genetically modified organisms and repealing Council Directive 90/220. EEC

Feng Z, Zhang B, Ding W, Liu X, Yang D-L, Wei P, Cao F, Zhu S, Zhang F, Mao Y (2013) Efficient genome editing in plants using a CRISPR/Cas system. Cell Research 23 (10):1229–1232

Funk C, Rainie L, Page D (2015) Public and scientists' views on science and society. Pew Research Center

Gaudelli NM, Komor AC, Rees HA, Packer MS, Badran AH, Bryson DI, Liu DR (2017) Programmable base editing of A·T to G·C in genomic DNA without DNA cleavage. Nature 551(7681):464

Georges F, Ray H (2017) Genome editing of crops: a renewed opportunity for food security. GM Crops Food 8(1):1–12

Glass Z, Lee M, Li Y, Xu Q (2018) Engineering the delivery system for CRISPR-based genome editing. Trends in biotechnology 36(2):173–185

Government of Canada (2015) Canadian patent database. http://www.ic.gc.ca/opiccipo/cpd/eng/search/results.html?query=crispr&start=1&num=50&type=basic_search&newSearch=0. Accessed 25 Nov 2018

Gratz SJ, Ukken FP, Rubinstein CD, Thiede G, Donohue LK, Cummings AM, O'Connor-Giles KM (2014) Highly specific and efficient CRISPR/Cas9-catalyzed homology-directed repair in Drosophila. Genetics 196(4):961–971

Guilinger JP, Thompson DB, Liu DR (2014) Fusion of catalytically inactive Cas9 to FokI nuclease improves the specificity of genome modification. Nat Biotechnol 32(6):577

Havlicek S, Shen Y, Alpagu Y (2017) Re-engineered RNA-guided FokI-nucleases for improved genome editing in human cells. Mol Ther 25(2):342–355

Heigwer F, Kerr G, Boutros M (2014) E-CRISP: fast CRISPR target site identification. Nature Methods 11(2):122

Hua K, Tao X, Yuan F, Wang D, Zhu J-K (2018) Precise A·T to G·C base editing in the rice genome. Mol Plant 11(4):627–630

Huang C-H, Lee K-C, Doudna JA (2018) Applications of CRISPR-Cas enzymes in cancer therapeutics and detection. Trends Cancer 4(7):499–512

Iqbal Z, Sattar MN, Shafiq M (2016) CRISPR/Cas9: a tool to circumscribe cotton leaf curl disease. Front Plant Sci 7:475

Ito Y, Nishizawa-Yokoi A, Endo M et al (2015) CRISPR/Cas9-mediated mutagenesis of the RIN locus that regulates tomato fruit ripening. Biochem Bioph Res Co 467(1):76–82

Jia H, Wang N (2014) Targeted genome editing of sweet orange using Cas9/sgRNA. PloS One 9(4): e93806

Jia J, Jin Y, Bian T, Wu D, Yang L, Terada N, Wu W, Jin S (2014) Bacterial delivery of TALEN proteins for human genome editing. PloS One 9(3):0091547

Jiang WZ, Henry IM, Lynagh PG et al (2017) Significant enhancement of fatty acid composition in seeds of the allohexaploid, Camelina sativa, using CRISPR/Cas9 gene editing. Plant Biotechnol J 15(5):648–657

Jinek M, Charpentier E, Chylinski K, Cate JH, Lim W, Qi L (2013) Methods and compositions for RNA-directed target DNA modification and for RNA-directed modulation of transcription. Google Patents

Kang BC, Yun JY, Kim ST, Shin Y, Ryu J, Choi M, Woo JW, Kim JS (2018) Precision genome engineering through adenine base editing in plants. Nat Plants 4:427–431

Kim H, Kim S-T, Ryu J, Kang B-C, Kim J-S, Kim S-G (2017) CRISPR/Cpf1-mediated DNA-free plant genome editing. Nature Communications 8(1):1–7

Komor AC, Kim YB, Packer MS, Zuris JA, Liu DR (2016) Programmable editing of a target base in genomic DNA without double-stranded DNA cleavage. Nature 533(7603):420

Lee K, Conboy M, Park HM, Jiang F, Kim HJ, Dewitt MA, Mackley VA, Chang K, Rao A, Skinner C (2017) Nanoparticle delivery of Cas9 ribonucleoprotein and donor DNA in vivo induces homology-directed DNA repair. Nature Biomedical Engineering 1(11):889

Leftwich PT, Edgington MP, Harvey-Samuel T, Carabajal Paladino LZ, Norman VC, Alphey L (2018) Recent advances in threshold-dependent gene drives for mosquitoes. Biochemical Society Transactions 46(5):1203–1212

Li J-F, Norville JE, Aach J, McCormack M, Zhang D, Bush J, Church GM, Sheen J (2013) Multiplex and homologous recombination–mediated genome editing in Arabidopsis and Nicotiana benthamiana using guide RNA and Cas9. Nature Biotechnology 31(8):688–691

Li J, Sun Y, Du J, Zhao Y, Xia L (2017) Generation of targeted point mutations in rice by a modified CRISPR/Cas9 system. Mol Plant 10(3):526–529

Li X, Wang Y, Chen S et al (2018) Lycopene is enriched in tomato fruit by CRISPR/Cas9-mediated multiplex genome editing. Front Plant Sci 9:559

Liang Z, Zhang K, Chen K, Gao C (2014) Targeted mutagenesis in Zea mays using TALENs and the CRISPR/Cas system. Journal of Genetics and Genomics 41(2):63–68

Liang Z, Chen K, Li T, Zhang Y, Wang Y, Zhao Q, Liu J, Zhang H, Liu C, Ran Y (2017) Efficient DNA-free genome editing of bread wheat using CRISPR/Cas9 ribonucleoprotein complexes. Nature communications 8(1):1–5

Liu J, Gaj T, Patterson JT, Sirk SJ, Barbas CF III (2014) Cell-penetrating peptide-mediated delivery of TALEN proteins via bioconjugation for genome engineering. PloS One 9(1):e0085755

Liu H, Ding Y, Zhou Y, Jin W, Xie K, Chen L-L (2017) CRISPR-P 2.0: an improved CRISPR-Cas9 tool for genome editing in plants. Molecular Plant 10(3):530–532

Lu J, Zhao C, Zhao Y et al (2018) Multimode drug inducible CRISPR/Cas9 devices for transcriptional activation and genome editing. Nucleic Acids Res 46(5):e25–e25

Ma X, Zhang Q, Zhu et al (2015) A robust CRISPR/Cas9 system for convenient, high-efficiency multiplex genome editing in monocot and dicot plants. Mol Plant 8(8):1274–1284

Macovei A, Sevilla NR, Cantos C, Jonson GB, Slamet-Loedin I, Čermák T, Voytas DF, Choi IR, Chadha-Mohanty P (2018) Novel alleles of rice eIF4G generated by CRISPR/Cas9-targeted mutagenesis confer resistance to Rice tungro spherical virus. Plant Biotechnol J 16 (11):1918–1927

Malnoy M, Viola R, Jung M-H, Koo O-J, Kim S, Kim J-S, Velasco R, Nagamangala Kanchiswamy C (2016) DNA-free genetically edited grapevine and apple protoplast using CRISPR/Cas9 ribonucleoproteins. Frontiers in Plant Science 7:1904

Miao J, Guo D, Zhang J, Huang Q, Qin G, Zhang X, Wan J, Gu H, Qu L-J (2013) Targeted mutagenesis in rice using CRISPR-Cas system. Cell research 23(10):1233–1236

Mikami M, Toki S, Endo M (2016) Precision targeted mutagenesis via Cas9 paired nickases in rice. Plant Cell Physiol 57(5):1058–1068

Misra S, Mohanty D (2019) Psychobiotics: a new approach for treating mental illness? Critical Reviews in Food Science and Nutrition 59(8):1230–1236

Montague TG, Cruz JM, Gagnon JA, Church GM, Valen E (2014) CHOPCHOP: a CRISPR/Cas9 and TALEN web tool for genome editing. Nucleic Acids Research 42(1):401–407

Morineau C, Bellec Y, Tellier F, Gissot L, Kelemen Z, Nogué F, Faure JD (2017) Selective gene dosage by CRISPR-Cas9 genome editing in hexaploid Camelina sativa. Plant Biotechnol J 15 (6):729–739

Mout R, Ray M, Yesilbag Tonga G, Lee Y-W, Tay T, Sasaki K, Rotello VM (2017) Direct cytosolic delivery of CRISPR/Cas9-ribonucleoprotein for efficient gene editing. ACS Nano 11 (3):2452–2458

Naito Y, Hino K, Bono H, Ui-Tei K (2015) CRISPRdirect: software for designing CRISPR/Cas guide RNA with reduced off-target sites. Bioinformatics 31(7):1120–1123

Nakayasu M, Akiyama R, Lee HJ, Osakabe K, Osakabe Y, Watanabe B, Sugimoto Y, Umemoto N, Saito K, Muranaka T, Mizutani M (2018) Generation of α-solanine-free hairy roots of potato by CRISPR/Cas9 mediated genome editing of the St16DOX gene. Plant Physiol Biochem 131:70–77

Nekrasov V, Staskawicz B, Weigel D, Jones JD, Kamoun S (2013) Targeted mutagenesis in the model plant Nicotiana benthamiana using Cas9 RNA-guided endonuclease. Nature Biotechnology 31(8):691

Neve P (2018) Gene drive systems: do they have a place in agricultural weed management? Pest Manag Sci 74(12):2671–2679

Nieves-Cordones M, Mohamed S, Tanoi K, Kobayashi NI, Takagi K, Vernet A, Guiderdoni E, Périn C, Sentenac H, Véry AA (2017) Production of low-Cs+ rice plants by inactivation of the K+ transporter Os HAK 1 with the CRISPR-Cas system. The Plant J 92(1):43–56

Nonaka S, Someya T, Zhou S et al (2017) An Agrobacterium tumefaciens strain with gamma-aminobutyric acid transaminase activity shows an enhanced genetic transformation ability in plants. Sci Rep 7:42649

Okuzaki A, Ogawa T, Koizuka C et al (2018) CRISPR/Cas9-mediated genome editing of the fatty acid desaturase 2 gene in Brassica napus. Plant Physiol Biochem 131:63–69

Ren B, Yan F, Kuang Y et al (2018) Improved base editor for efficiently inducing genetic variations in rice with CRISPR/Cas9-guided hyperactive hAID mutant. Mol Plant 11(4):623–626

Roitman LA, Fisher D, Theriot J (2017) Playing God: eradicating malaria and mosquitoes in the developing world with gene drives and CRISPR/Cas9. J Nanomed Nanosci JNAN-127. https://doi.org/10.29011/JNAN-127.100027

Sander JD, Maeder ML, Reyon D, Voytas DF, Joung JK, Dobbs D (2010) ZiFiT (Zinc Finger Targeter): an updated zinc finger engineering tool. Nucleic Acids Research 38 (suppl_2):462–468

Schnell J, Steele M, Bean J, Neuspiel M, Girard C, Dormann N, Pearson C, Savoie A, Bourbonniere L, Macdonald P (2015) A comparative analysis of insertional effects in genetically engineered plants: considerations for pre-market assessments. Transgenic Research 24 (1):1–17

Shan Q, Zhang Y, Chen K, Zhang K, Gao C (2015) Creation of fragrant rice by targeted knockout of the Os BADH 2 gene using TALEN technology. Plant Biotechnol J 13(6):791–800

Sherkow JS (2017) Patent protection for CRISPR: an ELSI review. Journal of Law and the Biosciences 4(3):565–576

Shimatani Z, Kashojiya S, Takayama M, Terada R, Arazoe T, Ishii H, Teramura H, Yamamoto T, Komatsu H, Miura K (2017) Targeted base editing in rice and tomato using a CRISPRCas9 cytidine deaminase fusion. Nat Biotechnol 35(5):441–443

Silva G, Poirot L, Galetto R, Smith J, Montoya G, Duchateau P, Pâques F (2011) Meganucleases and other tools for targeted genome engineering: perspectives and challenges for gene therapy. Current Gene Therapy 11(1):11–27

Sun J, Wang Q, Jiang Y, Wen Z, Yang L, Wu J, Yang S (2018) Genome editing and transcriptional repression in Pseudomonas putida KT2440 via the type II CRISPR system. Microb Cell Factories 17(1):1–17

Standage-Beier K, Zhang Q, Wang X (2015) Targeted large-scale deletion of bacterial genomes using CRISPR-nickases. ACS Synth Biol 4(11):1217–1225

Tang L, Mao B, Li Y, Lv Q, Zhang L, Chen C, He H, Wang W, Zeng X, Shao Y (2017) Knockout of OsNramp5 using the CRISPR/Cas9 system produces low Cd-accumulating indica rice without compromising yield. Sci Rep-UK 7(1):14438

Terns MP (2018) CRISPR-based technologies: impact of RNA-targeting systems. Mol Cell 72 (3):404–412

Tesson L, Usal C, Ménoret S, Leung E, Niles BJ, Remy S, Santiago Y, Vincent AI, Meng X, Zhang L (2011) Knockout rats generated by embryo microinjection of TALENs. Nature biotechnology 29(8):695–696

Tong C, Huang G, Ashton C, Wu H, Yan H, Ying Q-L (2012) Rapid and cost-effective gene targeting in rat embryonic stem cells by TALENs. Journal of Genetics and Genomics 39 (6):275–280

US Patent Collection (2018). http://patft.uspto.gov/netacgi/nph-Parser?Sect1=PTO2&Sect2=HITOFF&p=1&u=%2Fnetahtml%2FPTO%2Fsearchbool.html&r=0&f=S&l=50&TERM1=crispr&FIELD1=&co1=AND&TERM2=&FIELD2=&d=PTXT. Accessed 25 Nov 2018

USDA (2018) U.S. Department of Agriculture Secretary Perdue Issues USDA statement on plant breeding innovation. https://www.usda.gov/media/pressreleases/2018/03/28/secretary-perdue-issues-usda-statement-plant-breedinginnovation. Accessed 24 Nov 2018

Vigouroux A, Oldewurtel E, Cui L, Bikard D, van Teeffelen S (2018) Tuning dCas9's ability to block transcription enables robust, noiseless knockdown of bacterial genes. Mol Syst Biol 14(3): e7899

Waltz E (2016) CRISPR-edited crops free to enter market, skip regulation, p 582
Wang Y, Cheng X, Shan Q, Zhang Y, Liu J, Gao C, Qiu J-L (2014) Simultaneous editing of three homoeoalleles in hexaploid bread wheat confers heritable resistance to powdery mildew. Nature Biotechnology 32(9):947
Wang S, Zhang S, Wang W, Xiong X, Meng F, Cui X (2015) Efficient targeted mutagenesis in potato by the CRISPR/Cas9 system. Plant cell reports 34(9):1473–1476
Weeks DP, Spalding MH, Yang B (2016) Use of designer nucleases for targeted gene and genome editing in plants. Plant Biotechnology Journal 14(2):483–495
Wefers B, Panda SK, Ortiz O, Brandl C, Hensler S, Hansen J, Wurst W, Kühn R (2013) Generation of targeted mouse mutants by embryo microinjection of TALEN mRNA. Nature protocols 8 (12):2355
Widom JR, Rai V, Rohlman CE, Walter NG (2019) Versatile transcription control based on reversible dCas9 binding. RNA 25(11):1457–1469
Woo JW, Kim J, Kwon SI, Corvalán C, Cho SW, Kim H, Kim S-G, Kim S-T, Choe S, Kim J-S (2015) DNA-free genome editing in plants with preassembled CRISPR-Cas9 ribonucleoproteins. Nat Biotechnol 33(11):1162
Xu L, Park KH, Zhao L et al (2016) CRISPR-mediated genome editing restores dystrophin expression and function in mdx mice. Mol Ther 24(3):564–569
Yan F, Kuang Y, Ren B, Wang J, Zhang D, Lin H, Yang B, Zhou X, Zhou H (2018) Highly efficient A·T to G·C base editing by Cas9n-guided tRNA adenosine deaminase in rice. Mol Plant 11 (4):631–634
Yin X, Biswal AK, Dionora J (2017) CRISPR-Cas9 and CRISPR-Cpf1 mediated targeting of a stomatal developmental gene EPFL9 in rice. Plant Cell Rep 36(5):745–757
Zentner GE, Wade MJ (2017) The promise and peril of CRISPR gene drives: genetic variation and inbreeding may impede the propagation of gene drives based on the CRISPR genome editing technology. BioEssays 39(10):1700109
Zhang Y, Liang Z, Zong Y et al (2016) Efficient and transgene-free genome editing in wheat through transient expression of CRISPR/Cas9 DNA or RNA. Nat Commun 7(1):1–8
Zhang Y, Massel K, Godwin ID, Gao C (2018) Applications and potential of genome editing in crop improvement. Genome Biol 19(1):210
Zischewski J, Fischer R, Bortesi L (2017) Detection of on-target and off-target mutations generated by CRISPR/Cas9 and other sequence-specific nucleases. Biotechnology Advances 35 (1):95–104
Zong Y, Wang Y, Li C, Zhang R, Chen K, Ran Y, Qiu J-L, Wang D, Gao C (2017) Precise base editing in rice, wheat and maize with a Cas9-cytidine deaminase fusion. Nat Biotechnol 35 (5):438
Zou J, Huss M, Abid A, Mohammadi P, Torkamani A, Telenti A (2019) A primer on deep learning in genomics. Nature Genetics 51(1):12–18

The manufacturer's authorised representative in the EU is Springer Nature Customer Service Centre GmbH, Europaplatz 3, 69115 Heidelberg, Germany. If you have any concerns regarding our products, please contact ProductSafety@springernature.com

Printed and bound by CPI Group (UK) Ltd, Croydon, CR0 4YY
30/09/2026
02225290-0014